Additive Manufacturing

Kun Zhou
Editor

Additive Manufacturing

Materials, Functionalities and Applications

Springer

Editor
Kun Zhou
School of Mechanical and Aerospace
Engineering
Nanyang Technological University
Singapore, Singapore

ISBN 978-3-031-04720-6 ISBN 978-3-031-04721-3 (eBook)
https://doi.org/10.1007/978-3-031-04721-3

This Springer imprint is published by the registered company Springer Nature Switzerland AG
The registered company address is: Gewerbestrasse 11, 6330 Cham, Switzerland

Preface

Additive manufacturing (AM), trendily known as three-dimensional (3D) printing, is an emerging technology capable of converting computer-aided designs into physical objects through the layer-by-layer accumulation of feedstock without molding, machining, or tooling. With the great design freedom and high production efficiency that AM techniques offer, their development has become a key area of research in various fields. Since their introduction, AM applications have gradually shifted from rapid prototyping to the direct manufacturing of end-use products. The selection of available AM techniques has also expanded at an astonishing rate, with each technique designated for a specific set of materials, functionalities, and applications.

The development of AM has sparked a new wave of innovation and ingenuity among the scientific community, driving the invention of novel materials with unique functionalities and applications at an increasing pace. In recent years, rapid advances in AM have brought about significant improvements in several niche applications, such as flexible electronics, energy storage, drug delivery, biomedical implants and devices, technologies for four-dimensional (4D) printing, large-scale printing, and ceramics printing. This book reviews and discusses the recent progress, current state, application trends, and future outlook of AM in such research fields.

Chapter 1 reports the recent advances in AM pertaining to soft wearable electronics, in particular, their related techniques, materials, and applications. Representative AM device applications, including sensors, actuators, heaters, interconnects, and antennas, are also highlighted.

Chapter 2 covers the fundamentals and applications of AM in advanced energy storage. A wide range of AM materials are explored with an emphasis on the production of energy storage devices. Additionally, an in-depth review of AM-fabricated rechargeable batteries and electrochemical capacitors is presented.

Chapter 3 expounds upon the fundamental concepts and applications of 4D printing. A variety of 4D-printed stimuli-responsive materials, including polymers, metals, ceramics, and their composites, are reviewed extensively. In addition, a detailed discussion of the recent progress in the 4D printing of stimuli-responsive materials is conducted.

Chapter 4 presents an evaluation of AM as a production method for personalized medicine. The fundamentals of selecting excipient materials and active pharmaceutical ingredients that are compatible with current AM techniques are introduced. The key advantages and limitations of the AM techniques for small-scale and large-scale drug delivery device production are reviewed. Finally, an outlook of AM as a future production technology for the pharmaceutical industry is provided.

Chapter 5 outlines the fundamentals and typical applications of AM in fabricating metal implants and surgical plates. Popular metal materials selected for printing implants, including iron alloys, titanium and titanium alloys, cobalt alloys, tantalum, and degradable magnesium and zinc alloys, are analyzed. In particular, representative application cases for various metal implants and surgical plates are presented.

Chapter 6 presents a review of the recent development of wire arc AM (WAAM) intended for large-scale printing. Specifically, WAAM systems and their fundamentals, the prevalent engineering materials employed, existing challenges, quality control methods, and associated printing modeling techniques are specified. An outlook based on the contemporary status quo and trends of WAAM research is also provided.

Chapter 7 discusses the current state of the AM of advanced ceramics, providing a comprehensive evaluation of both advantages and limitations of each AM process. The advantages of advanced ceramics and ceramic matrix composites fabricated through AM are emphasized in terms of characterizations and applications. Guidelines for optimizing processes and material selection are also presented.

I would like to express my deepest appreciation to the contributing authors of all the chapters and Mayra Castro, Yogesh Padmanaban, and Kowsalya Raghunathan at Springer Nature. I would also like to thank the members of my research group, with whom my discussions have always been fruitful. In particular, I appreciate the efforts from Wei Zhu, Pengfei Tan, Wei Shian Tey, Boyuan Li, Yung Zhen Lek, Devesh Kripalani, Yujia Tian, Raj Kiran, Priyanka Viveganantham, Han Zheng, Liming You, and Asker Jarlöv.

Singapore Kun Zhou

Contents

1 3D-Printed Soft Wearable Electronics: Techniques, Materials, and Applications 1
Yuxuan Liu and Yong Zhu

2 Additive Manufacturing of Energy Storage Devices 51
Xiaocong Tian and Kun Zhou

3 4D Printing of Stimuli-Responsive Materials 85
Chunze Yan, Xiao Yang, and Hongzhi Wu

4 Personalized Medicine: Manufacturing Oral Solid Dosage Forms Through Additive Manufacturing 113
Yinfeng He, Maria Inês Evangelista Barreiros, and Hatim Cader

5 Additive Manufacturing of Metal Implants and Surgical Plates 151
Di Wang, Yongqiang Yang, and Changjun Han

6 Wire Arc Additive Manufacturing: Systems, Microstructure, Defects, Quality Control, and Modelling 205
Zhe Chen and Gim Song Soh

7 Additive Manufacturing of Ceramics: Materials, Characterization and Applications 245
Jiaming Bai, Jinxing Sun, and Jon Binner

Chapter 1
3D-Printed Soft Wearable Electronics: Techniques, Materials, and Applications

Yuxuan Liu and Yong Zhu

1.1 Background

Wearable electronics with soft features are emerging in recent years, enabling long-term applications on the human body. For example, wearable sensors can provide accurate monitoring of human physiological signals; [1–6] wearable drug delivery systems and wearable heaters can ensure precise and timely therapy; [7–10] and wearable displays with the visual interface can clearly inform the monitoring status [11]. Not only for human health, wearable devices can also be used on plants to monitor plant health for smart agriculture [12, 13]. Wearable electronics that feature gas permeability [14] and recyclability [15] have emerged, which further facilitate practical application of these wearable devices. Relatively simple and cost-effective fabrication of soft wearable electronics can greatly facilitate their deployment. Top-down microfabrication, e.g., deposition/evaporation, lithography, and etching, are conventional approaches for device fabrication [16–20]. They are widely used in the fabrication of electronics with high resolution and high integration levels [21–24]. However, they are typically complicated and expensive. Instead of top-down approaches, bottom-up approaches including coating and printing techniques are widely used to process nanomaterials with high throughput and low costs [25–28]. However, coating techniques and conventional printing techniques (e.g., inkjet printing, screen printing, and gravure printing) are typically limited to 2D planar fabrication; hence, most of the devices, including wearable soft sensors, actuators,

Y. Liu · Y. Zhu (✉)
Department of Mechanical and Aerospace Engineering, North Carolina State University, Raleigh, NC 27695, USA
e-mail: yong_zhu@ncsu.edu

Y. Liu
e-mail: yliu235@ncsu.edu

K. Zhou (ed.), *Additive Manufacturing*,
https://doi.org/10.1007/978-3-031-04721-3_1

and transistors, are relatively simple in architecture. To adapt to complex wearable electronics systems, 3D printing technique has been used to fabricate 3D structured devices. There are generally two approaches to constructing 3D structures by printing: (1) directly 3D printing functional materials or functional/structural composite materials onto a planar substrate and (2) printing 2D traces or 3D structures on a preformed 3D structure to enable the function.

Applying 3D printing to the fabrication of soft wearable electronics can offer the following advantages: (1) enabling rapid prototyping due to the simple setup and continuous fabrication process; (2) providing low-cost and large-scale fabrication for industrial applications; [3] being compatible with printing of multi-materials to reduce the time and cost; and [4] making flexible hybrid electronics including multi-materials and multi-structures possible. In this chapter, soft wearable electronics fully or partially fabricated by 3D printing techniques will be introduced. Wearable electronics are typically composited of several parts: a soft substrate for wearability, functional materials-enabled electrical components, and highly conductive interconnects.

1.2 3D Printing Techniques Used in Soft Wearable Electronics

1.2.1 Photopolymerization-Based Printing Approaches

Photopolymerization is among the mechanisms of the first-generation 3D printing techniques. Typically, ultraviolet (UV) light is used to solidify the photo-hardening polymer resin locally to pattern the structure layer-by-layer until a final 3D object is accomplished. The exposure of light over the material generates the desired pattern according to the G-code or CAD model. Photopolymerization-based 3D printing techniques, including stereo lithography appearance (SLA), digital light processing (DLP), and continuous direct light processing (CDLP) are representative approaches using the photopolymerization mechanism to construct 3D polymer structures. Materials that can be printed using photopolymerization-based printing approaches are usually photosensitive polymer resins that can be solidified by radiation (e.g., UV or visible light) [29]. The chemical, optical, and rheological properties need to be carefully considered when choosing proper materials for printing. Photosensitive polymers with functional groups such as acrylates, epoxides, thiolenes, and fumarates are commonly used. Acrylate oligomer, UV curable polydimethylsiloxane (PDMS), UV curable polyurethane acrylate, and commercially available photopolymer TangoPlus FLX930 have been 3D-printed as soft substrates or structural materials for soft electronics [30–33].

SLA and DLP are two most widely used photopolymerization-based printing approaches. In the conventional SLA process, a laser source across the polymer surface is utilized for performing the direct laser writing locally, as shown in Fig. 1.1a.

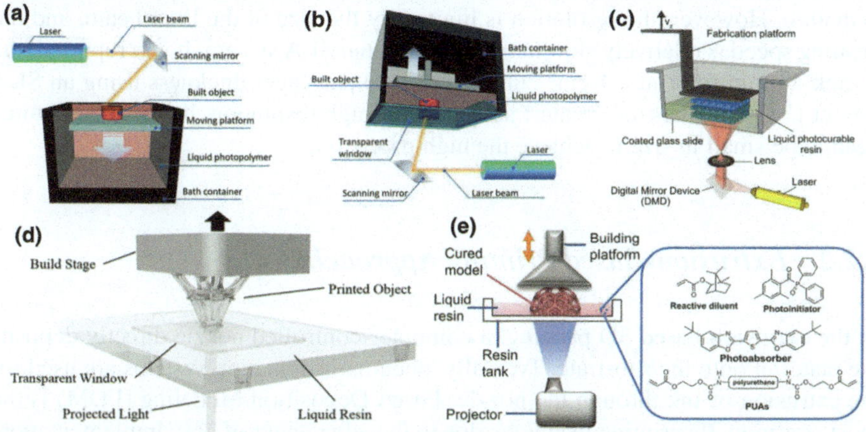

Fig. 1.1 Schematic representations of **a** conventional SLA printing [35], **b** modified SLA printing [35] and **c** DLP printing system [36]. **d** A fast fabrication method of high-resolution features using SLA printing [34]. **e** A DLP printer for the fabrication of a high-resolution 3D solid object and the composition of a photocurable resin [37]

Recently, the projection type SLA printers are developed, where the pattern is formed when the UV light goes through a transparent window and solidifies the resin layer-by-layer, as shown in Fig. 1.1b. In the DLP process, a digital mirror device is used to project the UV light on a whole layer to solidify the design pattern in one shot, as shown in Fig. 1.1c. Attributed to the capability of printing polymer materials, these methods have been used to fabricate the substrate or structural frameworks for wearable electronics in recent years. Zarek et al. demonstrated the fabrication of complex shape memory structures with a viscous melt (≈30 Pa s) using a commercial SLA printer (Picoplus39, Asiga) and a customized heated resin bath. The printing setup is shown in Fig. 1.1c. For each layer, the light source projects a cross-section of the part on a thin layer of resin in contact with the print platform. The platform then exits the photopolymer resin and restarts the process for another layer. Odent et al. used an Ember (Autodesk, Inc.) digital mask projection SLA printer to fabricate ionic conductors with high stretchability (up to 425%) and high toughness (up to 53.5 kJ m^{-3}) [34]. The printer employs a bottom-up process, where the photo-pattern is projected through a transparent, oxygen permeable window at the base of a vat of liquid resin (Fig. 1.1d). Peng et al. printed 3D polyurethane acrylate oligomer (PUA) elastomers structures by a homemade DLP 3D printer, as shown in Fig. 1.1e. Because the DLP printers share a similar mechanism with the video projector, a commercial video projector (1920 × 1080 pixels) with 405 nm UV light was used as the light source and the digital mirror device. The printed elastomer was then combined with the stretchable and conductive hydrogels to fabricate a piezoresistive strain sensor and a wearable finger motion monitoring sensor.

Advantages and drawbacks exist when these techniques are used in fabricating wearable electronics. SLA is capable of printing large-size models and the technique is mature. However, the resolution is limited by the size of the laser beam, and the printing speed is relatively slow in the conventional SLA system. It was reported that it took 44 min to print a 1 cm^3 cube with a 100 μm layer thickness using an SLA printer [38]. DLP has the greatest advantage of high resolution, however, structures need to be small in size to achieve the high precision.

1.2.2 Extrusion-Based Printing Approaches

In the extrusion-based 3D printing, a computer-controlled nozzle directly deposits the material onto the substrate. Typically, pneumatic actuation systems are used for the extrusion of ink through the nozzle. Fused Deposition Modeling (FDM) is the most common 3D printing technique due to the advantages of easy implementation, simplicity, and low costs. In an FDM printing process, the thermoplastic materials are molten in the heated nozzle and deposited on the substrate in a layer-by-layer manner to obtain the desired 3D shape, as shown in Fig. 1.2a. FDM technique has been widely used to fabricate the substrate or structural framework for wearable electronics [39–41]. Liang et al. developed an FDM method to fabricate a mouthguard type wearable oral delivery device, which can deliver a preloaded compound in the oral cavity to treat local oral disease [41]. Poly(L-lactic acid) (PLA) and poly(vinyl alcohol)

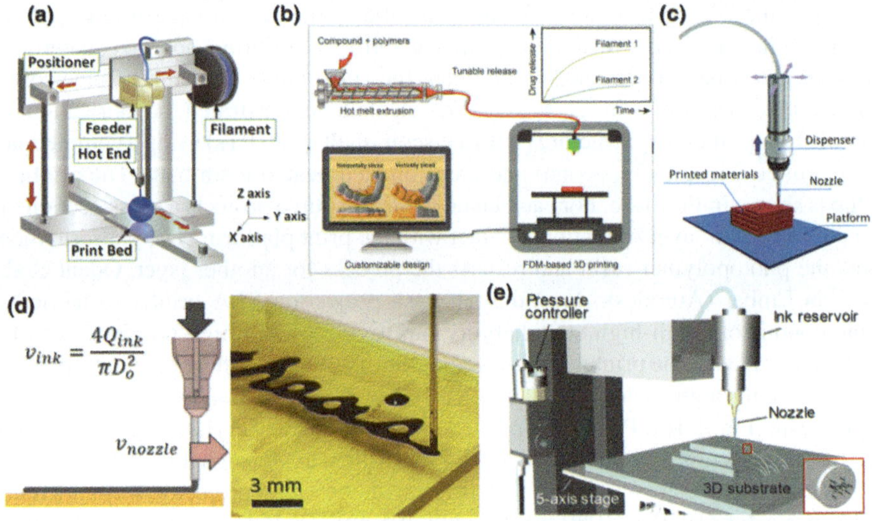

Fig. 1.2 **a** Schematic representation of the FDM printing system [42]. **b** Manufacture process of wearable oral delivery mouthguards using FDM 3D printing [41] **c** Schematic representation of the DIW printing system [35]. **d** Printed letters using CNT ink by the DIW printing technique [43]. **e** Schematic illustration of the DIW system for printing liquid metal/carbon nanotube composites [44]

(PVA) are used as thermoplastic printing materials to construct the device due to their nontoxicity and biocompatibility. The schematic setup is shown in Fig. 1.2b. The manufacture of the oral delivery mouthguards by FDM involved two stages. First, the maxillary anatomy of the subject was scanned and used as the template for 3D printing. Second, the structural materials, PLA and PVA loaded with drugs are hot-melted and extruded into the customizable designed 3D structure based on the scanned template.

Direct ink writing (DIW) is another popular method for 3D printing of both the substrate materials and the functional materials for wearable electronics (Fig. 1.2c) [45–47]. The process of the DIW printing starts with precisely controlling the position of the nozzle following a 3D model and hence the extrusion of fluidic inks to print the designed structures. Unlike the FDM printing, DIW printing does not need a hot nozzle to melt the printing materials for extrusion but requires the ink to be a liquid phase with shear-thinning rheology property and suitable viscosity. Non-Newtonian fluids whose viscosity is a function of the shear rate are the most widely used types of ink for DIW printing. The viscosity of the inks ranges from 10^5 (under a low shear rate of 10^{-2} s^{-1}) to 10^1 Pa s (under a high shear rate of to 10^2 s^{-1}) [46, 48]. Typically, after the inks are extruded and settled down on the substrate, the solidification is achieved by the evaporation of solvents in the ink naturally in the air or by post treatment. Owens et al. used an aqueous carbon nanotube (CNT)-based ink in DIW to fabricate conductive and flexible conductors on polymer and paper substrates [43]. However, the patterns are mostly 2D structures although they used a 3D printer for the fabrication (Fig. 1.2d). Park et al. used a 3D printer for direct writing of liquid metal/CNT composite ink [44]. The printing system and the printed structures are shown in Fig. 1.2e. Benefitting from the oxidation layer outside the liquid metal after printing, the printed structure could be self-standing with a high aspect ratio to maintain a 3D form. The printed 3D traces are demonstrated to be potentially used as interconnects for electronics.

Extrusion-based printing techniques have unique advantages compared to other 3D printing methods. They are more precise in controlling the deposition of materials into 3D structures. The substrates can be versatile. Furthermore, specific techniques such as DIW are compatible with emerging functional materials such as nanomaterials. However, the hot-melting-based FDM methods require thermoplastic polymers as the printing materials; the DIW based methods require the ink to have unique rheological properties, but a clear link between the rheological properties of inks and the device performance is not established. To apply these printing methods, the printing materials and the ink properties need to be tuned carefully to meet the required thermal melting properties, viscosity, and rheological properties.

1.2.3 Powder Bed-Based Printing Approaches

Powder bed-based 3D printing uses powder materials as the feeding materials during the printing. After sintering or binding the designed structure on one layer, another layer of powder will be fed, and the patterning of this layer continues. Typical powder bed-based 3D printing approaches include binder jetting and selective laser sintering (SLS). Based on the design from a digital file, binder jetting uses binder materials to solidify the designed structure and SLS uses laser to fuse the powder into the designed pattern. SLS has been used to fabricate components for wearable electronics, due to its capabilities to print both metal-based materials and polymers. The powders used for the SLS process should be able to absorb energy at the laser wavelength efficiently [29]. When choosing the printed materials, a suitable laser with the corresponding wavelength should be used. Zacharatos et al. studied the printing performance of silver nanoparticles (AgNPs) using SLS (wavelength of the laser: 532 nm) with high printing speed and high repetition rate (the number of pulses that occur per unit time at a particular point), as shown in Fig. 1.3a [49]. High-quality laser sintering of AgNP patterns with sub 100 μm width (Fig. 1.3b) were achieved with the scanning speed as high as 1 m/s and resistivity down to $8.91 \pm 0.9\ \mu\ \Omega$ cm. By tuning the laser pulse width and power, different printing performances such as resistivity of the printed NPs can be tailored. Considering that AgNPs are one of the most widely used conductive materials in soft wearable devices, this approach is promising for fabricating fine conducting paths for wearable electronics.

Considering that the feeding materials are in powders, SLS also shows great capability to print 3D structured polymer/conducting filler composites, in which the feeding materials are polymer powders with conducing materials wrapping these powders [49–52]. Li et al. have developed a process to construct a 3D electrically conductive network in a polymer matrix by SLS, as shown in Fig. 1.3c. The feeding powder was CNT-wrapped thermoplastic polyurethane (TPU) powder. Electrical conductivity of the printed TPU/CNT composite reached 10^{-1} S m^{-1}, which is seven orders of magnitude higher than that of the injected one. The cyclic stretching test of the printed TPU/CNT composite showed a 40% increase after 1000 stretching cycles, while no considerable change was observed after 1000 bending cycles. The same group also printed PDMS/CNT composites using SLS [50]. Covalent adaptable PDMS network with dynamic steric-hindrance pyrazole urea bond was first developed and fabricated into powders wrapped by CNTs, as shown in Fig. 1.3c. Furthermore, the printed PDMS/CNT composite possesses self-healing capability due to the presence of the dynamic reversible bonds.

When applying SLS printing to fabricate components for wearable devices, one of the typical issues that needs to be addressed is that the printing materials need to be manufactured into the powder phase in order to be compatible with conventional SLS printing systems, which makes it difficult to print liquid- or solvent-based materials such as liquid metal and nanomaterials suspended in solvents. To address this issue, a modified SLS was developed to combine the sintering process with an inkjet printing system so that liquid phased inks can be printed by the SLS techniques [53].

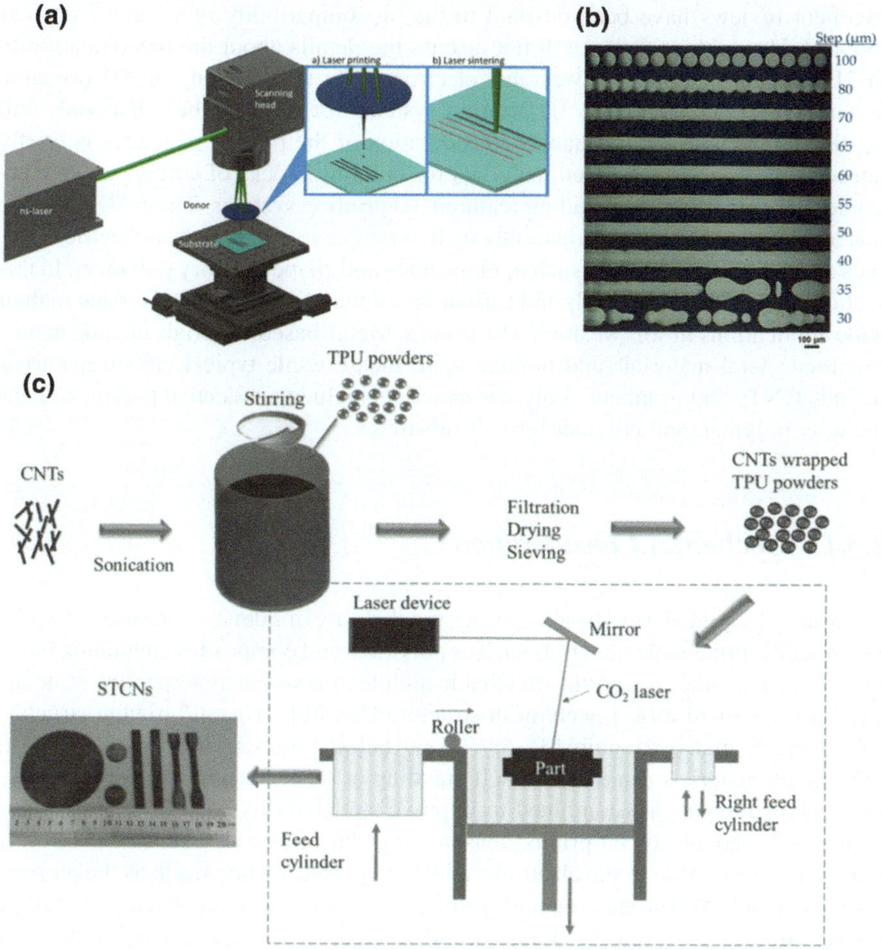

Fig. 1.3 **a** High speed laser printing and laser sintering set-up schematic representation [49]. **b** Microscope images of SLS printed AgNPs. **c** Schematic illustration of the procedures of preparing TPU/CNTs composites by SLS [50]

1.3 3D Printable Materials for Soft Wearable Electronics

Using traditional materials for soft wearable electronics is challenging due to the fact that most metallic and semiconducting materials are lack of flexibility and conformability [54, 55]. However, large strain (up to 100%) needs to be accommodated when being worn on the human body due to the daily motion of humans. To make the 3D-printed electronics deformable for wearable applications, both the printed materials and the structure design are crucial. The biocompatibility of these materials is also of great importance for the practical application of wearable electronics. Some excellent reviews have been devoted to the biocompatibility of wearable devices [56–63]. Hence this section will not discuss the details about the biocompatibility of 3D-printed materials for wearable electronics. In this section, the 3D printable materials and the associated 3D printing systems for printing these materials will be discussed. First, the mechanics consideration of the printed structures is briefly introduced. Then, widely used materials for building blocks of soft wearable electronics are summarized, including featured 3D printing systems. These 3D printable materials include functional materials such as conductive materials, dielectric materials and structural materials such as elastomers and shape memory polymers. In this section, metal-based materials and carbon-based materials are discussed due to their wide applications in soft wearable electronics. Metal-based materials include nanostructured metal materials and molten/liquid metal, while typical carbon materials include CNTs and graphene. Polymer materials include conductive polymer, shape memory polymer, and polymer for soft substrates.

1.3.1 Mechanics Consideration

There are two typical ways to achieve the stretchability of a device. One way is to print intrinsically stretchable materials such as polymers and composites containing nanomaterials. The other is to print stretchable architectures such as serpentine structure [64], wavy structure [47], web/mesh structure [65, 66], origami/kirigami structure [67], scaffold/matrix structure [37, 68–70] and helix/spiral structure [64, 71, 72]. The 3D-printed materials can either be self-standing or on a stretchable substrate. Often these two strategies are combined to improve stretchability/compressibility. Wang et al. developed a fully 3D-printed sensor using elastomer as the substrate, TPU/Ag nanoflakes composite as the electrode, and TPU/porous carbon black as the piezoresistive sensor [73]. The electrodes are printed into a helix structure to reduce the strain interference, as shown in Fig. 1.4a. The electrode layer with an 85% Ag content can be stretched up to 120% without losing conductivity; the resistance changes only 7% when the electrode is stretched by 50% strain, attributed to the inherent stretchability of TPU and the helix structure. Wong et al. developed a 3D printable shear-thinning ionogel with ionic conductivity [74]. By printing the ionogel into auxetic geometry (as shown in Fig. 1.4b), the fabricated sensor exhibits 310% higher stretchability compared to a continuous bulk film. The extremely high stretchability is due to the auxetic structure that exhibits in-plane and out-of-plane deformation to reduce the local stress during tensile stretching. Wei et al. printed 3D wavy structured conductors using PDMS/CNT composite ink [47]. The authors studied the effect of joining angles (θ) on the stretchability of the wavy conductor (shown in Fig. 1.4c, d), and found that a 45° joining angle results in the best mechanical stretchability. With the help of the wavy structure, the 3D-printed wavy structure can be stretched up to 315%; the resistance change is only 5% when stretched by 100% strain. These structural design strategies for improving the stretchability of the printed materials present a great opportunity for 3D printing due to its facile manufacturing process for these

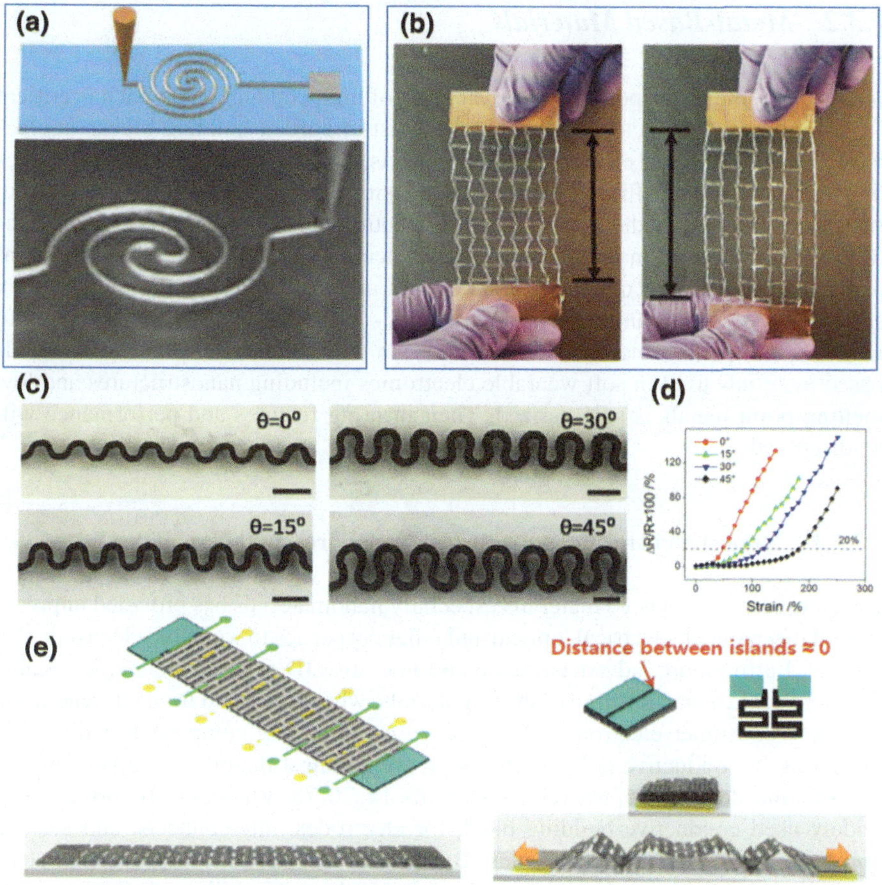

Fig. 1.4 **a** Helix structured TPU/Ag nanoflake electrode [73]. **b** The printed and crosslinked auxetic structure expands in the direction perpendicularly to the loading direction [74]. **c** The cross-sectional images of the 3D-printed wavy electrodes with different joining angles. Scale bar 5 mm [47]. **d** Relative resistance change versus applied strain on the electrodes with different joining angles [47]. The illustration of combining kirigami and origami design for minimizing the device size and enhancing the stretchability [67]

complex structures, which may not be applicable to other fabrication approaches. Jo et al. demonstrated a 3D printing-assisted approach to print TPU structures on a AgNW network film by FDM printing to fabricate AgNWs/TPU composite structures [67]. The printed structure can be used as stretchable conductor. To further enhance the stretchability, kirigami and origami structures were employed, which provides high stretchability and efficient space utilization, as shown in Fig. 1.4e.

1.3.2 Metal-Based Materials

Metal-based materials possess the advantages of high conductivity, which is critical for electronics. Conductors are one of the most important building blocks in electronic systems to act as electrodes, interconnects, and sensing parts. Higher conductivity leads to more efficient signal transmission and lower power consumption, which are the key to the performance of electronic systems. However, manufacturing of metal-based materials for wearable electronics needs further improvement to meet the requirement of low costs, scalability, and capability of patterning complex structures. Additive manufacturing approaches such as 3D printing are promising to fabricate and pattern metal-based materials. In this section, 3D printing of metal-based materials used in soft wearable electronics including nanostructures and low melting-point metals is summarized. Their printing features and performance will be discussed.

1.3.2.1 Nanostructures (Nanoparticles, Nanowires, and Nanoflakes)

Development of advanced materials especially nanomaterials has provided unprecedented mechanical, electrical, optical and other properties to wearable electronics [5, 75–79]. Furthermore, advanced facile and low-cost 3D printing techniques enable the large-scale manufacturing of these materials, which is critical for next-generation wearable consumer electronics. Conductive nanomaterial composites, with either insulating or conductive polymer matrices, are amongst the most promising candidates for the electrodes of wearable sensors [5, 14, 76, 78, 80]. Metal NPs are the most widely used conductive building block for electrodes, interconnects, and sensing components in wearable electronics. By mixing the NPs into a polymer matrix, moderate electrical conductivity and mechanical stretchability can be achieved attributed to the percolation structure of NP conductive traces. With proper mechanical design for the printed structure, the electrical and mechanical properties can be enhanced greatly. For example, Zhang et. al. printed serpentine AgNP structures and compared them with straight lines in terms of stretchability. They found that the serpentine structure with an arc diameter of 1600 μm exhibits much higher stretchability (>25% strain with resistance change ~10 folds) than the straight line (<2% strain with resistance change ~ 10^{10} folds) [81]. Among all metals, silver has the highest thermal conductivity of 427 W m^{-1} K^{-1} at 298 K and electrical conductivity of 62.9×10^6 S m^{-1} at 293 K [82]. Commercially available single element metallic NP inks contain AgNPs/flakes are the most commonly used for 3D printing. These inks typically mix AgNPs with polymer additives to adjust the rheology properties and the viscosity to adapt to 3D printers such as inkjet and aerosol jet printers. Specific rheology enables the ink to transit between fluidic behavior and solid behavior, which ensures the 3D shape retention after printing due to the increased rigidity of ink coming from the shear thinning effect of the polymer. After the printing of AgNP patterns, sintering with a temperature ranging from 100 to 300 °C is usually needed

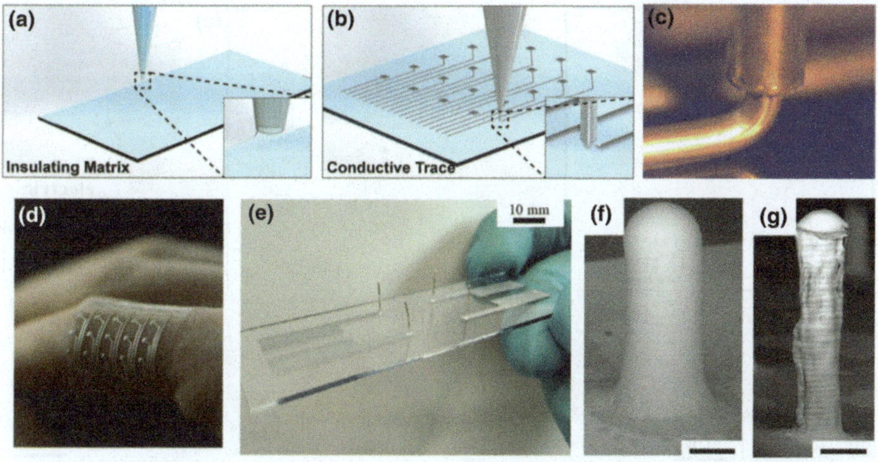

Fig. 1.5 **a** Direct ink writing of insulating matrix for the device body and B) conductive Ag/TPU traces for the sensing part [86]. **c** Image of direct writing of Ag/TPU ink through a 200 μm nozzle (scale bar 200 μm) [86]. **d** Image of a soft sensor array [86]. **e** Four printed and sintered silver pillars using the silver NP ink [87]. **f** A printed pillar with 48 layers when as-printed (not sintered) and **g** after sintered [87]

to enhance the conductivity and robustness [83–85]. Valentine et al. used DIW-based 3D printing to print both the insulating layer and the conductive layer to fabricate sensors and interconnects in a wearable electronics system, as shown in Fig. 1.5a, b [86]. The conductive ink in this work is the composite of Ag flakes and TPU, as shown in Fig. 1.5c. After the printing and the evaporation of the solvent, conductive electrodes made of a percolating network of silver flakes with a linewidth as small as 100 μm can be realized, as shown in Fig. 1.5d. Sowade et al. printed vertical 3D multilayer pillars as interconnectors by inkjet printing to achieve vertical integration of the circuit [87]. The maximum aspect ratio of the 3D-printed pillars can be up to 50:1, as shown in Fig. 1.5e. The as-printed patterns need post-process to dry and sinter the NPs, which will cause shrinkage of the pillars, as seen in Fig. 1.5f, g.

Metal nanowires (NWs) such as AgNWs [88–90], AuNWs [91], and CuNWs [92] are frequently used in electrodes for wearable electronics. Compared to NPs, NWs are more effective in forming conductive pathways according to the percolation theory [89, 93]. These percolated conductive pathways can maintain the electrical stability when the electronics are deformed [94–96]. However, 3D printing of 1D nanomaterials such as NWs is more challenging than NPs due to the severer agglomeration problem. The 3D printing technique that is applicable to NWs is extrusive-based printing. It requires the inks to have proper rheological properties and viscosity to achieve printability. The viscosity of a printable ink for extrusive-based printing such as inkjet printing, aerosol jet printing, and EHD printing ranges from 10^{-3} Pa·s to 10^{1} Pa·s. A large number of polymer additives need to be added to the ink to stabilize the NWs (i.e., reducing the agglomeration) and to increase the viscosity, which

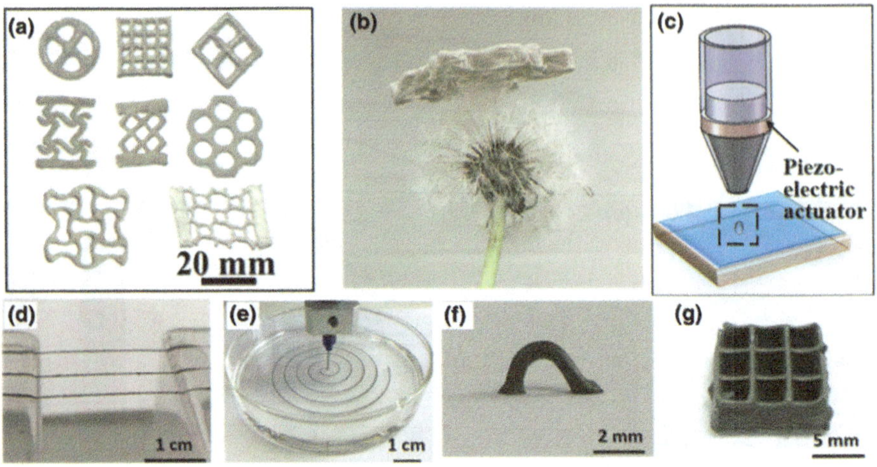

Fig. 1.6 **a** 3D-printed silver NW aerogels architectures [97]. **b** Printed silver NW aerogels standing on a dandelion [97]. **c** 3D printing setup for silver NW aerogels printing [97]. Photographs of **d** suspended microfibers, **e** coiled microfiber, **f** arch-like structure **e** and **g** scaffold structure printed using the NW gel [98]

may need harsh post-treatment including high temperature and repeated washing to remove the additives and improve the conductivity.

3D-printed AgNW aerogel was developed by Yan et al., allowing printing of ultralight rasterization structures, as shown in Fig. 1.6a, b [97]. In the printing system, an inkjet printer in a drop-on-demand mode with a piezoelectric actuator, as shown in Fig. 1.6c, is integrated with a freeze casting environment, which enables the in-situ freezing solidification of printed structures. The temperature of the freezing process highly influences the microstructure of the aerogel, where −10 to −50 °C is a proper temperature range to form tunable, uniformly thin AgNW compartments. Then, a further freezing process under a temperature of −70 °C is carried out to remove the water in the ink by sublimation. The printed AgNW aerogel is cross-linked by PVA additive in the ink, which contributes to the stable mechanical properties during bending, while sacrifices the conductivity of the printed structure. The aerogel AgNW with a density of 1.3 mg cm^{-3} only exhibited a conductivity of 1.3 S cm^{-1}, which is orders of magnitudes lower than that of the bulk silver. Instead of using polymer to crosslink AgNWs for reliable structure retention, Liu et al. developed a general strategy to make NW inks printable by gelling the NWs by 2D nanosheets including single-layer graphene oxide (GO) and titanium carbide $Ti_3C_2T_x$ (MXene) [98]. AgNW, manganese oxide (MnO_2) NWs, and zinc oxide (ZnO) NWs are successfully gelled with sufficiently high viscosity for 3D printing. Several printed structures including suspended microfibers, coiled microfiber, arch structure, and matrix structure are demonstrated, as shown in Fig. 1.6d–g. With this strategy, the conductivity of the self-standing structure can be improved to 10,000 S cm^{-1}.

1.3.2.2 Molten Metal and Liquid Metal

To achieve the high electrical conductivity needed for wearable electronics, 3D printing of molten metal with a low melting point [99] (e.g., Field's alloy and Wood's alloy) and liquid metal with a melting point at room temperature [100] (e.g., EGaIn) are emerging. Field's alloy has a melting point of 62 °C and Wood's alloy has a melting point of 70 °C. With a heater included in the 3D printing system, these alloys can be melted in the nozzle and printed out into the ambient environment, and then solidify at room temperature. Han et al. used an electrohydrodynamic (EHD) printing system (Fig. 1.7a) to successfully print molten metals (e.g., Field's metal, Wood's metal, and solder) with 3D structures [101]. The resolution can be as low as 50 μm. This is attributed to the Taylor cone formed at the tip of the nozzle under the electric field, which makes the diameter of the flow during printing much smaller than that of the nozzle. Metal trace, two-layer scaffold structure, vertical metal wire, thin wall, and metal bridge structure are printed to demonstrate the promising application of 3D fabrication for wearable electronics, as shown in Fig. 1.7b. Furthermore, the molten metal-based conductive traces encapsulated in an elastomer show self-healing properties by simply heating the device to the melting temperature of the molten alloy, as illustrated in Fig. 1.7c [99]. The required temperature for self-healing is as low as 75 °C so that they would not result in damages to the other components in the devices. However, molten metals are solid at room temperature, meaning that the

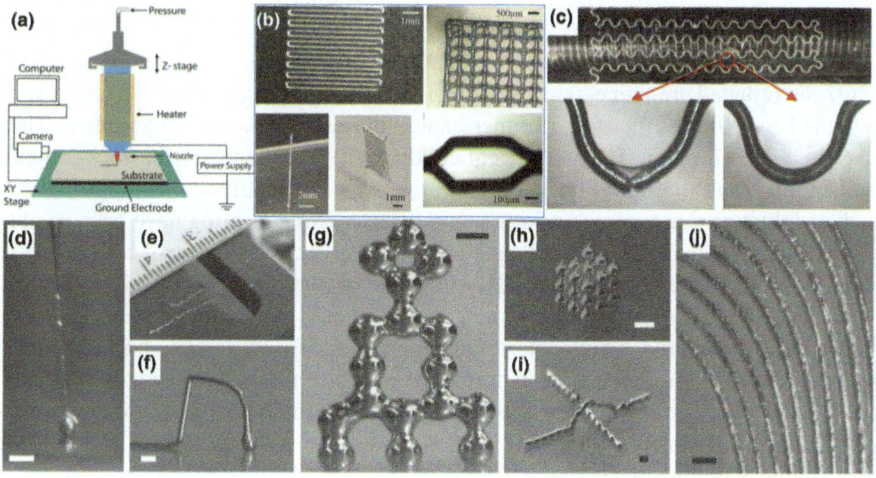

Fig. 1.7 **a** Schematic of the EHD printing system with a heater. **b** EHD printed molten metal 3D structures [101]. **c** Optical images of a local failure before and after healing [102]. The optical images of 3D-printed liquid metal with **d** thin wire shape, e free-standing wires between the gaps, **f** a free-standing liquid metal arch, **g** a tower of liquid metal droplets, **h** a 3D cubic array of stacked droplets, **i** a metal wire and an arch composed of liquid metal droplets, and **j** an array of in-plane lines. Scale bars 500 μm [45]

printed conductive traces are not stretchable. To fabricate stretchable wearable electronics, the printed patterns need to be carefully designed by a structural strategy so that the strain could be accommodated by the stretchable structures.

On the other hand, liquid metals with lower melting points remain at a liquid phase at room temperature, making them strain insensitive. So, the mechanical properties depend on the substrate material or encapsulation material but not the liquid metal itself. By choosing the proper stretchable substrate and encapsulation layer, the whole device can be stretchable. Liquid metal has been widely used as electrodes or interconnects for wearable electronics. However, facile and scalable fabrication is still needed. 3D printing technology has proven feasible to pattern free-standing 3D structured liquid metal at room temperature [45, 103]. A passive oxidation layer forms on the surface of the liquid metal pattern when the ink is exposed to air, which is key to maintaining the printed structure. This oxidation layer would not affect the conductivity of the liquid metal structure due to its thin feature on the scale of 1 nm. Extrusion, fast expelling, stacking, and sacrificial microchannel are four typical ways to 3D print liquid metal. Structures including self-standing vertical wires, stacked droplets, and bridges have been printed successfully using these methods, as shown in Fig. 1.7d–i. By controlling the diameter of the printing nozzle and the printing speed, the resolution of the printed trace can be controlled, as low as <2 μm [45].

1.3.3 Carbon-Based Materials

Carbon-based materials such as CNTs [22, 104–109], graphene (i.e. graphene oxide (GO) and reduced graphene oxide (rGO)) [110–114], graphite [115, 116], and carbon black [117–119] are emerging as functional materials in wearable electronics. They possess outstanding merits in electrical conductivity, mechanical robustness, chemical stability, and thermal properties, and are easy to be modified. Printing methods such as inkjet printing, screen printing, and direct writing printing have been used to print 2D patterns using these carbon materials. However, 3D printing of carbon-based materials is still challenging especially in terms of electrical conductivity, because typically polymer additives need to be added to the ink to maintain the structure of the printed 3D patterns, which deteriorates their conductivity. How to print stable 3D structures with high-concentration conductive carbon fillers is an urgent issue that needs to be addressed. In this section, 3D printing techniques for fabricating wearable electronics using representative carbon materials will be reviewed. The printing performances will be discussed.

1.3.3.1 CNTs

CNTs have been widely used in soft wearable electronics due to their unique 1D microstructure. Their high conductivity and robust mechanical properties make them emerge as a promising conductor or semiconductor material for soft wearable devices. The elastic modulus and strength of CNTs are on the order of 1.0 TPa and 50 GPa, respectively [120, 121]. Similar to metal NWs, the random or aligned percolation network formed by CNTs can provide robust conductive paths, even when stretched. This makes CNTs a promising candidate as the functional parts in wearable electronics where stretchability is needed. The electrical conductivity of single-walled CNTs can be as high as 10^6 S cm^{-1}, while it is only on the order of 10^4 S cm^{-1} for multi-walled CNTs [122–127]. This level of conductivity is usually several orders of magnitude lower than that of metals; it is desirable to increase the conductivity of the printed CNT structures in order to have them as substitutes for metal electrodes.

CNTs are usually fabricated into conductive traces, bridges, matrices, and other structures to act as conductors for the electrodes and interconnect in wearable electronics [104, 105, 128–132]. 3D printing technology provides a way to improve the printability of complex 3D structures used in emerging wearable electronics, further expanding the application of these devices in the area of healthcare [22, 107], human–machine interface [133, 134], and body motion detection [106]. To print 3D-structured CNTs instead of 2D patterns, a variety of polymer materials such as polyvinylpyrrolidone (PVP) [129], PVA [135], poly(glycerol sebacate) (PGS), photopolymer [33], cellulose [105], and elastomer [47, 68, 136] have been used as the matrix or dispersion agent. These additives can improve the printability and structural stability of the CNTs for 3D printing but decrease the conductivity of the printed structure. The most straightforward method to improve the conductivity of printed CNTs/polymer composite is to remove the polymer and the solvent in the printed patterns via post-processing. Kim et al. used PVP to disperse multi-walled CNTs in water uniformly to achieve a no-back-pressure 3D printing of complex conductive structures, including freestanding pillars, bridges, and wall architectures, as shown in Fig. 1.8a–c [129]. The conductivity is improved by a thermal annealing process with a temperature of 450 °C to remove the PVP. The printing is enabled by glass micronozzles with diameters from 8 to 30 μm, and the extrusion of the ink is achieved by capillary force instead of back pressure because of the low viscosity of the ink. Accurate three-axis stepping motors with a resolution of 250 nm are used to control the printing process, which allows the printer to print structures not only on planar but also on the curved surface, as shown in Fig. 1.8d. They demonstrated a gas sensor, a point emitter, and an RF inductor printed by this system with 3D structures, which can be potential used as components in wearable electronics.

While this thermal annealing process can enhance the conductivity of the printed patterns, the conductivity of the above CNT structures can only achieve ~2540 S m^{-1} with 75% CNTs content, which is much lower than metal-based materials and the pure CNTs. Furthermore, the high temperature used to remove the polymer may destroy the other components in the soft devices such as polymer substrates. The improvement of conductivity of the 3D-printed CNTs still needs to be addressed for the large-scale applications of 3D CNT structures as conductors in wearable electronics. Nevertheless, 3D-printed CNT structures can be widely used in scenarios

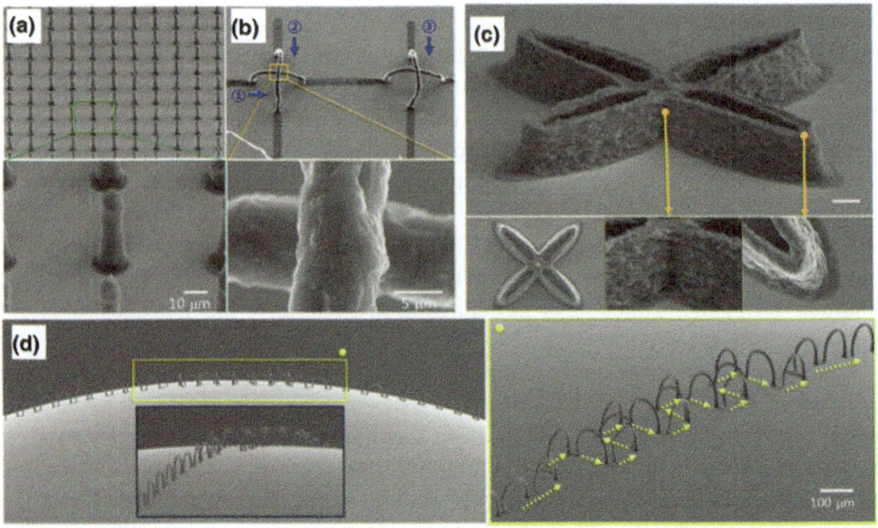

Fig. 1.8 **a** FE-SEM image of an array of free-standing CNT pillars. **b** FE-SEM image of a suspended junction pattern consisting of CNT bridges and straight lines. **c** 3D cross-shaped wall structure. All scale bars are 20 μm. **d** FE-SEM images of a concatenated CNT bridge-line structure on a curved glass substrate with an 8.4 mm curvature radius [129]

that do not request very high conductivity, such as pressure sensors [33, 68], strain sensors [136], supercapacitors [135], and triboelectric nanogenerators [137].

1.3.3.2 Graphene-Based Materials

3D printing of graphene-based materials such as GO and rGO has been attracting more and more attention recently, attributed to the increasing exploration of the applications of graphene in various areas. Soft wearable electronics is one of the most promising application that may take advantage of the excellent electrical, mechanical, and thermal properties of grapheme [138, 139]. As a typical 2D material, graphene can form percolation networks for the conductor or semiconductor building blocks in wearable electronics, such as sensors [140, 141]. Interconnects [142]. Transistors [143, 144], optical devices [145, 146], energy devices [147], etc. The development of 3D printing technology makes the manufacturing of customized structures possible for graphene-based electronics, facilitating the application of 3D-printed graphene in wearable electronics.

Similar to other nanomaterials, one common strategy to fabricate graphene-based electrodes is to mix graphene materials with a polymer matrix and print desired structures with direct writing-based 3D printing methods. By mixing graphene with PDMS elastomer with controlled weight ratio to adjust the rheological properties, a 3D printable ink was developed to fabricate scaffold structure for ultrasensitive soft strain sensors, as shown in Fig. 1.9a, b [69]. Higher graphene ratio leads to high viscosity, which may make the printer hard to extrude the ink, while the 3D structure cannot be maintained well after printing if the graphene ratio is too low. Apart from the application for wearable soft sensors, 3D-printed graphene scaffold was also used as bioelectronics for tissue regenerative engineering [148]. The device was fabricated by extrusion-based layer-by-layer printing of graphene/polylactide-co-glycolide (PLG) composite ink, which is conductive, mechanically robust, and biocompatible, as shown in Fig. 1.9c. Different annealing temperatures after the printing were proven to have a significant influence on the mechanical properties of the printed structure. With the increase in the annealing temperature, the printed scaffold becomes more

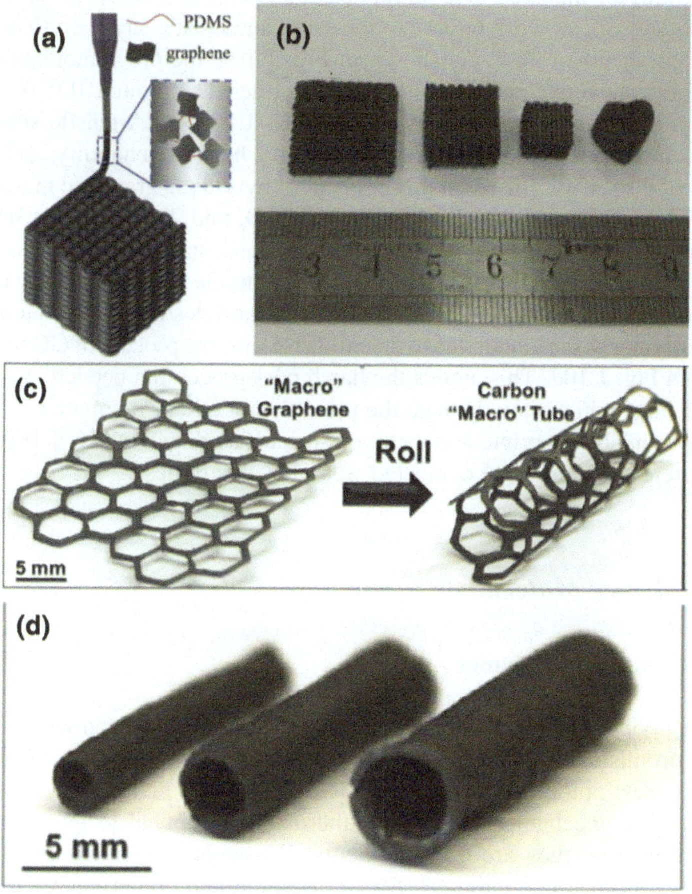

Fig. 1.9 **a** Schematic showing preparation of 3D graphene-PDMS. **b** Various 3D graphene-PDMS samples with different shapes and sizes [69]. **c** Photographs depicting that thin three-dimensional graphene-PDMS sheets are flexible and can be rolled into more complex 3D forms. **d** Various sizes that could serve as custom-sized nerve graft conduits [148]

and more brittle and undergoes brittle failure upon compression when the temperature exceeds 150 °C, due to the decomposition of PLG. In vitro and in vivo experiments found that the printed graphene scaffold can support tissue regeneration without an inflammatory response. The high programmability and resolution (~100 μm) also make it potentially applicable for precision surgical procedures, as demonstrated in Fig. 1.9d.

Graphene-based materials are also widely used as the building block of hybrid inks to enhance the mechanical, electrical, and thermal properties. Commonly used materials that are incorporated with graphene materials for 3D printable inks include conducting NPs [139, 149]. dielectric particles [150], CNTs [139], 2D MoS_2[139], and so on. For example, GO NPs were added into NP inks with PU as the binder for 3D printing of piezoresistive strain sensors [149]. Attributed to the 2D nature of GO, this sensor can work under compression and tensile strain up to 160% with a high gauge factor of 48.2. As shown in Fig. 1.10a, the GO nanoplate networks can maintain conductive paths under large stretching. The printed 3D sensor is used to detect pressure from touching of fingers (Fig. 1.10b, c). A tensile strain sensor was also demonstrated to show the high sensitivity, high stretchability, and excellent durability of the sensor. Based on this concept of hybrid 2D graphene materials with other low-dimensional materials including 0D, 1D, and 2D materials, 3D printing of GO-based aerogel hybrid with mixed-dimensional nanomaterials was achieved using a DIW printing method [139]. Benefitted from the unique structure of 2D GO nanosheets, these inks composited of GO and nanoparticles/nanotubes/nanosheets do not need polymer as the binder due to the mild crosslinking process of GO nanosheets, as shown in Fig. 1.10d. This avoids the harsh post-processing needed in most other composite inks, which may damage the printed structure or the matrix. The hybrid composites can also be printed on a curved surface, as shown in Fig. 1.10e, which is of great interest to conformal contacted wearable electronics.

1.3.4 Polymer Materials

1.3.4.1 Conductive Polymer

Conducting polymers are intrinsically conductive. They have been widely used in a variety of promising applications including wearable electronics, biosensors, energy storage devices, optoelectronics, etc. [151–153]. These materials possess unique properties (e.g., mechanical, electrical, thermal, and optical properties) of polymers. Furthermore, they are usually chemically stable and biocompatible, which makes them promising candidates for implantable and epidermal wearable electronics. It is of great importance to fabricate conductive polymers into diverse structures to meet the mechanical and electrical requirements for different applications. Conventional manufacturing of conductive polymers mainly includes drop casting, spin coating, ink-jet printing, screen printing, aerosol printing, and lithography. However, limitations exist when using these techniques to fabricate complex

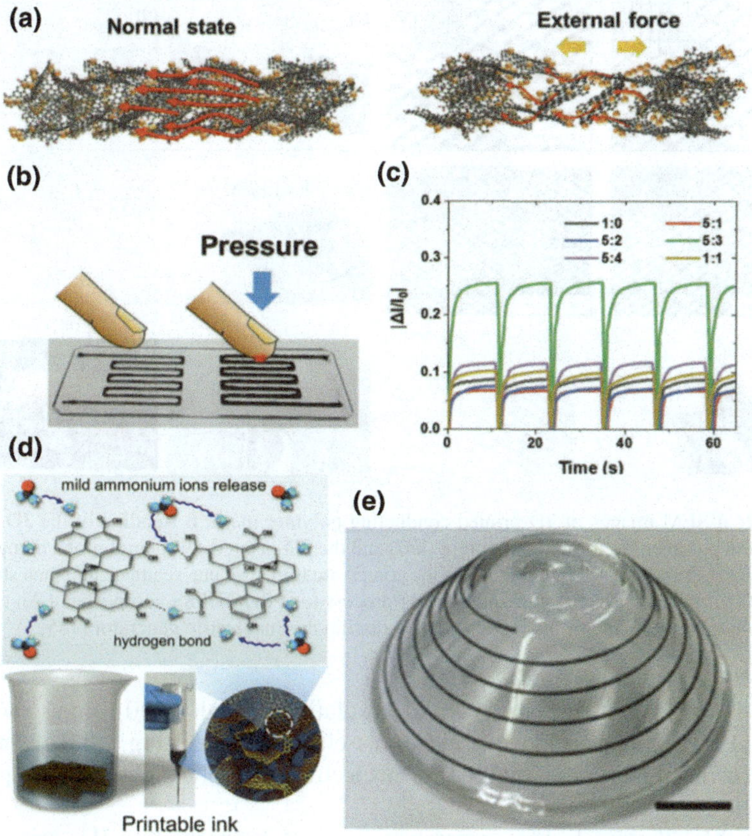

Fig. 1.10 **a** Schematic illustration of the sensing mechanism for pressure sensor fabricated by GO and NPs. **b** Schematic showing pressure sensing. **c** The current change when the pressure sensor is pressed with various weight ratios between GO and NPs [149]. **d** Schematic diagram of printable cross-linked graphene ink. **e** 3D-printed hemispherical spiral using the GO/Ag hybrid ink [139]

structures with high resolution, high aspect ratio, or curved surfaces. 3D printing techniques can potentially address these issues. Poly [3, 4-ethylenedioxythiophene]: polystyrene sulfonate (PEDOT:PSS), as one of the most widely used conductive polymers, has been used in 3D printing to fabricate wearable optoelectronic [154], thermoelectric [155], and bioelectronic [154] devices. To realize the printability of PEDOT:PSS, Yuk et al. developed a high-performance 3D printable ink [154]. The paste-based conducting polymer ink is prepared by cryogenic freezing of aqueous PEDOT:PSS solution followed by lyophilization and controlled re-dispersion in water and dimethyl sulfoxide (DMSO) mixture. To make the ink with suitable rheological properties, relatively high concentration PEDOT:PSS (5–7 wt%) was added to the solvent to transform the ink from low viscosity liquid to gels with the shear-thinning property. Conducting meshes fabricated by multilayer printing

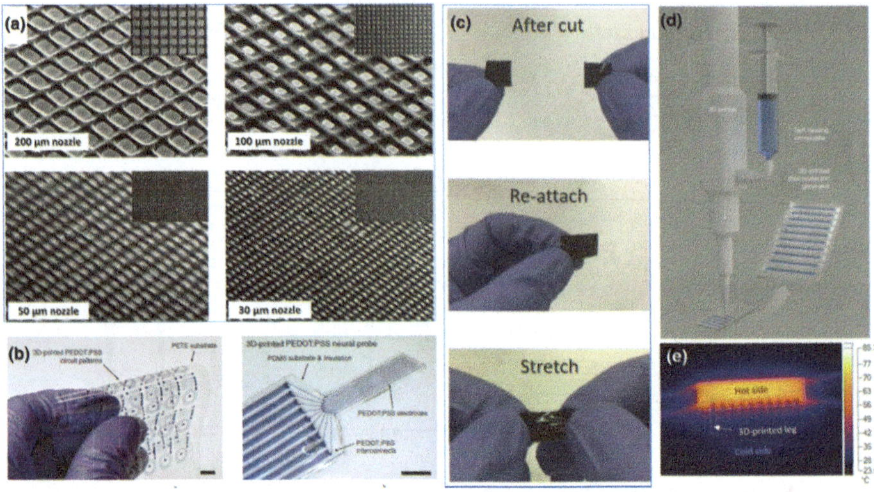

Fig. 1.11 **a** SEM images of 3D-printed conducting polymer mesh. **b** Bending of the 3D-printed conducting polymer circuit without failure (left) and the 3D-printed soft neural probe in magnified view (right) [154]. **c** Photographs showing the composite film after cut, reattach, and then stretch. **d** Schematic of the 3D-printer setup and printing process for PEDOT:PSS composite. **e** An example of infrared thermal image of the ten-leg composite film thermoelectric generator [155]

of PEDOT:PSS was demonstrated with resolution as high as 30 μm, as shown in Fig. 1.11a. High-density flexible electronic circuit patterns (Fig. 1.11b) were 3D-printed using this ink and applied to bioelectronics applications as an implanted probe.

Wearable thermoelectric generators were fabricated by 3D printing of PEDOT:PSS composite to harvest low-grade heat from the human body and convert it into electricity [155]. With polymeric surfactant Triton X-100 added as a healing agent, the printed composite structure can not only bear a 35% strain without decrease in electrical conductivity and Seebeck coefficient, but also self-heal after being cut, as shown in Fig. 1.11c. A ten-leg composite film electrically connected in series and thermally in parallel was printed to investigate the power generation of the thermoelectric generator (Fig. 1.11d and e). The thermoelectric generator provides a maximum output power of 12.2 nW and can retain over 85% power output after self-healing from a mechanical cut. The 3D printing technique makes the fabrication of composite structures all solution-based and facile for large-scale manufacturing.

Apart from printing 3D structures directly, another category of 3D printing is printing patterns or structures on a 3D surface such as concave/convex hemispherical surface and step-shaped surface. Polymer photodetector arrays were successfully printed on both the planar surface and the inner face of a hemispherical surface, as shown in Fig. 1.12a and b [156]. The single photodetector is fabricated by 3D printing of different materials including poly [3-hexylthiophene] (P3HT) photoactive layer, PEDOT:PSS transparent anode, EGaIn liquid metal cathode, and AgNP interconnect layer-by-layer (Fig. 1.12c). The one-pot 3D printing platform makes the fabrication

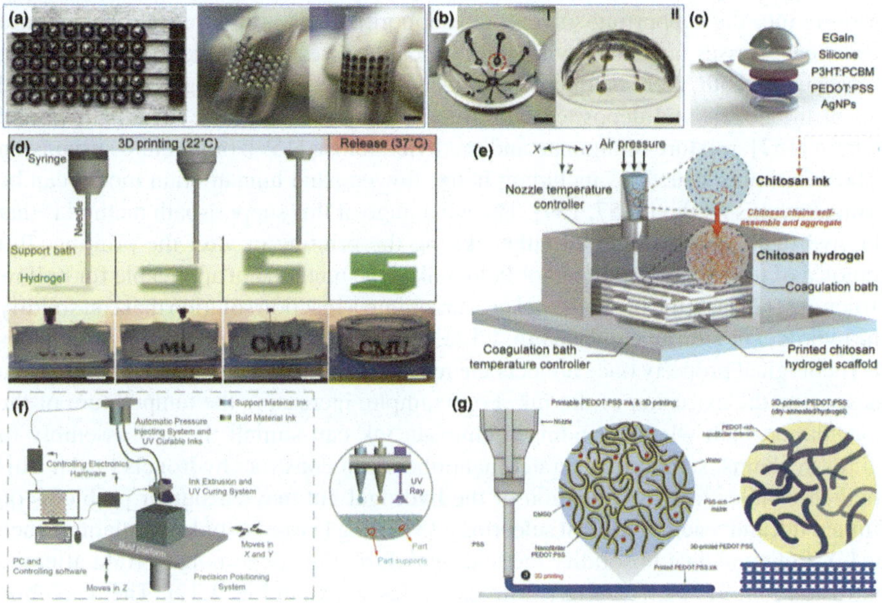

Fig. 1.12 **a** photodetector array printed on PET when flat (left) and bent (right). **b** Photographs of the photodetector array printed onto the inner surface of a hemispherical glass dome. **c** Schematic diagram of the structure of the photodetector [156]. **d** Schematic of printing the hydrogel precursor into a supporting solvent [157]. **e** Schematic of printing the hydrogel precursor into a reaction solution that can solidify the hydrogel in-situ [161]. **f** Schematic of 3D printing process of UV curable ink [30]. **g** Schematic of the set up for directly printing PEDOT:PSS hydrogel without external assistance [154]

of multi-material systems more efficient and facile. Different nozzles were used on the same printer to print different inks, with proper printing parameters, for layer-by-layer fabrication of the photodetectors. The capability of printing photodetectors on a hemispherical surface makes it possible to be integrated into a "bionic eye" vision system.

Besides the intrinsically electrical conductive polymers, hydrogels, a type of ionic conducting polymers consisting of a large amount of water, salt ions, and cross-linked polymer frameworks, are emerging for applications in soft wearable electronics. 3D printing of these hydrogels enables fast prototyping of soft biomedical devices that can be mounted on the skin or inside the body for applications like portable patient monitoring, wound treatment, and tissue/organ engineering. Due to the solidification requirement of the hydrogel during or after printing, there are generally four strategies for printing hydrogel structures. (1) Printing the hydrogel precursor into a supporting solvent such as hydrophobic fluids, thermoreversible fluids, and thixotropic materials, or supporting field such as electric field, to fix the shape of the hydrogel precursor first [157, 158]. After the printing, the hydrogel precursor is solidified by stimulation and the supporting solvents or field can be removed, as shown in Fig. 1.12d. When printing into the supporting solvent, the support bath should have rigid behavior at low shear stresses but can flow at higher shear stresses. When the extrusion nozzle moves in the bath, there is little mechanical resistance; the hydrogel being extruded out of the nozzle and deposited within the bath is held in place. The bath can be gelatin [157]. mixture of alginate and methylcellulose [159]. or laponite suspension [160]. Complex structures including helix, flowers, and human brain model can be printed by this method [157, 159]. The advantage of this support-bath method is that the hydrogel can be solidified either during the printing or after the printing. But removal of the supporting liquid or field makes the method not applicable for hollow structures. (2) Printing the hydrogel precursor into a reaction solution that can solidify the hydrogel in-situ, as shown in Fig. 1.12e. [161] The hydrogel precursor ink changes its rheological property once it meets the reaction solution so that the structures form as soon as the extrusion of the ink. For example, increasing the temperature of the coagulation bath when extruding a chitosan ink can stimuli the self-assembly of chitosan chains, so that regenerated nanofibers can construct hydrogels [161]. This method enables fast solidification of the hydrogel but may encounter problems on limited ink and reaction solvent selections. (3) Using fast cross-link stimulations such as UV light and heat to solidify the hydrogel during the patterning or right after the extrusion [30, 31, 34, 162–164], with examples shown in Fig. 1.12f. The resolution and quality of the printed structure highly rely on the curing speed. With fast curing, resolution of the printed features can reach 140 μm [162]. More complex structures with the shape of snowflake and butterfly can be printed. But these procedures usually have the problem of clogging nozzles when being used in a direct writing printing system. (4) Directly printing the self-standing hydrogel structures without external assistance [64, 65, 154, 165]. such as the process shown in Fig. 1.12g. For example, PAAm-based hydrogels were modified by Carbomer to enable self-standing after printing, where Carbomer acted as a rheology modifier in the ink [65]. This method can address the issue of structure fidelity and nozzle clogging. But challenges have risen in developing hydrogel inks and tuning printing parameters.

1.3.4.2 Shape Memory Polymer

Shape memory polymers (SMPs) are smart stimuli-responsive polymeric materials that can memorize permanent shapes and recover from temporary shapes to permanent shapes under external stimuli, such as temperature, electric potential, light, etc. [166]. The thermally induced shape memory behavior requires two critical temperature: a lower temperature where chemical or physical crosslinks occurs to set the permanent 3D shape, and a transition temperature to control the molecular switching segments that fix the temporary shape. When the material is deformed above the transition temperature, the temporary shape is obtained. Then the material will recover to its permanent shape when the temperature is reduced below the transition temperature [38]. When the SMPs are used as the functional materials in soft wearable electronics, they not only satisfy the material requirements of conventional soft electronics, but also bring some special features that traditional polymers do not possess, such as shape memory effect, self-healing capability, and variable elastic modulus. These unique properties along with the high stretchability and low cost of SMPs make them more and more popular in the application of wearable electronics. 3D printing of SMPs further benefits the application of the materials in various fields by reducing the fabrication steps/time and increasing the programmability of the structure.

One of the widely used SMPs is polycaprolactone (PCL) based semicrystalline polymer. Zarek et al. developed a general 3D printing method that can print PCL macromonomer through stereolithography 3D digital light processing (DLP) printer with UV light as the solidification stimuli [38]. Attributed to the shape memory property of the printed structure, the system was used to print a 3D structural electric circuit with the shape memory parts as switches. As shown in Fig. 1.13a, the

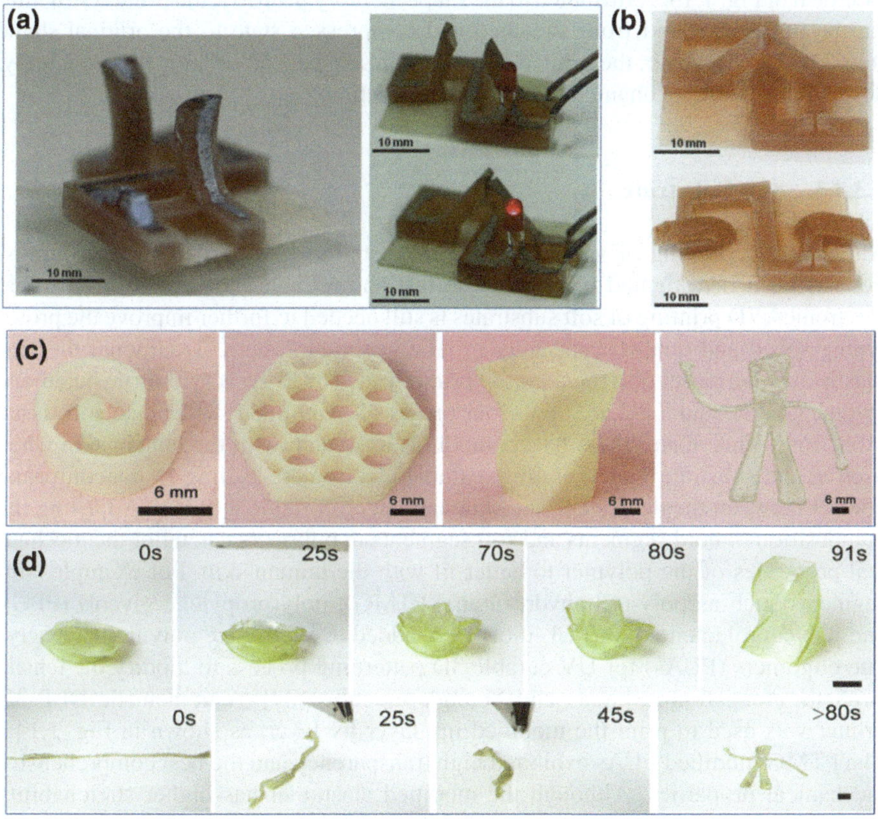

Fig. 1.13 **a** 3D-printed shape memory construct (left) in its off state (right top) and on state (right bottom). **b** A 3D-printed construct (top) programmed into its temporary state to enable inkjet printing on a 2D surface (bottom) [38]. **c** 3D structures by 3D printing of self-healing SMP. **d** Recovery of the printed SMP structure after compressed and stretched. All the scale bars are 6 mm [167]

temporary shape is an open electrical circuit, and when heated above the transition temperature, the circuit closes and turns on an LED. Furthermore, the circuit is printed at a temperature above the transition temperature, which may transform the printing of the 3D structure into a simpler 2D structure, shown in Fig. 1.13b. This strategy reduces the size of the unused devices and avoids using a complex 5-axis printing system. Such responsive printed structures have the potential to be used in soft robotics, minimally invasive medical devices, sensors, and wearable electronics [38].

Kuang et al. use 3D printing to print a self-healable shape memory polymer [167]. A highly stretchable semi-interpenetrating polymer network elastomer with an embedded PCL was printed using the DIW approach. The PCL provides both the self-healing and shape memory properties for the printed structure. Archimedean spiral, honeycomb, hollow vase, and Gumby model were printed using this system, as shown in Fig. 1.13c. Attributed to the shape memory property, the printed structure can be fully recovered from stretched and compressed state to the original shape (Fig. 1.13d). Moreover, the printed structure can self-heal after being cut repeatedly, despite with some mechanical property deterioration.

1.3.4.3 Soft Substrate

Although 3D printing of various functional materials and even polymer composites has been investigated to achieve the fast and reliable fabrication of wearable electronics, 3D printing of soft substrates is still needed to further improve the prototyping speed and quality, especially in an all-printed system. Traditional thermoplastic and thermoset polymers for 3D printing such as PLA, acrylonitrile butadiene styrene (ABS), and TPU are up to four orders of magnitude stiffer than soft tissue [168, 169]. This makes them have a modulus mismatch with the human skin when used as the substrate materials for wearable electronics, leading to nonconformal contact between the device and the skin and uncomfortable wearing. Adjusting the composition of hard segments and soft segments in polymers can tailor the mechanical properties of the polymer to better fit with the human skin. For example, soft segments such as poly-(tetrahydrofuran) (PTMG), poly(propylene glycol) (PPG), and polycaprolactone diol (PCL-diol) were added to synthesize polyurethane acrylate oligomers (PUAs) for UV curable 3D patterning process to modify the tensile strength, elongation at break, and Young's modulus [37]. A homemade DLP 3D printer was used to print the modified ink layer by layer, as shown in Fig. 1.14a. The PTMG modified PUAs exhibited high transparency and the best comprehensive mechanical properties. Although the modified elastomer has higher stretchability and maintains its mechanical strength without sacrificing the elongation at break compared with original polymers, the Young's Modulus is around 6.7 MPa, much higher than that of human skin (2–600 kPa).

To address this problem, one approach is to apply 3D printing to printing soft silicone elastomer materials such as PDMS to fabricate soft substrates for wearable devices. Zheng et al. developed PDMS composite elastomer inks for DIW 3D

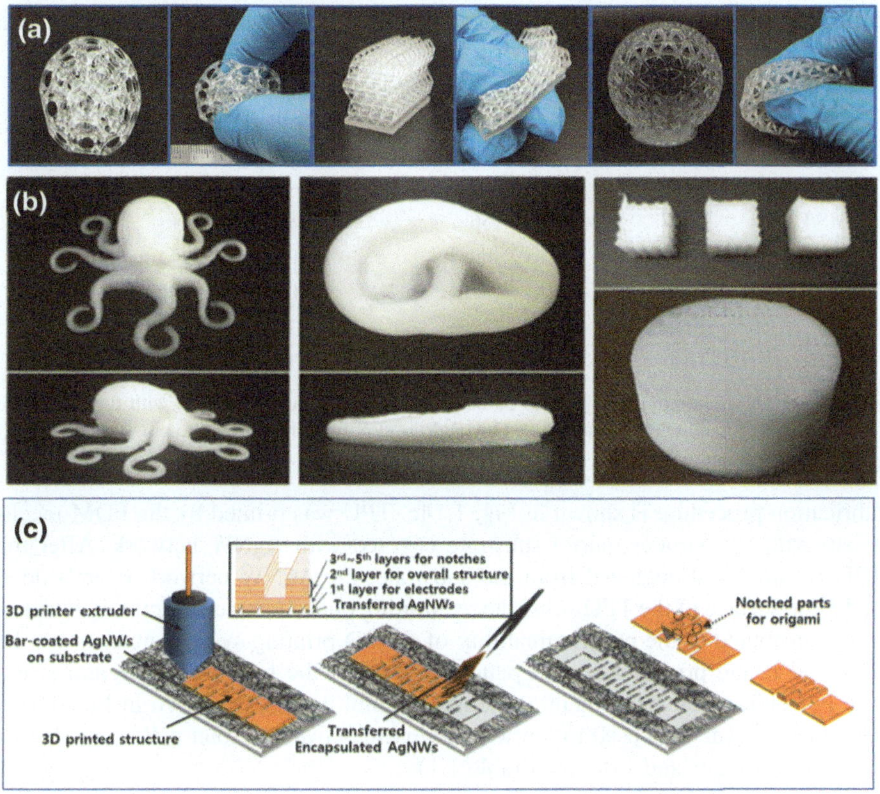

Fig. 1.14 **a** Photographs of the 3D-printed hollow PUAs 3D structure and their shape deformation with finger pressing [37]. **b** Photographs of 3D-printed PDMS/PTFE structure [170]. **c** Schematic of the 3D printer-based kirigami/origami approach combined with silver nanowires (AgNWs) [67]

printing by adding PTFE micro powder as a thixotropic agent [170]. The mechanical properties of the printed structures can be tailored by adjusting the ratio of PDMS and PTFE. An octopus model, an ear model, cubic cellular lattices with different filament gaps, and cylinder structures were printed using this approach, as shown in Fig. 1.14b. Benefiting from the increased electron affinity of PDMS by adding PTFE, a 3D-printed triboelectric nanogenerator (TENG) with a PDMS composite as the supporting structure and friction layer was fabricated. TENG works by cyclic compressing-releasing and is self-recoverable. Light-emitting diode (LED) arrays were successfully lighted by manually pressing TENG, demonstrating the good output performance of the fabricated TENG.

Another approach to enhance the wearability of the device is to print structurally designed patterns with high stretchability that can adapt to the movement of the human body. A 3D printing strategy was developed to fabricate kirigami/origami-based 3D structure with extreme stretchability and high areal coverage [67]. The

Table 1.1 Mechanical properties of representative 3D-printed soft substrates compared to human skin

Material	Printing technique	Young's modulus (kPa)	Stretchability (%)	Reference
Human skin	–	2–600	100	[5]
TPU	FDM	–	440	[67]
PDMS/PTFE	DIW	156	483	[170]
SMP	DIW	400	500–600	[167]
PBSF-FA-BMI	DIW	1870 ± 80	300–400	[171]
PDMS	FDM	1200 ± 100	130	[169]
PUAs	DLP	6700 ± 700	414.3 ± 7.6	[37]

PBSF-FA-BMI: poly(butylene sebacate-co-butylene fumarate) functionalized with furfurylamine (FA) and cross-linked with bismaleimide (BMI); PUAs: polyurethane acrylate oligomers

fabrication procedure is shown in Fig. 1.14c. TPU was printed by the FDM mechanism with an extruder onto a substrate covered with AgNW network. After the TPU is cured and detached from the substrate, the AgNW network is embedded under the surface of the TPU substrate, exhibiting outstanding conductivity. Furthermore, attributed to the programmability of the 3D printing system, notches can be designed in the printed kirigami pattern and assist the forming of origami structure by folding the 2D kirigami patterns into multilayers, as shown in Fig. 1.14c. This method was then used to fabricate high-area-coverage solar cell arrays with a designed bridge-island structure (Table 1.1).

1.4 Application of 3D-Printed Wearable Electronics

3D printing technology provides reliable solutions to facile fabrication of wearable devices. Due to the capability of printing solution-based materials, bottom-up fabrication of micro/nano-materials can be achieved by a 3D printing system. Moreover, polymer materials with proper rheology properties, thermoset/thermoplastic properties, and optical solidification response can be printed into customized structures more easily with higher product quality using 3D printing than conventional approaches such as laser cutting, injection molding, and soft lithography. Advances in functional materials have facilitated the ever-increasing performance of wearable sensors. However, the fabrication processes adapted to these functional materials still need further investigation. Considering that the most important building blocks of soft wearable electronics are functional materials and polymer substrates/encapsulations, 3D printing possesses a great potential in fabricating such electronics for novel applications including sensors, actuators, heaters, antennas, and interconnects, among others. This section will briefly summarize 3D-printed wearable systems in these application areas.

1.4.1 *Wearable Sensors*

In the past decade, development of soft wearable sensors has seen tremendous progress due to the increasing interest in electronic skin, human–machine interface, health monitoring, and other applications [80, 172, 173]. In this section, applications of using 3D printing techniques to fabricate representative wearable sensors will be discussed.

1.4.1.1 Strain Sensors

Wearable strain sensors can evaluate the skin deformations and body movements. Parameters to evaluate the performance of a strain sensor include sensitivity, linearity, hysteresis, and dynamic durability. To effectively capture the deformation level or motion features, a strain sensor should meet several requirements. (1) The sensor should be highly stretchable with low Young's modulus. The high stretchability will ensure the working function under human's daily movements, which may introduce a local strain over 75% [174, 175]. The low Young's modulus is to match the mechanical property of human skin (the modulus of epidermis is 140–600 kPa, and that of dermis is 2–80 kPa) and keep conformal contacts, which is vital for signal accuracy and quality. (2) For capturing of subtle strains, the sensor should have a high sensitivity. For example, by mounting a highly sensitive strain sensor on throat, the subtle motions of throat including swallow and speaking can be detected. (3) The fabrication process is scalable for mass production. 3D printing provides reliable solutions to meet these requirements. 3D printing is compatible with complex structural designs such as kirigami, origami, serpentine shape, and island-bridge are widely used to accommodate the large strain. In addition, 3D printing of advanced materials such as NPs, NWs, and graphene can easily integrate these materials into wearable sensors and increase the sensor sensitivity. Furthermore, the facile and mature 3D printing processes are amenable to scalable manufacturing of wearable sensors with high throughput at low cost.

Strain sensors can be mainly divided into capacitive and resistive types. Capacitive strain sensors typically exhibit low sensitivity, high linearity, and negligible hysteresis. In contrast, resistive strain sensors typically show high sensitivity but suffer from large nonlinearity and hysteresis. Typical capacitive wearable strain sensors have sandwich structures with conductors as electrodes and polymer as dielectric layer sandwiched between electrodes. The gauge factors (GFs) can be calculated by

$$GFs = \frac{\Delta C}{\varepsilon C_0} = \frac{1 - \varepsilon \nu_e + \nu_d - \nu_e}{1 - \varepsilon \nu_d} \tag{1.1}$$

where ΔC corresponds to the capacitance change of sensors, ε refers to strain, C_0 is measured as the initial capacitance, ν_e and ν_d represent the Poisson ratio of the electrode and dielectric layers, respectively [5, 176]. The maximum GF is 1 according to this equation. Yin and co-workers use DLP printing technique to print copolymerize acrylamide monomers and PEGDA hydrogel [162]. $MgCl_2$ was added to hydrogel to provide ion conductivity. To fabricate a capacitive strain sensor, two printed copolymer films with multiple parallel grooves were sandwiched by three 3 M very-high-bond (VHB) tapes, as shown in Fig. 1.15a. The VHB tape acted as both encapsulation layer and the dielectric layer. The fabricated sensor can be used as both pressure senor and strain sensor, as shown in Fig. 1.15b. The pressure-dependent capacitance changes and sensitivity are demonstrated in Figure xc.

A straightforward method to fabricate resistive strain sensors using 3D printing is to print conductive materials or conductive filler/polymer composites with stretchability. When the printed structure is stretched, the elongation of the structure changes its resistance. A tough and conductive double network hydrogel was fabricated by 3D printing technology to be used as a strain sensor with rapid sensing response, high sensing stability, and high strain sensitivity [164]. A wearable resistive strain sensor was fabricated and mounted on a finger joint, which can rapidly and precisely detect the motions of finger bending, as shown in Fig. 1.15d. Gao et al. demonstrated an integrated wearable sensor array fabricated with coaxial extrusion of conductive fiber by 3D printing technology [177]. The sensor array enables strain detection and

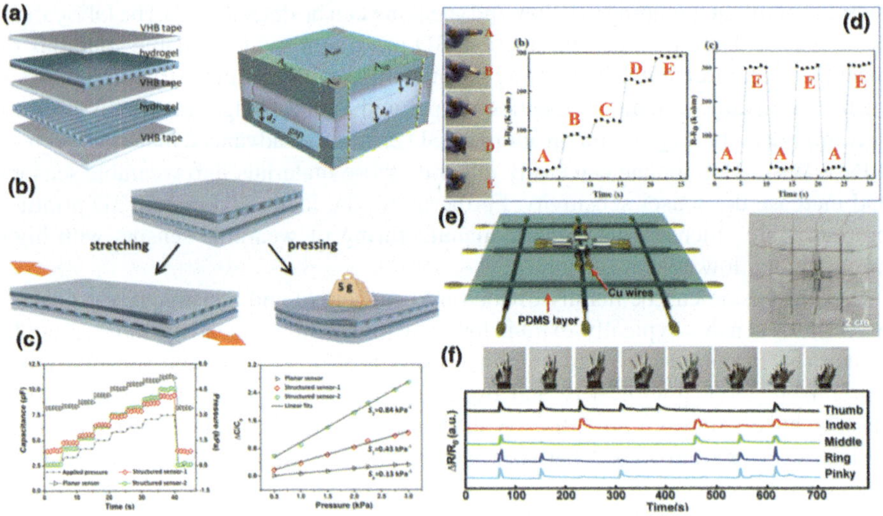

Fig. 1.15 **a** Schematical illustration of 3D-printed strain sensor by sandwiching a layer of stretchable dielectric (VHB 4905 tape, 3 M) between two layers of hydrogel film. **b** Schematic illustrations for the two sensing mechanisms of the capacitive sensor. **c** Pressure-dependent capacitance changes and sensitivity. **d** Photographs of finger bending (left) and associated resistance change of the resistive strain sensor [164]. **e** Schematic of the integrated sensor (left) and optical image of the integrated sensor (right). **f** Single-fiber strain sensors were used to identify the human hand gesture [177]

pressure detection on one platform, owing to the design of the sensor configuration. As shown in Fig. 1.15e, four resistive strain sensors in "U" shapes with a gauge factor of 11.8 were printed in the center of the array, where the gauge factor is calculated by

$$GFs = \frac{\Delta R_{\varepsilon}/\Delta R_{\varepsilon 0}}{\varepsilon} \tag{1.2}$$

where ε is the applied strain, ΔR_{ε} is the resistance at applied strain ε, $\Delta R_{\varepsilon 0}$ is the resistance when applied strain is zero, and $\Delta R_{\varepsilon}/\Delta R_{\varepsilon 0}$ is the relative resistance change. The strain sensor shows a small variation in resistance after repeat loading/unloading test for 10,000 cycles, exhibiting good durability for the strain sensor. As demonstrated, the strain sensors were applied to monitor human hand gestures, as shown in Fig. 1.15f.

A new method, embedded 3D printing (e-3DP), was developed for fabricating strain sensors by printing conductive carbon grease filler directly into an uncured elastomer reservoir, which was then cured to encapsulate the conductive path. The device maintains the stretchability of the elastomer matrix and exhibits high conformability. By printing the strain sensors within a pre-molded glove-shaped reservoir, an electronic glove that can capture the movement of finger joints was demonstrated, illustrating the potential of this method for fabricating soft functional devices for wearable electronics.

1.4.1.2 Pressure Sensors

Pressure sensors are a basic component for electronic skin, which can be mounted on human skin to evaluate the touch force/pressure or mounted on robotics to establish a human–machine interface. The piezoresistive type is widely used for wearable pressure sensors. The compression from an external force deforms the sensor and changes its resistance. Simply because larger deformation (under the same force) results in higher sensitivity, hollow or porous structures and scaffold structures are employed to enhance the performance of the sensor. 3D printing provides a reliable route to fabricate hollow structures, which shows promising potential to program the structure of pressure sensors. For example, Wang et al. printed 3D scaffold structured carbon black/TPU composite on silver electrodes to fabricate the pressure sensor, as shown in Fig. 1.16a [73]. Due to the unique configuration of the sensor, the tensile stress concentrates at the helix interdigital electrodes and substrate underneath the piezoresistive sensing element. The maximum stress in the sensing layer is only about 5% when the device is stretched up to 50%, which ensures the stable out-of-plane pressure response and minimized in-plane stretching interference.

On the other hand, if the Poisson's effect causes compression in the transverse direction when the sensor is stretched, the sensor can work as both the pressure sensor and strain sensor. Microstructured ionic conductive hydrogel electrodes were printed using a commercial DLP printer to fabricate a highly sensitive transparent

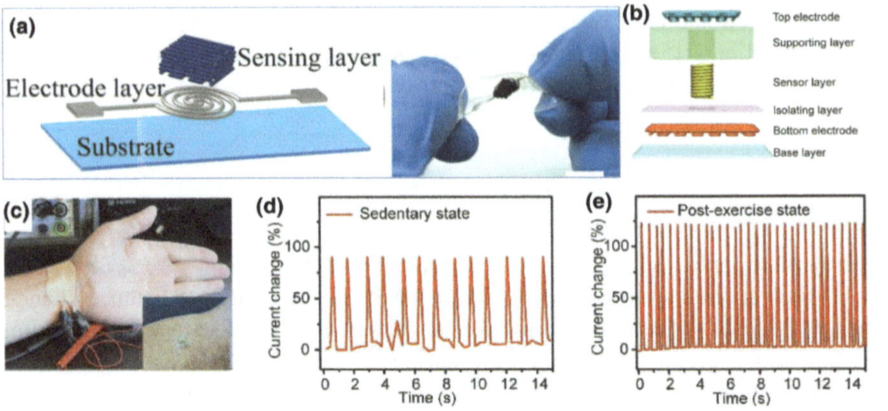

Fig. 1.16 **a** Schematic structure of the stretchable piezoresistive sensor device (left) and optical image of a representative stretchable piezoresistive sensor (right) [73]. b Schematic of the tactile sensor consisting of a base layer, top and bottom electrodes, an isolating layer, a sensor layer, and a supporting layer. **c** Photograph showing the tactile sensor mounted directly above the radial artery. Measurement signals of the radial pulse under **d** sedentary and **e** post-exercise states [72]

pressure/strain sensor [162]. The sensor is a capacitive type, where the thickness of the dielectric layer changes with the pressure, leading to a capacitance change. The microstructures on the printed electrodes introduce air gaps into the capacitor, which is critical to enhancing the sensitivity. When the sensor is pressed, the air gap and the dielectric layer shrink, the sensor works as a pressure sensor. When the sensor is stretched, the air gap and the dielectric layer are also compressed due to the Poisson's effect, the sensor works as a strain sensor [176]. 3D printing technology makes the fabrication of skin-like sensors more facile and programmable. By mounting the sensors at different locations of the human body, such as joints (e.g., finger knuckle, wrist, elbow, knee, and ankle), chest, and throat, the movement of the body, respiration, and speaking can be monitored, respectively.

Multi-material fabrication using one printing system is also a notable advantage of 3D printing. The conventional fabrication processes usually include several steps such as drop casting, spin coating, printing, and molding to complete the fabrication of one device. However, all-printed wearable electronics enabled by 3D printing of different materials by one setup have emerged. An all-printed piezoresistive pressure sensor was developed by Guo et al. with a one-step customized 3D printing process, where only the print nozzle needs to be adjusted to adapt different inks [72]. The sensor was composed of several layers including a base layer (silicone), a sensor layer (68% Ag/silicone), two electrode layers (75% Ag/silicone), an isolating layer (silicone), and a sacrificial supporting layer (40% Pluronic), as shown in Fig. 1.16b. Owing to the high flexibility, stretchability, and sensitivity of the 3D-printed pressure sensor, it was mounted at the wrist to monitor the pulse rate, as shown in Fig. 1.16c–e. Besides, pressing motion and bending motion of human fingers were also measured as a demonstration to show the potential of the all-printed pressure sensor for wearable sensors and human–machine interfaces.

1.4.1.3 Temperature Sensors

Wearable temperature sensors can be used to monitor human health in real time by evaluating temperature variation related to some diseases such as respiration disease, hypothyroidism, diabetes, and Parkinson's disease. Several types of temperature sensors exist including thermocouples, resistance temperature detectors (RTDs), thermistors, and semiconductor sensors. To get reliable body temperature data for precise diagnosis, wearable temperature sensors should meet two critical requirements. (1) The sensor should be conformally mounted on the human body to avoid impeding the heat transfer from the human body to the sensor and minimize the influence of the ambient environment. (2) The sensor should be strain-insensitive to reduce the interference from body movement. The first issue can be addressed by soft electrodes that can adapt to the curved surface of the human body, and the second issue can be addressed by designing the stretchable sensor structure, where strain does not change the resistance [85]. 3D printing is superior in fabricating soft electrodes and complex structures, which makes it an excellent fabrication approach for wearable temperature sensors. A DIW-based 3D printing method was used to print graphene/PDMS composites to fabricate wearable RTD type of temperature sensors [178]. Three different mesh-type structural designs including grid, triangular, and hexagonal structures are constructed, as shown in Fig. 1.17a–c. Compared with the solid-type temperature sensor, the gird structured sensor exhibits much higher stability when there is an external strain applied on the sensor, i.e., body movement. Grid sensors and solid sensors were mounted on the human wrist with the resistance signal being monitored. The drift of the measured temperature is as high as 0.5 °C due to the bending of the wrist in the case of the solid sensor, while a negligible drift value of only ~0.2% is observed for the grid sensor, under the same external deformation. 3D printing can be used to print, with high fidelity, not only temperature sensors but also other wearable sensors that need to eliminate strain artifacts with complex mechanical designs.

1.4.1.4 Pulse Oximeters

Pulse oximeter is clinically used to detect the vital signs such as heart rate and peripheral capillary oxygen saturation (SpO_2) for evaluating physiologic status of human objects, especially patients with neonates, postoperative, and cardiopulmonary diseases. With the COVID-19 pandemic spreading all over the world, wearable pulse oximeter becomes one of the most straightforward and reliable devices that can provide easy-accessed, off-clinical evaluation of the symptoms of suspected cases. Traditional pulse oximeters are bulky and rigid, causing a lot of interference when used. Lightweight soft pulse oximeter offers a great opportunity for continuous

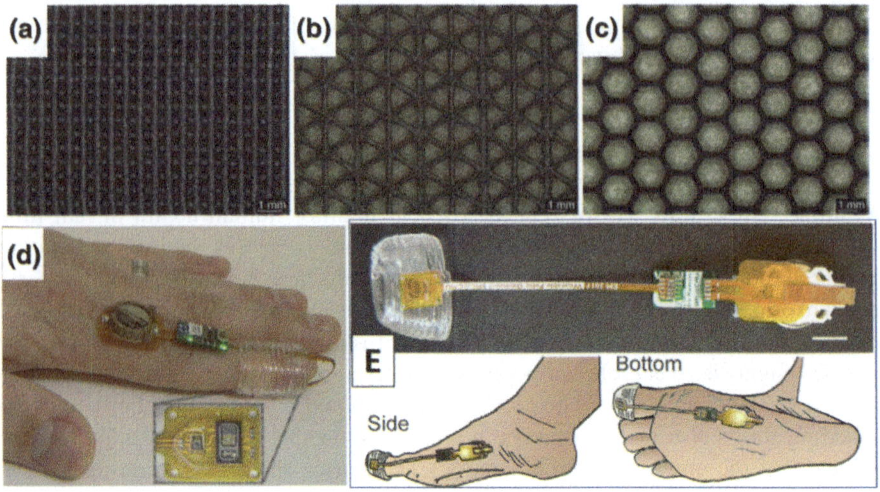

Fig. 1.17 Optical microscopy images of the 3D-printed graphene/PDMS composites with **a** grid structure; **b** triangular porous structure; and **c** hexagonal porous structure [178]. **d** Representative image of the customized wearable pulse oximeter tested on the index finger. **e** Complete wearable toe pulse oximeter and the schematic images of pulse oximeter placed on the side and bottom of the toe. Scale bar is 1 cm [169]

and remote heart rate and SpO_2 monitoring. There are typically four parts in a pulse oximeter system: light sources, photodetectors, holding frameworks, and data acquisition components. 3D printing techniques have been proven to be efficient and robust to fabricate these components. Abdollahi et al. applied freeform reversible embedding (FRE) 3D printing system to fabricate user-customizable soft pulse oximeter [169]. Like the portable pulse oximeter on the market, the soft pulse oximeter is mounted on the fingertip. However, 3D printing allows premeasure and on-demand printing of PDMS to fabricate the finger cuff for specific users, which make the soft pulse oximeter fit for fingers or toes, as shown in Fig. 1.17d, e. The toe-type sensor was tested at rest, while sitting and during walking. The human movement activity, pulse, and oxygen saturation can be monitored by the customized pulse oximeter. However, still some challenges remain in 3D printing of pulse oximeters.

1.4.2 Wearable Energy Devices

One of the key challenging issues for wearable systems is the wearable power sources. These power source devices should have a small footprint to make the system unobtrusive, and the energy they provide should be sustainable and continuous. Moreover, the power source devices should be biocompatible. This section will discuss the research efforts on 3D-printed wearable energy systems including storage and conversion systems.

1.4.2.1 Energy Storage Devices

Basically, energy storage devices can be categorized into supercapacitors and batteries. The main building blocks in these two devices are electrodes, electrolytes, and separators. They are different in their mechanisms of energy storage. A supercapacitor relies on the mechanisms of electrical-double-layer capacitance and/or pseudocapacitance, while a battery is based on relatively slow faradaic reactions [179]. Conventional fabrication of these storage devices mainly uses 2D casting methods such as doctor blade casting. These 2D devices are bulky and nonconformal when used as wearable power sources. 3D printing can not only fabricate soft wearable energy storage devices with good conformability, but also print complex electrodes that have higher surface area for electrolyte contacting, which is conducive for increasing the capacity of the device. For example, a bracelet-shaped Li–S battery with a scaffold structure as the electrode was designed and manufactured using the DIW- (Fig. 1.18a) and FDM-based 3D printing (Fig. 1.18b) [150]. The GO cathode is printed by the DIW technique, and the battery case is printed by the FDM technique. The printed wearable battery is shown in Fig. 1.18c–e. Graphene/phenol formaldehyde (PF) resin paste was modified with SiO_2 particles to achieve the desired rheology property for printing. Attributed to the large surface area provided by the printed scaffold electrode, an interpenetrating transmission path for electrons and ions is established to enhance the charge transfer at the interface between the electrodes and the electrolyte. And the porous structured electrodes also provide the transport channels for mass transportation of Li^+ and sufficient space for the electrolyte to diffuse to the area of the electrodes. The loading density of the active materials is also improved

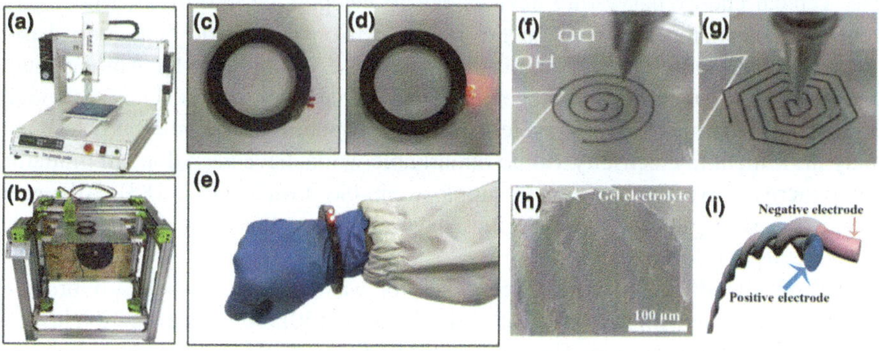

Fig. 1.18 **a** DIW 3D printer, and **b** FDM 3D printer. **c–e** Bracelet batteries with lights on and off, and bracelet battery on wrist, respectively [150]. **f, g** Optical images of the wet fiber during the printing process. **h** Cross-sectional SEM image of the gel electrolyte coating on the fiber. **i** Schematic of the assembled structure of the asymmetric supercapacitor [135]

by the 3D structured electrodes. An all-printed Li-ion battery was fabricated by 3D printing approaches [180]. Versatilely designed batteries can be fabricated owing to the flexibility of 3D printing. This work used well-established and commercially accepted $LiNi_{0.8}Co_{0.15}Al_{0.05}O_2$ (NCA) and natural graphite as the cathode and anode active materials, respectively, and the ink was modified to possess similar properties with slurries used in commercial coating fabrications. These make the fabrication procedure easily acceptable by industry. By integrating these 3D-printed batteries with other wearable electronics, a compact wearable system with a reduced size and weight can be developed.

Electrochemical supercapacitors have been widely used in powering wearable electronics due to their high power density, large capacity, low cost, and excellent reversibility. But the energy density is low, which impedes their practical application in wearable systems. Furthermore, the conventional planar rigid supercapacitors are not easy to be integrated into soft wearable systems. Advanced active materials can provide higher energy density and flexible configurations may make them fit for wearable applications. A fiber-shaped asymmetric supercapacitor was fabricated by 3D-printed single-walled CNT/V_2O_5 fiber cathodes and single-walled CNT/VN fiber anodes [135]. Electrodes are 3D-printed fibers, as shown in Fig. 1.18f, g. PVA/KOH gel electrolyte was printed on the surface of the as-printed positive and negative fibers (Fig. 1.18h). The two fiber-shaped electrodes are twisted to form a complete device, as shown in Fig. 1.18i. This asymmetric supercapacitor is fully printed with a maximum operating voltage of 1.6 V.

1.4.2.2 Energy Harvesting Systems

Energy harvesting technologies provide promising solutions to overcome the short lifetime of wearable energy storage devices. Unlike the batteries and supercapacitors, the energy harvesting devices do not need to charge repeatedly for continuous use. Typical energy harvesting devices can be categorized into the thermoelectric generator, oscillator, electromagnetic, piezoelectric triboelectric generators based on movement, and energy harvesting devices base on light [181]. Among them, thermoelectric generators and movement-enabled harvesting approaches are widely used in the wearable electronic system due to the capability to transfer body heat and body movement to electricity. Some remarkable works have been done in miniaturizing energy harvesting devices to supply wearable devices [182, 183]. Advanced fabrication strategies for these devices have also been investigated to reduce the manufacturing effort and cost.

An ultra-flexible 3D triboelectric generator was developed using a hybrid 3D printing technique with ink extrusion and UV curing [30]. The electrification layer is a printed UV curable composite resin structure, and the electrode is an ionic hydrogel, as shown in Fig. 1.19a. The triboelectric generator can achieve a peak power density of 10.98 W/m^3 and a transferred charge density of 0.65 mC/m^3 when working under a 1.3 Hz frequency. A self-powered buzzing SOS device was developed including printed ultra-flexible 3D triboelectric generators and external management circuits,

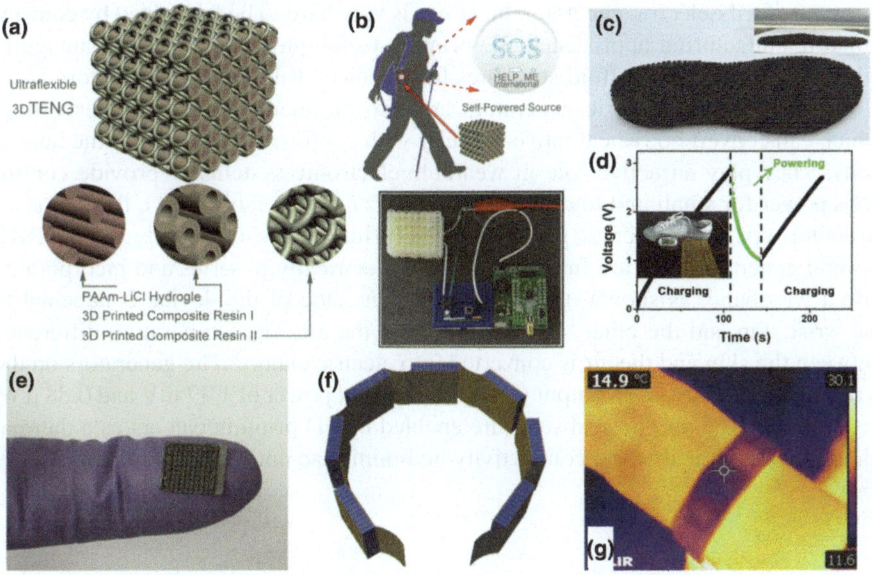

Fig. 1.19 **a** The design of an ultra-flexible 3D-printed triboelectric generator. **b** Schematic of an application as self-powered source for SOS distress signal system (top) and the picture of self-powered buzzing SOS distress system circuit (bottom) [30]. **c** Top-view images of 3D-printed triboelectric generator insole. **d** Voltage profile of a 22 μF capacitor being charged by the 3D-printed triboelectric generator insole and powering the electronic watch [137]. **e** 3D-printed flexible thermoelectric generator without the top electrode. **f** Schematic diagram of the thermoelectric generator connected in series. **g** Thermal infrared spectrum of the wristband wearable thermoelectric generator [184]

LEDs, and buzzers, as shown in Fig. 1.19b. It was also demonstrated that the triboelectric generators can be integrated with shoes to drive or charge low-power consuming electronics.

Another triboelectric generator with a tailorable sophisticated 3D architecture was fabricated using a one-step 3D printing process without additional assembling processes [137]. In this work, the electricity is generated by the compressive deformation of the scaffold electrodes consisting of CNTs and PGS. The exposed CNTs touch PGS on other filaments and become positively charged, while the PGS is negatively charged. The 3D-printed scaffold structure provides lots of open pores between the filaments, and these pores deformed/recover when the device is deformed by an external force, which will provide additional electricity due to the increase/decrease in the contact surface area. The hierarchical porous structure enabled by 3D printing results in a peak output voltage of 71 V at a 70% compression strain. A 3D-printed porous insole was fabricated and used to charge and drive electronics, as shown in Fig. 1.19c, d. And a ring-shaped triboelectric generator integrated with a LED indicator was mounted on the finger joint to monitor the finger motion. The hierarchical porous structure plays a critical role in the enhancement of performance of this type of triboelectric generator, however, is very hard to be fabricated by conventional manufacturing approaches. 3D printing technique shows great advantage in fabricating porous or scaffold structures for advanced triboelectric generators.

Wearable thermoelectric generators are also promising for energy harvesting, which can convert body heat into electricity with conformal contact with the human body. They play an active role in wearable electronic systems to provide continuous power for small and low-power devices. Zeng et al. fabricated a flexible grid-structure-type thermoelectric generator by 3D printing, as shown in Fig. 1.19e [184]. Several generator units are further connected electrically in series and incorporated into a wristband, as shown in Fig. 1.19f, g. One side of the device is attached to the wrist skin and the other side is exposed to the air. The temperature difference between the skin and the air is converted into electric energy. The generators on the wristband can generate an output voltage and output power of 1.47 mV and 0.88 μW, respectively. The porous grid structure enabled by 3D printing can act as a thermal blocker with a low thermal conductivity and minimize unnecessary thermal energy loss.

1.4.3 Interconnects for Wearable Systems

Stretchable interconnects are one of the most important passive components in wearable electronic systems, transmitting signals between other device components. For a wearable system, the interconnects should be flexible, stretchable, and conformable to nonplanar surfaces for reliable signal transmission. Stretchable conductors that exhibit nearly constant electrical properties when undergoing large deformation and repeated loading cycles can be excellent interconnects in soft wearable systems. To achieve high stretchability with stable electrical properties, strategies employing stretchable materials s and/or stretchable structures can be used. For the stretchable material strategy, soft polymers such as TPU, silicone, PVA are widely used as the matrix materials for the interconnects, with conductive fillers added as the conducting agents. For stretchable structure strategy, the conductors are patterned into stretchable structures such as serpentine, fractal pattern, helix, and wavy structure to reduce the local stresses in the conductors. 3D printing techniques enable the printing of advanced conducting materials such as NPs, NWs, nanotubes, and graphene. For complex stretchable structures, 3D printing can also be used for fast and precise fabrication. Furthermore, due to the increasing interests in vertically configured electronics to minimize the space for wearable devices and/or achieve more functions, vertical interconnects (so-called VIAs) are required to connect between layers. 3D printing provides reliable solution to fabricate VIAs, which cannot be fabricated by conventional 2D manufacturing methods [87].

As a popular intrinsic stretchable conductive filler, liquid metal is widely used as interconnects in wearable electronic systems. 3D printing of liquid metal renders unique advantages in fabricating sophisticated structures. A bridge structure of 3D-printed liquid metal enabled by fast oxidation of the liquid metal surface after extrusion can produce a freestanding conductive path that hangs over a previously printed line without touching it, as shown in Fig. 1.20a [45]. This makes the design of printed circuits more flexible and versatile. This method was used to fabricate a soft LED array with both transverse and longitudinal interconnects connected individually on the same plane, but no short circuit occurred between them, as shown in Fig. 1.20b. A dynamic switch circuit for LEDs enabled by the self-standing liquid metal and a reconfigurable antenna was also demonstrated, as shown in Fig. 1.20c, d. A similar strategy was reported by Park et al., with CNTs in the liquid metal to enhance the mechanical properties of the printed traces [44].

Proper structural design of the interconnects can increase the stretchability and decrease the strain sensitivity of the interconnects. A freestanding stretchable electrode based on the wavy shape was fabricated using a DIW 3D printing method [47]

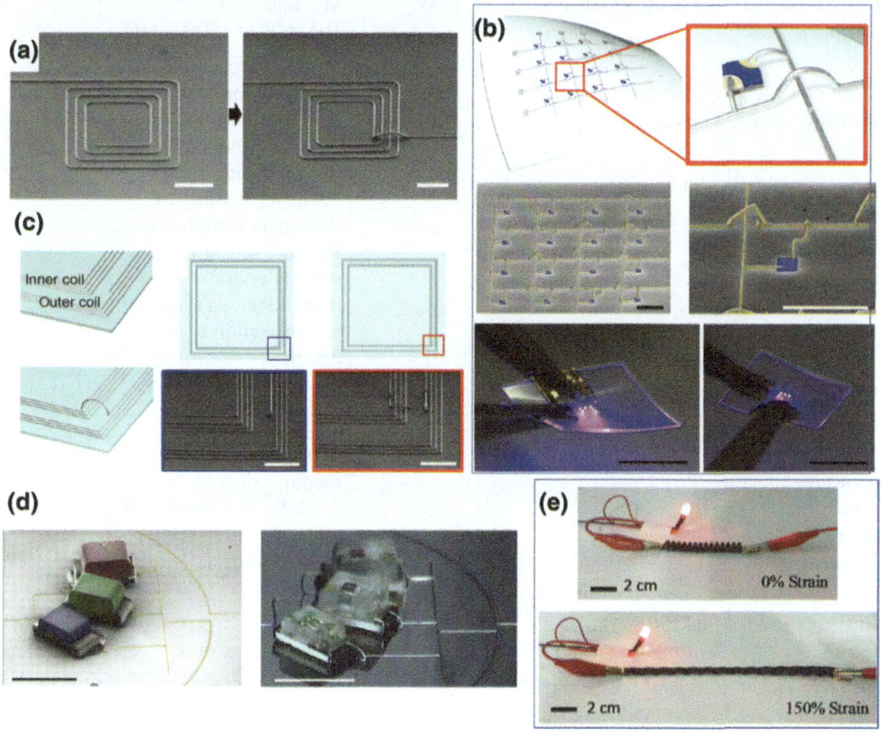

Fig. 1.20 **a** SEM images of the bridge-structured interconnects fabricated by 3D printing. Scale bars 200 mm. **b** An LED array with 3D liquid metal interconnects. **c** Schematic illustrations and SEM images of the reconfigurable antenna. **d** Colorized SEM image of three LED pixels and liquid meal interconnects, scale bar 1 mm (left) and photograph of the interconnects and LED, scale bar 1 mm (right) [45]. **e** An LED connected to the 3D-printed wavy shape interconnects under 0 (top) and 150% strain (bottom) [47]

The stretchability of the printed structure can be as high as over 300% for PDMS; and only <5% resistance change was observed at 100% strain. The 3D wavy structure was used to connect an LED light with a 12 V voltage applied, as shown in Fig. 1.20e. When the interconnect was stretched from 0 to 150%, the LED maintained a constant brightness (Table 1.2).

Table 1.2 Summary of 3D-printed wearable devices

Application	Material	Printing technique	Performance	Referecne
Capacitive strain/pressure sensor	PAAm–PEGDA hydrogel	DLP	GF: 0.92 Limit of detection: < 1.5% Strain range: 100% Stability: > 5000 cycles between 0–25% strain	[162]
Resistive strain sensor	PDMS/CNT fiber	DIW	Maximum GF: 2.5×10^6 at 90 to 150% strain Strain range: 150%	[185]
Strain sensor, Pressure sensor	Elastomer/carbon grease	coaxial extrusion	Strain sensor: GF of 11.8 for the strain less than 20% and 4.3 within the strain range of 20–80%, Stability of more than 10,000 cycles Pressure sensor: sensitivity of 0.562 kPa^{-1}, a response and relaxation speed of 230 ms	[177]
Resistive strain sensor	agar/calcium alginate (CA)/ PAAm hydrogel	UV curing + DIW	GF: 3.83	[164]
Capacitive pressure sensor, Resistive strain sensor	TPU/Silver flakes	DIW	Pressure sensor: elastic modulus of E ≈ 2.3 MPa; linear range of <3 MPa Strain sensor: GF of 13.3	[86]
Resistive strain sensor	PDMS/PAAm-LiCl hydrogel	UV curing + DIW	GF: 0.84 Linear range: 40%	[31]
Pressure sensor based on TENG	covalent crosslinked elastomer/CNT/AgNF/CB	DIW	Sensitivity: 0.147–0.528 MPa^{-1} Stability: 1.3% mass loss in 20 days	[171]
Resistive tactile sensors	AgNPs/silicone, AgNW/ silicone, Silicone, Pluronic	DIW	GF: 180 Stability: more than 100 cycles	[72]

(continued)

Table 1.2 (continued)

Application	Material	Printing technique	Performance	Referecne
Piezoresistive pressure sensor	PDMS, TPU/CB/NaCl , TPU/Ag microflakes	DIW	Sensitivity: 5.54 kPa^{-1} for <10 kPa, 0.123 kPa^{-1} for 10–100 kPa and 0.0048 kPa^{-1} for 100–800 kPa GF: 13.3 Stability: more than 10,000 cycles	[73]
Piezoresistive pressure sensor	Modified PUAs	DLP	Sensitivity: 2010.1 MPa^{-1} in the region of 0.09–0.23 MPa Stability: After 100 compression cycles at 80% strain, the maximum compressive stress remained above 93.3%	[37]
Piezoresistive pressure sensor	CNTs/silica NPs (SiNPs)/Ecoflex	DIW	Sensitivity: 0.096 kPa^{-1} Range: 0–175 kPa Stability: over 10,000 compression cycles	[68]
Piezoelectric air pressure sensor	Functionalized piezoelectric particles	SLA	Detectable low aire pressure: <50 Pa	[70]
Temperature sensor	Graphene/PDMS	DIW	Resolution: 0.5 °C Sensitivity: 0.008 $°C^{-1}$ Response time: 1.32–3.91 s Recovery time: 1.01–4.58 s	[178]
Photodetector	P3HT:PCBM, liquid metal, silicone, PEDOT:PSS, AgNPs	DIW	External quantum efficiency: 25.3%,	[156]
TENG	acrylate oligomer-based UV curable polymer, PAAm-LiCl hydrogel	UV curing + DIW	Power density: 10.98 W/m3 under a frequency of 1.3 Hz, Transferred charge density: 0.65 mC/m^{-2} under a frequency of 1.3 Hz,	[30]

(continued)

Table 1.2 (continued)

Application	Material	Printing technique	Performance	Referecne
TENG	PUA/silver flakes/liquid metal	DIW	Stretchability: 2500% power density: 40 μWcm^{-2} with load resistance of 1MΩ	[32]
TENG	poly(glycerol sebacate)/(CNTs)	DIW	Power density: 1.11 W m^{-3} Stability: more than 6000 cycles at a compression strain of 40%	[137]
Lithium-Sulfur Battery	GO/phenolformaldehyde (PF)/SiO_2	FDM	Specific capacity: 505.4 mAh g^{-1} after 500 cycles Active material loading: 10.2 mg cm^{-1}	[150]
Interconnect	PDMS/CNT composite	DIW	Stretchability: 315% Resistance change: 5% change at 100% strain and 35% change at 200% Stability: >120 cycles between 0–100% strain	[47]
Interconnect	Liquide metal	DIW	Minimum line width: 1.9 mm 60–90 Ω contact resistance between interconnect and Ag, Au, and Cu	[45]
Interconnect	TPU/AgNW	FDM	System Stretchability: 440% Interconnect stretchability: 25 000% Stability: over 1000 cycles under Interconnect stretchability	[67]

1.5 Summary and Outlook

As an advanced additive manufacturing approach, 3D printing has been applied to a wide variety of applications. Soft wearable electronics are thriving with the advances in functional materials and new fabrication approaches such as 3D printing. This chapter has provided a systematic perspective on the 3D printing techniques, 3D printable materials, and wearable electronic devices enabled by the 3D printing techniques. In our opinion, the following directions are promising in order to expand the applications of 3D printing in wearable electronics:

1. Investigating advanced ink formulation for printing more functional materials. More and more emerging nanomaterials are used to fabricate wearable electronics. Although a variety of widely used materials discussed in this chapter, e.g., nanoparticles, carbon-based nanomaterials, and liquid metal, have been proven to be printable using a 3D printing platform, the conducting fillers in the inks are still limited by the strict requirements on the thermal properties for FDM-based printing and rheology properties for DIW-based printing.
2. Developing and optimizing 3D printing systems that are applicable to multiple materials and structures. Using one platform to prepare different components in one device, i.e., all-printed devices, can further simplify the manufacturing procedure and lower the cost. This requires the 3D printer to be capable of printing inks with different properties into a variety of different structures. For instance, it should be compatible with structural materials like polymers and functional materials like metal NPs/NWs. It should be able to construct different types of structures such as in-pane traces, vertical structures, grid structures, and self-standing bridges.
3. Further increasing the resolution of the printed features. Minimizing the footprint and increasing the integration level of wearable devices are of great significance in practical applications. However, to date, most of the widely used 3D printing techniques such as SLA, DLP, and FDM have much lower resolution compared to conventional lithography manufacturing approaches.
4. Improving printing capabilities by incorporating advanced algorithms to control the 3D printer, such as predictive algorithms to anticipate future deformations and motions.

References

1. Trung TQ, Lee NE (2016) Flexible and stretchable physical sensor integrated platforms for wearable human-activity monitoringand personal healthcare. Adv Mater 28(22):4338–4372
2. Ma Y, Zhang Y, Cai S, Han Z, Liu X, Wang F et al (2019) Flexible hybrid electronics for digital healthcare. Adv Mater e1902062
3. Yao S, Yang J, Poblete FR, Hu X, Zhu Y (2019) Multifunctional electronic textiles using silver nanowire composites. ACS Appl Mater Interfaces 11(34):31028–31037
4. Yao S, Myers A, Malhotra A, Lin F, Bozkurt A, Muth JF et al (2017) A Wearable hydration sensor with conformal nanowire electrodes. Adv Healthc Mater 6(6):1601159
5. Yao S, Swetha P, Zhu Y (2018) Nanomaterial-enabled wearable sensors for healthcare. Adv Healthc Mater 7(1):1700889
6. Liu Y, Shukla D, Newman H, Zhu Y (2021) Soft wearable sensors for monitoring symptoms of COVID-19 and other respiratory diseases: a review. Progress Biomed Eng 4(1):012001
7. Lee H, Song C, Hong YS, Kim MS, Cho HR, Kang T et al (2017) Wearable/disposable sweat-based glucose monitoring device with multistage transdermal drug delivery module. Sci Adv 3(3):1601314

8. Wu S, Cui Z, Baker GL, Mahendran S, Xie Z, Zhu Y (2021) A biaxially stretchable and self-sensing textile heater using silver nanowire composite. ACS Appl Mater Interfaces 13(49):59085–59091
9. Soto F, Mishra RK, Chrostowski R, Martin A, Wang J (2017) Epidermal tattoo patch for ultrasound-based transdermal microballistic delivery. Adv Mater Technol 2(12):1700210
10. Amjadi M, Sheykhansari S, Nelson BJ, Sitti M (2018) Recent advances in wearable transdermal delivery systems. Adv Mater 30(7):1704530
11. Kim J, Shim HJ, Yang J, Choi MK, Kim DC, Kim J et al (2017) Ultrathin quantum dot display integrated with wearable electronics. Adv Mater 29(38):1700217
12. Zhang T, Li Z, Li K, Yang X (2019) Flexible pressure sensors with wide linearity range and high sensitivity based on selective laser sintering 3D printing. Adv Mater Technol 4(12):1900679
13. Li Z, Liu Y, Hossain O, Paul R, Yao S, Wu S et al (2021) Real-time monitoring of plant stresses via chemiresistive profiling of leaf volatiles by a wearable sensor. Matter 4(7):2553–2570
14. Zhou W, Yao S, Wang H, Du Q, Ma Y, Zhu Y (2020) Gas-permeable, ultrathin, stretchable epidermal electronics with porous electrodes. ACS Nano 14(5):5798–5805
15. Liu Y, Wang H, Zhu Y (2021) Recycling of nanowire percolation network for sustainable soft electronics. Adv Electron Mater 7(9):2100588
16. Xue Z, Song H, Rogers JA, Zhang Y, Huang Y (2019) Mechanically-guided structural designs in stretchable inorganic electronics. Adv Mater e1902254
17. Zhang Y, Xu S, Fu H, Lee J, Su J, Hwang KC et al (2013) Buckling in serpentine microstructures and applications in elastomer-supported ultra-stretchable electronics with high areal coverage. Soft Matter 9(33):8062–8070
18. Chang JK, Chang HP, Guo Q, Koo J, Wu CI, Rogers JA (2018) Biodegradable electronic systems in 3D, heterogeneously integrated formats. Adv Mater 30(11):1704955
19. Pan T, Pharr M, Ma Y, Ning R, Yan Z, Xu R et al (2017) Experimental and theoretical studies of serpentine interconnects on ultrathin elastomers for stretchable electronics. Adv Func Mater 27(37):1702589
20. Hong YJ, Jeong H, Cho KW, Lu N, Kim DH (2019) Wearable and implantable devices for cardiovascular healthcare: from monitoring to therapy based on flexible and stretchable electronics. Adv Func Mater 29(19):1808247
21. Huang Z, Hao Y, Li Y, Hu H, Wang C, Nomoto A et al (2018) Three-dimensional integrated stretchable electronics. Nat Electron 1(8):473–480
22. Koo JH, Jeong S, Shim HJ, Son D, Kim J, Kim DC et al (2017) Wearable electrocardiogram monitor using carbon nanotube electronics and color-tunable organic light-emitting diodes. ACS Nano 11(10):10032–10041
23. Park M, Do K, Kim J, Son D, Koo JH, Park J et al (2015) Oxide nanomembrane hybrids with enhanced mechano- and thermo-sensitivity for semitransparent epidermal electronics. Adv Healthcare Mater 4(7):992–997
24. Zou Z, Zhu C, Li Y, Lei X, Zhang W, Xiao J (2018) Rehealable, fully recyclable, and malleable electronic skin enabled by dynamic covalent thermoset nanocomposite. Sci Adv 4(2):eaaq0508
25. Huang Q, Zhu Y (2019) Printing conductive nanomaterials for flexible and stretchable electronics: a review of materials, processes, and applications. Adv Mater Technol 4(5):1800546
26. Cui Z, Han Y, Huang Q, Dong J, Zhu Y (2018) Electrohydrodynamic printing of silver nanowires for flexible and stretchable electronics. Nanoscale 10(15):6806–6811
27. Huang Q, Zhu Y (2018) Gravure printing of water-based silver nanowire ink on plastic substrate for flexible electronics. Sci Rep 8(1):15167
28. Huang Q, Zhu Y (2021) Patterning of metal nanowire networks: methods and applications. ACS Appl Mater Interfaces 13(51):60736–60762
29. Tan LJ, Zhu W, Zhou K (2020) Recent progress on polymer materials for additive manufacturing. Adv Func Mater 30(43):2003062
30. Chen B, Tang W, Jiang T, Zhu L, Chen X, He C et al (2018) Three-dimensional ultraflexible triboelectric nanogenerator made by 3D printing. Nano Energy 45:380–389

31. Tian K, Bae J, Bakarich SE, Yang C, Gately RD, Spinks GM et al (2017) 3D printing of transparent and conductive heterogeneous hydrogel-elastomer systems. Adv Mater 29(10):1604827
32. Parida K, Thangavel G, Cai G, Zhou X, Park S, Xiong J et al (2019) Extremely stretchable and self-healing conductor based on thermoplastic elastomer for all-three-dimensional printed triboelectric nanogenerator. Nat Commun 10(1):2158
33. Emon MOF, Alkadi F, Philip DG, Kim D-H, Lee K-C, Choi J-W (2019) Multi-material 3D printing of a soft pressure sensor. Addit Manuf 28:629–638
34. Odent J, Wallin TJ, Pan WY, Kruemplestaedter K, Shepherd RF, Giannelis EP (2017) Highly elastic, transparent, and conductive 3d-printed ionic composite hydrogels. Adv Func Mater 27(33):1701807
35. Ambrosi A, Pumera M (2016) 3D-printing technologies for electrochemical applications. Chem Soc Rev 45(10):2740–2755
36. Billiet T, Vandenhaute M, Schelfhout J, Van Vlierberghe S, Dubruel P (2012) A review of trends and limitations in hydrogel-rapid prototyping for tissue engineering. Biomaterials 33(26):6020–6041
37. Peng S, Li Y, Wu L, Zhong J, Weng Z, Zheng L et al (2020) 3D Printing mechanically robust and transparent polyurethane elastomers for stretchable electronic sensors. ACS Appl Mater Interfaces 12(5):6479–6488
38. Zarek M, Layani M, Cooperstein I, Sachyani E, Cohn D, Magdassi S (2016) 3D printing of shape memory polymers for flexible electronic devices. Adv Mater 28(22):4449–4454
39. Cheng-Yu H, Ahmed Abro Z, Yi-Fan Z, Ahmed LR (2019) An FBG-based smart wearable ring fabricated using FDM for monitoring body joint motion. J Ind Text 50(10):1660–1673
40. Yang W, Zhao W, Li Q, Li H, Wang Y, Li Y et al (2020) Fabrication of smart components by 3D printing and laser-scribing technologies. ACS Appl Mater Interfaces 12(3):3928–3935
41. Liang K, Carmone S, Brambilla D, Leroux JC (2018) 3D printing of a wearable personalized oral delivery device: a first-in-human study. Sci Adv 4(5):eaat2544
42. Li Z, Rathore AS, Song C, Wei S, Wang Y, Xu W (2018) PrinTracker: Fingerprinting 3D printers using commodity scanners. In: Proceedings of the 2018 ACM SIGSAC conference on computer and communications security, pp 1306–1323
43. Owens CE, Headrick RJ, Williams SM, Fike AJ, Pasquali M, McKinley GH et al (2021) Substrate-versatile direct-write printing of carbon nanotube-based flexible conductors, circuits, and sensors. Adv Func Mater 31(25):2100245
44. Park YG, Min H, Kim H, Zhexembekova A, Lee CY, Park JU (2019) Three-dimensional, high-resolution printing of carbon nanotube/liquid metal composites with mechanical and electrical reinforcement. Nano Lett 19(8):4866–4872
45. Park YG, An HS, Kim JY, Park JU (2019) High-resolution, reconfigurable printing of liquid metals with three-dimensional structures. Sci Adv 5(6):eaaw2844
46. Zhang Y, Shi G, Qin J, Lowe SE, Zhang S, Zhao H et al (2019) Recent progress of direct ink writing of electronic components for advanced wearable devices. ACS Appl Electron Mater 1(9):1718–1734
47. Wei H, Li K, Liu WG, Meng H, Zhang PX, Yan CY (2017) 3D Printing of free-standing stretchable electrodes with tunable structure and stretchability. Adv Eng Mater 19(11):1700341
48. Li H, Tan YJ, Leong KF, Li L (2017) 3D Bioprinting of highly thixotropic alginate/methylcellulose hydrogel with strong interface bonding. ACS Appl Mater Interfaces 9(23):20086–20097
49. Zacharatos F, Theodorakos I, Karvounis P, Tuohy S, Braz N, Melamed S et al (2018) Selective laser sintering of laser printed Ag nanoparticle micropatterns at high repetition rates. Materials (Basel). 11(11):2142
50. Wang J, Sun S, Li X, Fei G, Wang Z, Xia H (2021) Selective laser sintering of polydimethylsiloxane composites. 3D Printing Addit Manuf
51. Ronca A, Rollo G, Cerruti P, Fei G, Gan X, Buonocore G et al (2019) Selective laser sintering fabricated thermoplastic polyurethane/graphene cellular structures with tailorable properties and high strain sensitivity. Appl Sci 9(5):864

52. Li Z, Wang Z, Gan X, Fu D, Fei G, Xia H (2017) Selective laser sintering 3D printing: a way to construct 3D electrically conductive segregated network in polymer matrix. Macromol Mater Eng 302(11):1700211
53. Agarwala S, Goh GL, Le Dinh TS, An J, Peh ZK, Yeong WY et al (2019) Wearable bandage-based strain sensor for home healthcare: combining 3D aerosol jet printing and laser sintering. ACS Sens 4(1):218–226
54. Choi S, Lee H, Ghaffari R, Hyeon T, Kim DH (2016) Recent advances in flexible and stretchable bio-electronic devices integrated with nanomaterials. Adv Mater 28(22):4203–4218
55. Park S, Kim H, Vosgueritchian M, Cheon S, Kim H, Koo JH et al (2014) Stretchable energy-harvesting tactile electronic skin capable of differentiating multiple mechanical stimuli modes. Adv Mater 26(43):7324–7332
56. Irimia-Vladu M (2014) "Green" electronics: biodegradable and biocompatible materials and devices for sustainable future. Chem Soc Rev 43(2):588–610
57. Ganesh RS, Yoon HJ, Kim SW (2020) Recent trends of biocompatible triboelectric nanogenerators toward self-powered e-skin. EcoMat. 2(4):12065
58. Lee HR, Kim CC, Sun JY (2018) Stretchable Ionics—a promising candidate for upcoming wearable devices. Adv Mater 30(42):e1704403
59. Song Y, Min J, Gao W (2019) Wearable and implantable electronics: moving toward precision therapy. ACS Nano 13(11):12280–12286
60. Choi S, Han SI, Jung D, Hwang HJ, Lim C, Bae S et al (2018) Highly conductive, stretchable and biocompatible Ag–Au core-sheath nanowire composite for wearable and implantable bioelectronics. Nat Nanotechnol 13(11):1048–1056
61. Sekitani T, Yokota T, Kuribara K, Kaltenbrunner M, Fukushima T, Inoue Y et al (2016) Ultraflexible organic amplifier with biocompatible gel electrodes. Nat Commun 7:11425
62. Wang C, Yokota T, Someya T (2021) Natural biopolymer-based biocompatible conductors for stretchable bioelectronics. Chem Rev 121(4):2109–2146
63. Oh JY, Bao Z (2019) Second skin enabled by advanced electronics. Adv Sci (Weinh) 6(11):1900186
64. Wei H, Cauchy X, Navas IO, Abderrafai Y, Chizari K, Sundararaj U et al (2019) Direct 3D printing of hybrid nanofiber-based nanocomposites for highly conductive and shape memory applications. ACS Appl Mater Interfaces 11(27):24523–24532
65. Chen Z, Zhao D, Liu B, Nian G, Li X, Yin J et al (2019) 3D printing of multifunctional hydrogels. Adv Func Mater 29(20):1900971
66. Lei ZY, Wang QK, Wu PY (2017) A multifunctional skin-like sensor based on a 3D printed thermo-responsive hydrogel. Mater Horiz 4(4):694–700
67. Jo M, Bae S, Oh I, Jeong JH, Kang B, Hwang SJ et al (2019) 3D Printer-based encapsulated origami electronics for extreme system stretchability and high areal coverage. ACS Nano 13(11):12500–12510
68. Tang Z, Jia S, Zhou C, Li B (2020) 3D printing of highly sensitive and large-measurement-range flexible pressure sensors with a positive piezoresistive effect. ACS Appl Mater Interfaces 12(25):28669–28680
69. Huang K, Dong S, Yang J, Yan J, Xue Y, You X et al (2019) Three-dimensional printing of a tunable graphene-based elastomer for strain sensors with ultrahigh sensitivity. Carbon 143:63–72
70. Yao D, Cui H, Hensleigh R, Smith P, Alford S, Bernero D et al (2019) Achieving the upper bound of piezoelectric response in tunable, wearable 3D printed nanocomposites. advanced functional materials 29(42)
71. Zhou LY, Gao Q, Zhan JF, Xie CQ, Fu JZ, He Y (2018) Three-dimensional printed wearable sensors with liquid metals for detecting the pose of snakelike soft robots. ACS Appl Mater Interfaces 10(27):23208–23217
72. Guo SZ, Qiu K, Meng F, Park SH, McAlpine MC (2017) 3D printed stretchable tactile sensors. Adv Mater 29(27):1701218
73. Wang Z, Guan X, Huang H, Wang H, Lin W, Peng Z (2019) Full 3D printing of stretchable piezoresistive sensor with hierarchical porosity and multimodulus architecture. Adv Func Mater 29(11):1807569

74. Wong J, Gong AT, Defnet PA, Meabe L, Beauchamp B, Sweet RM et al (2019) 3D Printing Ionogel auxetic frameworks for stretchable sensors. Adv Mater Technol 4(9):1900452
75. Yao S, Ren P, Song R, Liu Y, Huang Q, Dong J et al (2020) Nanomaterial-enabled flexible and stretchable sensing systems: processing, integration, and applications. Adv Mater 32(15):e1902343
76. Yao S, Zhu Y (2015) Nanomaterial-enabled stretchable conductors: strategies, materials and devices. Adv Mater 27(9):1480–1511
77. Jayathilaka W, Qi K, Qin Y, Chinnappan A, Serrano-Garcia W, Baskar C et al (2019) Significance of nanomaterials in wearables: a review on wearable actuators and sensors. Adv Mater 31(7):e1805921
78. Son D, Bao Z (2018) Nanomaterials in skin-inspired electronics: toward soft and robust skin-like electronic nanosystems. ACS Nano 12(12):11731–11739
79. Kamyshny A, Magdassi S (2019) Conductive nanomaterials for 2D and 3D printed flexible electronics. Chem Soc Rev 48(6):1712–1740
80. Cui Z, Poblete FR, Cheng G, Yao S, Jiang X, Zhu Y (2014) Design and operation of silver nanowire based flexible and stretchable touch sensors. J Mater Res 30(1):79–85
81. Zhang S, Cai L, Li W, Miao J, Wang T, Yeom J et al (2017) Fully printed silver-nanoparticle-based strain gauges with record high sensitivity. Adv Electron Mater 3(7):1700067
82. Serway RA, Jewett JW (2018) Physics for scientists and engineers. Cengage learning
83. Gaspar C, Passoja S, Olkkonen J, Smolander M (2016) IR-sintering efficiency on inkjet-printed conductive structures on paper substrates. Microelectron Eng 149:135–140
84. Lee HH, Chou KS, Huang KC (2005) Inkjet printing of nanosized silver colloids. Nanotechnology 16(10):2436–2441
85. Cui Z, Poblete FR, Zhu Y (2019) Tailoring the temperature coefficient of resistance of silver nanowire nanocomposites and their application as stretchable temperature sensors. ACS Appl Mater Interfaces 11(19):17836–17842
86. Valentine AD, Busbee TA, Boley JW, Raney JR, Chortos A, Kotikian A et al (2017) Hybrid 3D printing of soft electronics. Adv Mater 29(40):1703817
87. Sowade E, Polomoshnov M, Willert A, Baumann RR (2019) Toward 3D-printed electronics: inkjet-printed vertical metal wire interconnects and screen-printed batteries. Adv Eng Mater 21(10):1900568
88. Yu Z, Zhang Q, Li L, Chen Q, Niu X, Liu J et al (2011) Highly flexible silver nanowire electrodes for shape-memory polymer light-emitting diodes. Adv Mater 23(5):664–668
89. Lee P, Lee J, Lee H, Yeo J, Hong S, Nam KH et al (2012) Highly stretchable and highly conductive metal electrode by very long metal nanowire percolation network. Adv Mater 24(25):3326–3332
90. Jiang ZH, Cui Z, Yue T, Zhu Y, Werner DH (2017) Compact, highly efficient, and fully flexible circularly polarized antenna enabled by silver nanowires for wireless body-area networks. IEEE Trans Biomed Circ Syst 11(4):920–932
91. Zhu B, Ling Y, Yap LW, Yang M, Lin F, Gong S et al (2019) Hierarchically structured vertical gold nanowire array-based wearable pressure sensors for wireless health monitoring. ACS Appl Mater Interfaces 11(32):29014–29021
92. Han S, Kim MK, Wang B, Wie DS, Wang S, Lee CH (2016) Mechanically reinforced skin-electronics with networked nanocomposite elastomer. Adv Mater 28(46):10257–10265
93. Liang J, Li L, Tong K, Ren Z, Hu W, Niu X et al (2014) Silver nanowire percolation network soldered with graphene oxide at room temperature and its application for fully stretchable polymer light-emitting diodes. ACS Nano 8(2):1590–1600
94. Xu F, Zhu Y (2012) Highly conductive and stretchable silver nanowire conductors. Adv Mater 24(37):5117–5122
95. Song L, Myers AC, Adams JJ, Zhu Y (2014) Stretchable and reversibly deformable radio frequency antennas based on silver nanowires. ACS Appl Mater Interfaces 6(6):4248–4253
96. Yao S, Cui J, Cui Z, Zhu Y (2017) Soft electrothermal actuators using silver nanowire heaters. Nanoscale 9(11):3797–3805

97. Yan P, Brown E, Su Q, Li J, Wang J, Xu C et al (2017) 3D printing hierarchical silver nanowire aerogel with highly compressive resilience and tensile elongation through tunable poisson's ratio. Small 13(38):1701756
98. Liu S, Shi X, Li X, Sun Y, Zhu J, Pei Q et al (2018) A general gelation strategy for 1D nanowires: dynamically stable functional gels for 3D printing flexible electronics. Nanoscale 10(43):20096–20107
99. Han Y, Dong J (2018) Electrohydrodynamic (EHD) printing of molten metal ink for flexible and stretchable conductor with self-healing capability. Adv Mater Technol 3(3):1700268
100. Neumann TV, Dickey MD (2020) Liquid metal direct write and 3D printing: a review. Adv Mater Technol 2000070
101. Han Y, Dong J (2017) High-resolution direct printing of molten-metal using electrohydrodynamic jet plotting. Manufact Lett 12:6–9
102. Kang L, Chen H, Yang Z-J, Yuan Y, Huang H, Yang B et al (2018) Seesaw-like polarized transmission behavior of silver nanowire arrays aligned by off-center spin-coating. J Appl Phys 123(20):205110
103. Ladd C, So JH, Muth J, Dickey MD (2013) 3D printing of free standing liquid metal microstructures. Adv Mater 25(36):5081–5085
104. Lipomi DJ, Vosgueritchian M, Tee BC, Hellstrom SL, Lee JA, Fox CH et al (2011) Skin-like pressure and strain sensors based on transparent elastic films of carbon nanotubes. Nat Nanotechnol 6(12):788–792
105. Li Y, Zhu H, Wang Y, Ray U, Zhu S, Dai J et al (2017) Cellulose-Nanofiber-enabled 3D printing of a carbon-nanotube microfiber network. Small Methods 1(10):1700222
106. Yamada T, Hayamizu Y, Yamamoto Y, Yomogida Y, Izadi-Najafabadi A, Futaba DN et al (2011) A stretchable carbon nanotube strain sensor for human-motion detection. Nat Nanotechnol 6(5):296–301
107. Dorozhkin PS, Tovstonog SV, Golberg D, Zhan J, Ishikawa Y, Shiozawa M et al (2005) A liquid-Ga-filled carbon nanotube: a miniaturized temperature sensor and electrical switch. Small 1(11):1088–1093
108. Wang K, Chang Y-H, Zhang C, Wang B (2016) Conductive-on-demand: tailorable polyimide/carbon nanotube nanocomposite thin film by dual-material aerosol jet printing. Carbon 98:397–403
109. Goh GL, Agarwala S, Yeong WY (2019) Aerosol-jet-printed preferentially aligned carbon nanotube twin-lines for printed electronics. ACS Appl Mater Interfaces 11(46):43719–43730
110. Sun S, Guo L, Chang X, Liu Y, Niu S, Lei Y et al (2019) A wearable strain sensor based on the ZnO/graphene nanoplatelets nanocomposite with large linear working range. J Mater Sci 54(9):7048–7061
111. Trung TQ, Ramasundaram S, Hong SW, Lee N-E (2014) Flexible and transparent nanocomposite of reduced graphene oxide and P(VDF-TrFE) copolymer for high thermal responsivity in a field-effect transistor. Adv Func Mater 24(22):3438–3445
112. Wei Y, Qiao Y, Jiang G, Wang Y, Wang F, Li M et al (2019) A Wearable skinlike ultra-sensitive artificial graphene throat. ACS Nano 13(8):8639–8647
113. Kabiri Ameri S, Ho R, Jang H, Tao L, Wang Y, Wang L et al (2017) Graphene electronic tattoo sensors. ACS Nano 11(8):7634–7641
114. Yan C, Wang J, Lee PS (2015) Stretchable graphene thermistor with tunable thermal index. ACS Nano 9(2):2130–2137
115. Zhu YS, Li JW, Cai HB, Wua YM, Ding HY, Pan N et al (2018) Highly sensitive and skin-like pressure sensor based on asymmetric double-layered structures of reduced graphite oxide. Sens Actuators B-Chem 255:1262–1267
116. Jason NN, Wang SJ, Bhanushali S, Cheng W (2016) Skin inspired fractal strain sensors using a copper nanowire and graphite microflake hybrid conductive network. Nanoscale 8(37):16596–16605
117. Arduini F, Zanardi C, Cinti S, Terzi F, Moscone D, Palleschi G et al (2015) Effective electrochemical sensor based on screen-printed electrodes modified with a carbon black-Au nanoparticles composite. Sens Actuators B-Chem 212:536–543

118. Kim J, Sempionatto JR, Imani S, Hartel MC, Barfidokht A, Tang G et al (2018) Simultaneous monitoring of sweat and interstitial fluid using a single wearable biosensor platform. Adv Sci (Weinh). 5(10):1800880
119. Mohan AMV, Kim N, Gu Y, Bandodkar AJ, You JM, Kumar R et al (2017) Merging of thin- and thick-film fabrication technologies: toward soft stretchable "Island–Bridge" devices. Adv Mater Technol. 2(4):1600284
120. Salvetat JP, Briggs GAD, Bonard JM, Bącsa RR, Kulik AJ, Stockli T et al (1999) Elastic and shear moduli of single-walled carbon nanotube ropes. Phys Rev Lett 82(5):944–947
121. Yu MF, Files BS, Arepalli S, Ruoff RS (2000) Tensile loading of ropes of single wall carbon nanotubes and their mechanical properties. Phys Rev Lett 84(24):5552–5555
122. Cao Q, Yu Q, Connell DW, Yu G (2013) Titania/carbon nanotube composite (TiO_2/CNT) and its application for removal of organic pollutants. Clean Technol Environ Policy 15(6):871–880
123. Sreekumar TV, Liu T, Kumar S, Ericson LM, Hauge RH, Smalley RE (2002) Single-wall carbon nanotube films. Chem Mater 15(1):175–178
124. Slobodian P, Riha P, Lengalova A, Saha P (2010) Compressive stress-electrical conductivity characteristics of multiwall carbon nanotube networks. J Mater Sci 46(9):3186–3190
125. Lamela H, Dadrasnia E (2012) Terahertz analysis reveals nanostructural dependence of carbon thin-film properties. SPIE Newsroom
126. Dai H, Wong EW, Lieber CM (1996) Probing electrical transport in nanomaterials: conductivity of individual carbon nanotubes. Science 272(5261):523–526
127. Bachtold A, Henny M, Terrier C, Strunk C, Schönenberger C, Salvetat JP et al (1998) Contacting carbon nanotubes selectively with low-ohmic contacts for four-probe electric measurements. Appl Phys Lett 73(2):274–276
128. Xu F, Wang X, Zhu Y, Zhu Y (2012) Wavy ribbons of carbon nanotubes for stretchable conductors. Adv Func Mater 22(6):1279–1283
129. Kim JH, Lee S, Wajahat M, Jeong H, Chang WS, Jeong HJ et al (2016) Three-dimensional printing of highly conductive carbon nanotube microarchitectures with fluid ink. ACS Nano 10(9):8879–8887
130. Goh GL, Agarwala S, Yeong WY (2019) Directed and on-demand alignment of carbon nanotube: a review toward 3D printing of electronics. Adv Mater Interfaces 6(4):1801318
131. Ribeiro B, Botelho EC, Costa ML, Bandeira CF (2017) Carbon nanotube buckypaper reinforced polymer composites: a review. Polímeros 27(3):247–255
132. Zhu Y, Xu F (2012) Buckling of aligned carbon nanotubes as stretchable conductors: a new manufacturing strategy. Adv Mater 24(8):1073–1077
133. Roh E, Hwang BU, Kim D, Kim BY, Lee NE (2015) Stretchable, transparent, ultrasensitive, and patchable strain sensor for human-machine interfaces comprising a nanohybrid of carbon nanotubes and conductive elastomers. ACS Nano 9(6):6252–6261
134. Doganay D, Cicek MO, Durukan MB, Altuntas B, Agbahca E, Coskun S et al (2021) Fabric based wearable triboelectric nanogenerators for human machine interface. Nano Energy
135. Zhao J, Zhang Y, Huang Y, Xie J, Zhao X, Li C et al (2018) 3D Printing fiber electrodes for an all-fiber integrated electronic device via hybridization of an asymmetric supercapacitor and a temperature sensor. Adv Sci (Weinh) 5(11):1801114
136. Abshirini M, Charara M, Marashizadeh P, Saha MC, Altan MC, Liu Y (2019) Functional nanocomposites for 3D printing of stretchable and wearable sensors. Appl Nanosci 9(8):2071–2083
137. Chen S, Huang T, Zuo H, Qian S, Guo Y, Sun L et al (2018) A Single integrated 3D-printing process customizes elastic and sustainable triboelectric nanogenerators for wearable electronics. Adv Func Mater 28(46):1805108
138. Castro Neto AH, Guinea F, Peres NMR, Novoselov KS, Geim AK (2009) The electronic properties of graphene. Rev Mod Phys 81(1):109–162
139. Tang X, Zhou H, Cai Z, Cheng D, He P, Xie P et al (2018) Generalized 3D printing of graphene-based mixed-dimensional hybrid aerogels. ACS Nano 12(4):3502–3511
140. Wang Y, Wang L, Yang T, Li X, Zang X, Zhu M et al (2014) Wearable and highly sensitive graphene strain sensors for human motion monitoring. Adv Func Mater 24(29):4666–4670

141. Wang Y, Yang R, Shi Z, Zhang L, Shi D, Wang E et al (2011) Super-elastic graphene ripples for flexible strain sensors. ACS Nano 5(5):3645–3650
142. Wang Z, Liu X, Shen X, Han NM, Wu Y, Zheng Q et al (2018) An Ultralight graphene honeycomb sandwich for stretchable light-emitting displays. Adv Func Mater 28(19):1707043
143. Bay HH, Vo R, Dai X, Hsu HH, Mo Z, Cao S et al (2019) Hydrogel Gate graphene field-effect transistors as multiplexed biosensors. Nano Lett 19(4):2620–2626
144. Sun Q, Kim DH, Park SS, Lee NY, Zhang Y, Lee JH et al (2014) Transparent, low-power pressure sensor matrix based on coplanar-gate graphene transistors. Adv Mater 26(27):4735–4740
145. Chen X, Shehzad K, Gao L, Long M, Guo H, Qin S et al (2019) Graphene hybrid structures for integrated and flexible optoelectronics. Adv Mater e1902039
146. Choi MK, Park I, Kim DC, Joh E, Park OK, Kim J et al (2015) Thermally controlled, patterned graphene transfer printing for transparent and wearable electronic/optoelectronic system. Adv Func Mater 25(46):7109–7118
147. Bellani S, Petroni E, Del Rio Castillo AE, Curreli N, Martín-García B, Oropesa-Nuñez R et al (2019) Scalable production of graphene inks via wet-jet milling exfoliation for screen-printed micro-supercapacitors. Adv Func Mater 29(14):1807659
148. Jakus AE, Secor EB, Rutz AL, Jordan SW, Hersam MC, Shah RN (2015) Three-dimensional printing of high-content graphene scaffolds for electronic and biomedical applications. ACS Nano 9(4):4636–4648
149. Kwon SN, Kim SW, Kim IG, Hong YK, Na SI (2018) Direct 3D Printing of graphene nanoplatelet/silver nanoparticle-based nanocomposites for multiaxial piezoresistive sensor applications. Adv Mater Technol 1800500
150. Chen C, Jiang J, He W, Lei W, Hao Q, Zhang X (2020) 3D printed high-loading lithium-sulfur battery toward wearable energy storage. Adv Funct Mater 30(10)
151. Savagatrup S, Chan E, Renteria-Garcia SM, Printz AD, Zaretski AV, O'Connor TF et al (2015) Plasticization of PEDOT:PSS by common additives for mechanically robust organic solar cells and wearable sensors. Adv Func Mater 25(3):427–436
152. Nair NM, Pakkathillam JK, Kumar K, Arunachalam K, Ray D, Swaminathan P (2020) Printable silver nanowire and PEDOT:PSS nanocomposite ink for flexible transparent conducting applications. ACS Appl Electron Mater
153. Nair NM, Pakkathillam JK, Kumar K, Arunachalam K, Ray D, Swaminathan P (2020) Printable silver nanowire and PEDOT:PSS nanocomposite ink for flexible transparent conducting applications. ACS Appl Electron Mater 2(4):1000–1010
154. Yuk H, Lu B, Lin S, Qu K, Xu J, Luo J et al (2020) 3D printing of conducting polymers. Nat Commun 11(1):1604
155. Kee S, Haque MA, Corzo D, Alshareef HN, Baran D (2019) Self-healing and stretchable 3D-printed organic thermoelectrics. Adv Func Mater 29(51):1905426
156. Park SH, Su R, Jeong J, Guo SZ, Qiu K, Joung D et al (2018) 3D printed polymer photodetectors. Adv Mater e1803980
157. Hinton TJ, Jallerat Q, Palchesko RN, Park JH, Grodzicki MS, Shue HJ et al (2015) Three-dimensional printing of complex biological structures by freeform reversible embedding of suspended hydrogels. Sci Adv 1(9):e1500758
158. Wang J, Lu T, Yang M, Sun D, Xia Y, Wang T (2019) Hydrogel 3D printing with the capacitor edge effect. Sci Adv 5(3):eaau8769
159. Li H, Tan YJ, Kiran R, Tor SB, Zhou K (2021) Submerged and non-submerged 3D bioprinting approaches for the fabrication of complex structures with the hydrogel pair GelMA and alginate/methylcellulose. Addit Manuf 37:101640
160. Jin Y, Compaan A, Chai W, Huang Y (2017) Functional nanoclay suspension for printing-then-solidification of liquid materials. ACS Appl Mater Interfaces 9(23):20057–20066
161. Zhou L, Ramezani H, Sun M, Xie M, Nie J, Lv S et al (2020) 3D printing of high-strength chitosan hydrogel scaffolds without any organic solvents. Biomater Sci 8(18):5020–5028
162. Yin X-Y, Zhang Y, Cai X, Guo Q, Yang J, Wang ZL (2019) 3D printing of ionic conductors for high-sensitivity wearable sensors. Mater Horiz 6(4):767–780

163. Yin XY, Zhang Y, Xiao J, Moorlag C, Yang J (2019) Monolithic Dual-Material 3D Printing of Ionic Skins with Long-Term Performance Stability. Adv Func Mater 29(39):1904716
164. Wang J, Liu Y, Su S, Wei J, Rahman SE, Ning F et al (2019) Ultrasensitive wearable strain sensors of 3D printing tough and conductive hydrogels. Polymers (Basel). 11(11):1873
165. Zhu Z, Park HS, McAlpine MC (2020) 3D printed deformable sensors. Sci Adv 6(25):eaba5575
166. Gao H, Li J, Zhang F, Liu Y, Leng J (2019) The research status and challenges of shape memory polymer-based flexible electronics. Mater Horiz 6(5):931–944
167. Kuang X, Chen K, Dunn CK, Wu J, Li VCF, Qi HJ (2018) 3D printing of highly stretchable, shape-memory, and self-healing elastomer toward novel 4D printing. ACS Appl Mater Interfaces 10(8):7381–7388
168. Meyers MA, Chen P-Y, Lin AY-M, Seki Y (2008) Biological materials: structure and mechanical properties. Prog Mater Sci 53(1):1–206
169. Abdollahi S, Markvicka EJ, Majidi C, Feinberg AW (2020) 3D printing silicone elastomer for patient-specific wearable pulse oximeter. Adv Healthc Mater 9(15):e1901735
170. Zheng R, Chen Y, Chi H, Qiu H, Xue H, Bai H (2020) 3D printing of a polydimethylsiloxane/polytetrafluoroethylene composite elastomer and its application in a triboelectric nanogenerator. ACS Appl Mater Interfaces 12(51):57441–57449
171. Guo Y, Chen S, Sun L, Yang L, Zhang L, Lou J et al (2020) Degradable and fully recyclable dynamic thermoset elastomer for 3D-printed wearable electronics. Adv Func Mater 31(9):2009799
172. Yao S, Vargas L, Hu X, Zhu Y (2018) A novel finger kinematic tracking method based on skin-like wearable strain sensors. IEEE Sens J 18(7):3010–3015
173. Myers AC, Huang H, Zhu Y (2015) Wearable silver nanowire dry electrodes for electrophysiological sensing. RSC Adv 5(15):11627–11632
174. Chortos A, Liu J, Bao Z (2016) Pursuing prosthetic electronic skin. Nat Mater 15(9):937–950
175. Edwards C, Marks R (1995) Evaluation of biomechanical properties of human skin. Clin Dermatol 13(4):375–380
176. Yao S, Zhu Y (2014) Wearable multifunctional sensors using printed stretchable conductors made of silver nanowires. Nanoscale 6(4):2345–2352
177. Gao Y, Yu G, Shu T, Chen Y, Yang W, Liu Y et al (2019) 3D-printed coaxial fibers for integrated wearable sensor skin. Adv Mater Technol 4(10):1900504
178. Wang Z, Gao W, Zhang Q, Zheng K, Xu J, Xu W et al (2019) 3D-Printed graphene/polydimethylsiloxane composites for stretchable and strain-insensitive temperature sensors. ACS Appl Mater Interfaces 11(1):1344–1352
179. Liu Z, Mo F, Li H, Zhu M, Wang Z, Liang G et al (2018) Advances in flexible and wearable energy-storage textiles. Small Methods 2(11):1800124
180. Praveen S, Santhoshkumar P, Joe YC, Senthil C, Lee CW (2020) 3D-printed architecture of Li-ion batteries and its applications to smart wearable electronic devices. Appl Mater Today 20:100688
181. Magno M, Boyle D (2017) Wearable energy harvesting: from body to battery. In: 2017 12th IEEE international conference on design & technology of integrated systems in Nanoscale Era (Dtis 2017)
182. Khalid S, Raouf I, Khan A, Kim N, Kim HS (2019) A review of human-powered energy harvesting for smart electronics: recent progress and challenges. Int J Prec Eng Manufact-Green Technol 6(4):821–851
183. Zhu M, Zhu F, Schmidt OG (2021) Nano energy for miniaturized systems. Nano Mater Sci 3(2):107–112
184. Zeng W, Tao X-M, Lin S, Lee C, Shi D, Lam K-h, et al. Defect-engineered reduced graphene oxide sheets with high electric conductivity and controlled thermal conductivity for soft and flexible wearable thermoelectric generators. Nano Energy. 2018;54:163–74.
185. Tang Z, Jia S, Shi X, Li B, Zhou C (2019) Coaxial printing of silicone elastomer composite fibers for stretchable and wearable piezoresistive sensors. Polymers (Basel) 11(4):666

Chapter 2
Additive Manufacturing of Energy Storage Devices

Xiaocong Tian and Kun Zhou

2.1 Background

Manufacturing methods are of significance for various material fields ranging from mechanical materials to functional materials. To fabricate three-dimensional (3D) objects or parts with expected materials and structures, three major technologies, including subtractive, formative, and additive manufacturing (AM), have been created and evolved rapidly. Compared with the AM, formative and subtractive manufacturing have a much longer history and are characterized as traditional manufacturing [1, 2]. Formative manufacturing refers to the production of products by means of casting, forging, injection molding, etc., in which tools or molds are needed during the fabrication process. Subtractive manufacturing is a widely applied method that relies on material removal processes such as cutting, milling, drilling, and grinding. These processes are conducted by removing materials from a bulk monolith with cutting tools, which enable manufactured objects to exhibit specific shapes and perform specific functions. According to the manufacturing principles, the above two traditional manufacturing technologies show low capability to create highly complex multi-component objects.

Over the years, AM has emerged as a new but efficient method using the additive-based manufacturing principle. AM is a formalized term which is more colloquially known as rapid prototyping (RP) and 3D printing, and received unprecedented levels

X. Tian (✉)
Faculty of Material Science and Chemistry, China University of Geosciences, Wuhan 430074, P. R. China
e-mail: tianxc@cug.edu.cn

Zhejiang Institute, China University of Geosciences, Hangzhou 311305, P. R. China

K. Zhou
School of Mechanical and Aerospace Engineering, Nanyang Technological University, 50 Nanyang Avenue, Singapore 639798, Singapore
e-mail: kzhou@ntu.edu.sg

K. Zhou (ed.), *Additive Manufacturing*,
https://doi.org/10.1007/978-3-031-04721-3_2

of interest due to the high flexibility in fabricating 3D objects. AM uses the layer-by-layer manufacturing principle based on computer-aided design (CAD) model files rather than tools or molds. Thus, AM shows unique advantages such as high design freedom, rapid prototyping, and customized fabrication. These are significantly important for customized and complex material architectures.

With the rapid increase in energy consumption and environmental pollution, clean and sustainable energy-based technologies are in great demand in the twenty-first century. Advanced electrochemical energy storage techniques, including the rechargeable battery technique and electrochemical capacitor technique, are promising candidates in addressing energy and environmental issues due to their inherent advantages such as good environmental friendliness, high safety, and high comprehensive electrochemical performance. These techniques have achieved great successes and have been widely used for portable electronics, electric vehicles, hybrid electric vehicles, and energy stations. For energy storage devices, manufacturing methods are of significance for the structural configuration and eventually for the mechanical properties and electrochemical performance of obtained materials and devices. To access robust mechanical properties and high electrochemical performance, traditional manufacturing methods of electrochemical energy storage devices have been developed rapidly with the scientific and technological progress, in which subtractive-based and formative-based manufacturing act as the mainstream. In recent years, AM has been explored as well for advanced energy storage materials and devices, and a wide range of advanced 3D architectures were manufactured (Fig. 2.1). In the meantime, some key advantages have been recognised and listed as follows [4]:

(1) AM allows a freeform and cost-effective fabrication and RP of energy storage materials and components with customized geometries.
(2) Chemical formula, external shapes, and internal microstructure can be readily tuned via AM.
(3) The manufacturing of components and the full device can both be achieved.
(4) AM provides a convenient way in building on-chip energy storage devices, especially in asymmetric configurations.
(5) For energy storage devices, their integration with electronics can be achieved via AM.

It should be noted that the optimization in AM largely affects the quality and electrochemical performance of obtained objects, and manufacturing parameters and raw feeding materials are key elements towards state-of-the-art energy storage materials and devices. In this chapter, basic AM techniques are described first, followed by the summaries of various additively manufactured energy storage materials. Advances on AM of electrochemical energy storage devices including rechargeable batteries and electrochemical capacitors are subsequently indicated. Outlook regarding the future development of AM of energy storage devices is presented at the end of this chapter.

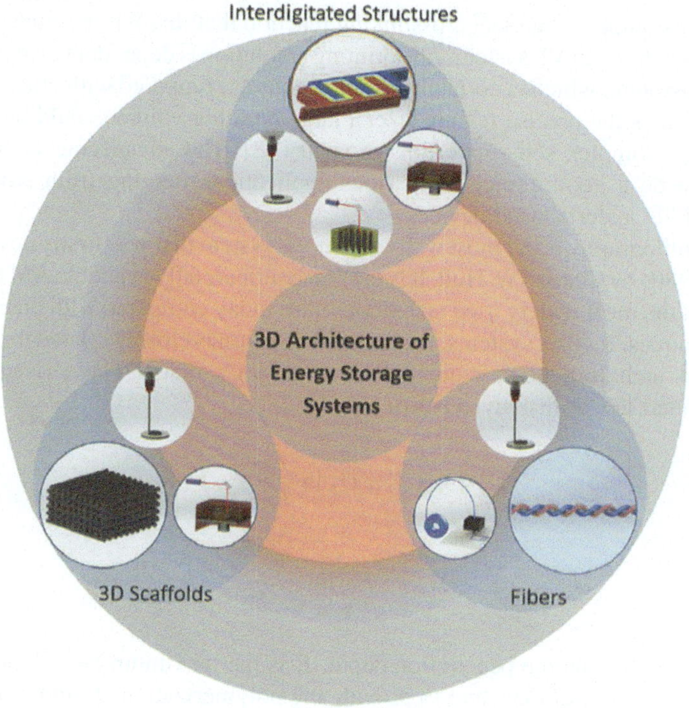

Fig. 2.1 Various 3D energy storage architectures fabricated by AM techniques [3]

2.2 Basic Additive Manufacturing Categories

According to the International Organization for Standardization/American Society of Testing and Materials (ISO/ASTM) 52,900 standard, AM can be classified into seven categories including vat photopolymerization (VP), material jetting (MJ), binder jetting (BJ), material extrusion (ME), powder bed fusion (PBF), directed energy deposition (DED), and sheet lamination (SL) [5]. Each specific technical category has its unique features such as printing sizes, printing resolution, material, and complexity of manufactured objects. During the manufacturing process, they also differ from each other in terms of the equipment cost, energy efficiency, and material utilization degree. For energy-related applications, these AM categories possess different advantages and disadvantages. To obtain desirable energy storage devices, a primary consideration is the selection of a specific AM manufacturing category that is appropriate for the entire manufacturing process.

Vat photopolymerization is the first-generation AM category that includes the stereolithography (SLA) and digital light processing (DLP) techniques. Both SLA and DLP use an ultraviolet (UV) light-based energy source to trigger the core photopolymerization process, where the liquid resins or photopolymers in a vat are selectively cured to form the solid 3D geometry. The photopolymerization process is usually a complex chemical crosslinking reaction that involves multiple chemical components [6, 7]. In VP systems, the printing resolution is dependent on the optics of the curing system, which is normally advantageous as compared with other AM categories. Besides the precise feature size, a large building volume could be expected with a high complex self-supporting geometry [8]. The advancements enable VP systems to be a good choice for various applications ranging from structural to functional 3D materials and devices.

Stereolithography is the most well-known VP system that was firstly developed in the mid-1980s by Charles W. Hull. It is the first commercially available AM technique and one of the most widely used techniques until today compared with other existing AM techniques. In SLA systems, the curing resolution is closely related to the curing parameters including the cure depth C_{d} and width C_{w} that can be derived from the Beer-Lambert law as displayed below [11–13]:

$$C_{\mathrm{d}} = D_p \ln \frac{E}{E_c} \tag{2.1}$$

$$C_{\mathrm{w}} = F\sqrt{\ln \frac{E}{\varphi \cdot E_c}} \tag{2.2}$$

where D_p represents the penetration depth, E is the maximum laser exposure, and E_c is the critical laser exposure to provide the polymerization. F and φ are related to the laser beam profile and resin nature.

Stereolithography processes can be classified according to the build-platform motion and laser movement [13]. Based on the build-platform motion, top-down and bottom-up approaches can be classified with different curing directions of the utilized laser. According to the laser movement, SLA can be classified into scanning-based stereolithography (SSL) and projection-based stereolithography (PSL). The SSL creates every layered two-dimensional (2D) pattern by scanning the layer surface, while the PSL utilizes a shot of masked laser light for every 2D fabrication layer [13]. In contrast to SSL, PSL possesses a lower resolution and is more suitable to manufacture small objects due to the limited sizes of the patterned laser light. A digital micromirror device has been used to combine SSL and PSL to form the scanning-projection-based stereolithography (SPSL), which makes it easier to do large-size printing with a high resolution.

With SLA techniques, polymer-based energy storage materials can be readily fabricated for favorable electrode or electrolyte components, templates or supports. He et al. used SLA to obtain 3D-Archimedean spiral-structured solid polymer electrolytes for all-solid-state lithium metal batteries (Fig. 2.2a–d) [9]. The rationally designed structure ensured the shortened Li-ion transport pathway, reinforced interfacial adhesion, and improved mass loading of active materials. A high ionic conductivity of 3.7×10^{-4} S cm^{-1} at 25 °C was achieved for the solid polymer electrolytes and a high specific capacity of 128 mAh g^{-1} was maintained after 250 cycles for Li|3D-printed solid polymer electrolyte|LFP cells. Zekoll et al. employed SLA to

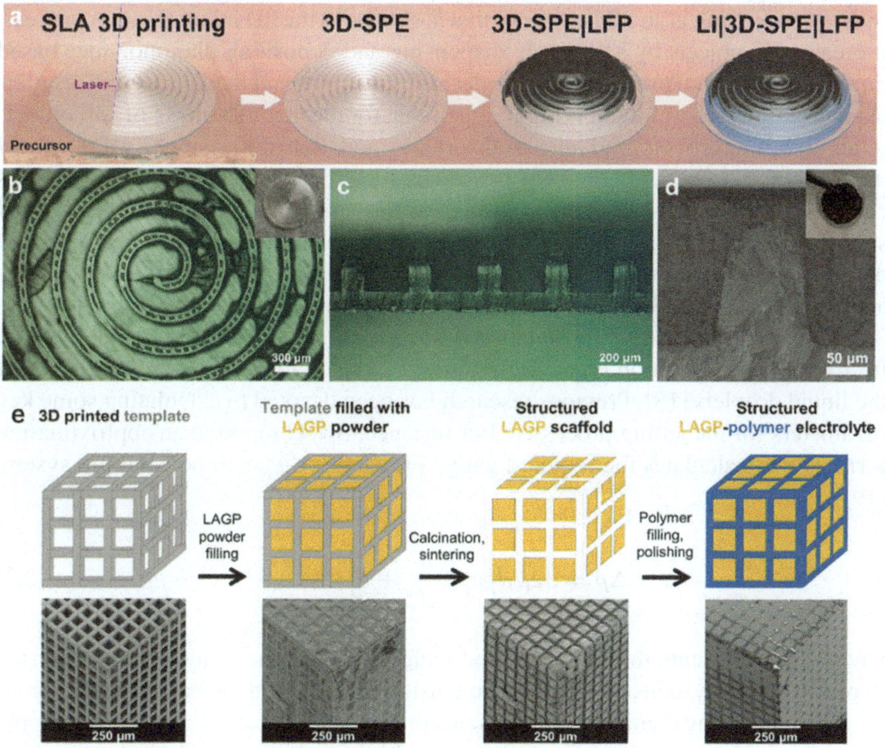

Fig. 2.2 **a** Schematic of fabrication process of all-solid-state lithium metal batteries based on 3D-printed solid polymer electrolytes. **b** Top view, and **c** Cross-sectional photographs of 3D-printed solid polymer electrolytes. **d** Cross-sectional SEM image showing interface between 3D-printed solid polymer electrolyte and electrodes [9]. **e** Schematic of the SLA-based templating procedure used for the fabrication of structured hybrid electrolytes with the example of the cube microarchitecture, and their corresponding SEM images of each fabrication stage [10]

design and construct cellular polymer templates. The filling of $Li_{1.4}Al_{0.4}Ge_{1.6}(PO_4)_3$ (LAGP), template removal, and filling of polymer were subsequently conducted to obtain hybrid LAGP-epoxy electrolytes with a 3D bicontinuous microstructure (Fig. 2.2e), where an ionic conductivity of 1.6×10^{-4} S cm^{-1} was measured at room temperature [10].

Digital light processing is another most common VP technique that uses photocurable resin for 3D printing [14]. The main difference between SLA and DLP is the energy source. In the case of DLP, a digital projector screen is applied to flash a detailed image of the entire layer, which speeds up the printing process. Similar to SLA, DLP can be classified into top-down and bottom-up manufacturing, in which the digital projector screen and resin vat are placed in different directions. The resolution in the *XY* plane is related to the pixel size of the projection, which is normally fixed in a DLP system. The resolution in the *Z*-direction is related to the cure depth of the resin, which represents the layer thickness in DLP printing.

Material jetting is an AM category that is similar to the 2D inkjet printing process. It creates 3D objects by selectively dispensing and depositing photopolymer-based liquid droplets from printingheads on desired substrates and then curing the droplets using UV light. Therefore, it is highly favorable for the manufacturing of soft objects (e.g., soft microrobots and micro absorbers) and four-dimensional (4D)-printed objects [15–17]. Nevertheless, MJ possesses some disadvantageous characteristics compared with other AM categories such as the less geometric complexity. In MJ systems, the liquid formulations play a key role in the jetting behavior of liquid droplets. To attain acceptable jetting behaviors, some substances such as surfactants are usually introduced to the liquid or ink in 2D ink-jet printing processes as well as 3D MJ. Besides the liquid formulations, some small changes to the liquid droplets, for example, the addition of small particles, can also tailor the jetting behaviors of the liquid droplets [18]. Previous research has been devoted to calculating some key parameters for the jetting processes. For instance, Sweet reported an approximation formula that calculates the required gauge pressure for a continuous jetting system [19]:

$$\Delta p = 32\mu d_j^2 v_j \int_{l_1}^{l_2} \frac{dl}{d_n^4} + \frac{2\sigma}{d_j} + \frac{\rho v_j^2}{2} \tag{2.3}$$

where Δp represents the total required gauge pressure; μ, ρ, and σ stand for the dynamic viscosity, density, and surface tension of the liquid, respectively. d_j means the resultant jetting diameter, d_n represents the inner diameter of the nozzle or supply tubing, v_j is the resultant jetting velocity, and l refers to the length of the nozzle or supply tubing. In this approximation, three pressure terms are included, in which the viscous friction in the nozzle and supply tubing, the surface tension, and the kinetic energy of the droplet or jet are corresponding relative elements.

To access the layer-by-layer assembly for MJ systems, the deposition of droplets on top of the previously printed layer should be further controlled with good bonding forces between the two adjacent layers. The liquid exits from the printing heads usually have two modes, the continuous stream (CS) mode and the drop-on-demand (DOD) mode, in which continuous liquid columns and discrete droplets are dispensed correspondingly. The DOD mode is widely applied with DOD printing heads, from which the droplets form only with the pressure pulses at specific times using piezoelectric, electrostatic, thermal, or other actuators. Currently, piezoelectric and thermal actuators are the dominant tools for DOD mode.

Binder jetting is an AM category that was developed in the early 1990s [20]. In contrast to MJ, BJ dispenses a liquid binding agent (binder droplets) to a powder bed to print a 2D pattern. Laser is not included in the manufacturing process, demonstrating that BJ is relatively inexpensive and fast. In BJ systems, the binder volume is related to binder saturation, which represents a significant input parameter by users and can be defined by Bai and Williams [21].

$$S = \frac{V_{\text{binder}}}{V_{\text{air}}} \tag{2.4}$$

where S is the binder saturation, V_{air} is the air space between powder particles, and V_{binder} is the binder volume. It is inferred that the binder saturation is an estimation of the powder bed density, and influences the quality (i.e., surface quality and geometric accuracy) of manufactured objects or parts [21]. Other parameters in BJ systems such as the raw feeding material formula, the formation of the powder bed, and post-processing treatments also affect the final properties of the objects or parts.

Material extrusion is a representative of the AM technology and has been widely used for industries, academic research, and even recreation. It involves the layer-by-layer deposition through a simple extrusion of feeding materials from a nozzle. ME facilities are simple, leading to relatively lower implementation complexity and lower cost compared with other AM categories. ME systems show a high capability in fabricating customized structures with reliable materials. The shapes and sizes of the extruded material filaments are highly related to the extrusion nozzles, of which diameter also determines the minimum feature size and printing resolution for ME systems. Besides, other parameters such as post-processing treatment also affect the final quality of additively manufactured objects or parts.

Fused deposition modeling (FDM) and direct ink writing (DIW) are two ME examples. FDM utilizes solid filaments or pellets, typically thermoplastic polymers such as polylactic acid and acrylonitrile–butadiene–styrene, as the feeding materials to be melted and then solidified immediately to enable the layer-by-layer deposition on target substrates. Solidification is basically dependent on the crystallization and chain entanglement of polymers [24]. It can be easily inferred that the FDM-based manufactured materials are polymers, while the addition of additive materials may bring a supplement to manufactured polymer-based composites.

Instead of solid filaments or pellets, DIW uses printable inks as the feeding materials. For this technique, the ink formula can be easily tuned during the ink preparation process, and the ink extrusion is conducted without the presence of photopolymerization. To meet the printability criteria, the ink rheological properties should be controlled at a suitable range to deliver a continuous and self-supporting behavior, which enables the 3D layer-by-layer stacking. Diverse driving forces for the ink dispensing have been demonstrated and investigated as shown in Fig. 2.3a [22]. The diversity of DIW-based additively manufactured materials is highly advantageous to other AM techniques, and the manufactured materials include polymer-based, ceramic-based, metal-based, oxide-based, and even nanomaterial-based ones. Besides, the chemical components of manufactured materials can also be tailored to optimize unique functions for various applications. Recently, a multimaterial multinozzle 3D printing system was developed (Fig. 2.3b), where multimaterial printing was brought to the next level [23]. It is worth noting that, similar to DIW, other ink-based printing techniques such as screen printing and ink-jet printing also need to prepare suitable ink, while some differences still lie in ink rheological behaviors. For example, higher viscosity is normally expected from the printable ink for DIW than for screen printing and ink-jet printing.

Powder bed fusion is an important member of the first commercialized AM categories. It uses focused energy sources to selectively fuse or melt the powder particles that are spread over the build platform and assemble the fused particles layer-by-layer,

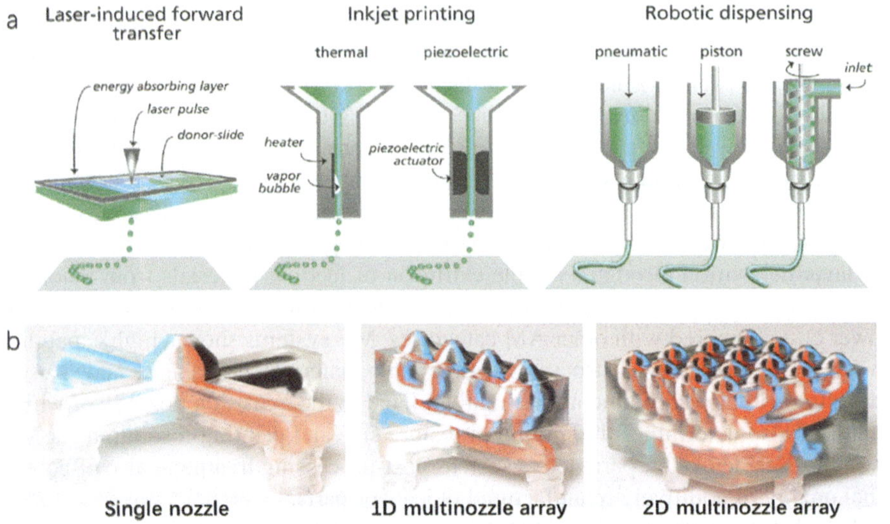

Fig. 2.3 **a** Schematic of diverse ink-based fabrication approaches [22]. **b** Multimaterial multinozzle 3D printing with zero-dimensional (0D), 1D, and 2D printing heads [23]

where polymer-based and metal-based powders are two major categories of powder particles. Frequently used focused energy sources include the laser beam, infrared ray, and electron beam. Selective laser sintering (SLS), selective laser melting (SLM), multi-jet fusion (MJF), and electron beam melting (EBM) are typical PBF processes.

Selective laser sintering is the first commercial PBF system, and has attracted extensive attention with the capability of producing large-scale polymer-based objects or parts. It utilizes high-power laser such as CO_2 laser as the focused energy source to conduct the selective sintering of solid-state polymer-based powders such as the polyamide 12 (PA12) powder. To obtain a desired result, the processing temperature is a key element. For commonly applied semicrystalline polymers, crystallization should be inhibited during the process, indicating that the processing temperature should be controlled precisely at a range between the melting temperature (T_m, the red line, Fig. 2.4) and the crystallization temperature (T_c, the blue line, Fig. 2.4) of the polymer. For a polymer, the sintering window of SLS processing represents the meta-stable thermodynamic region of the undercooled polymer melt. Figure 2.4 displays the differential scanning calorimetry-thermogram curve for the commercial PA12 powder, showing the sintering window region [25, 26].

In addition to SLS, SLM and EBM are also efficient and widely applied PBF techniques that are mainly used for metal-based 3D printing. SLM and EBM utilize high-power lasers such as Nd: YAG laser and electron beam, respectively, to selectively induce fusion between metal-based powders on the build platform [5]. They are suitable for manufacturing electrically conductive materials such as alloys and their composites. For PBF processes such as SLS, SLM, and EBM, the surface finish and accuracy (resolution) of additively manufactured materials are highly

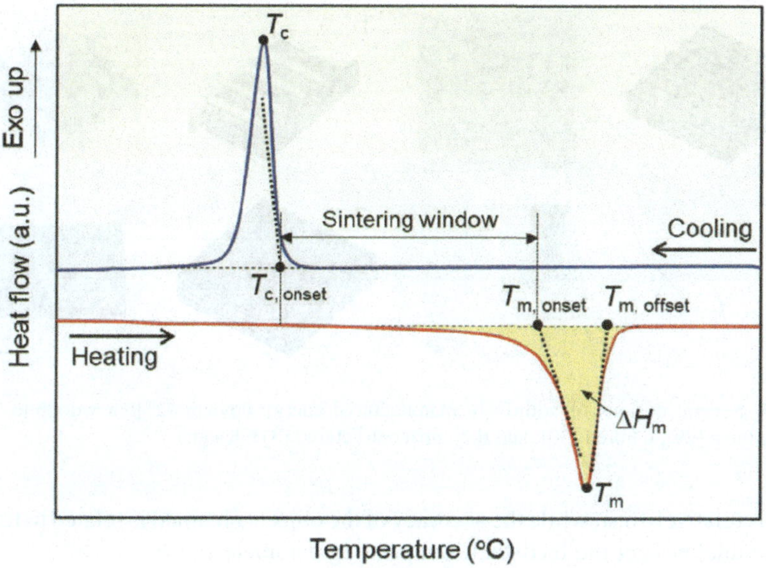

Fig. 2.4 Typical differential scanning calorimetry (DSC) curve of a polymer and the parameters for evaluating its sinterability [25]

related to the features such as the powder particle size and the operating conditions during the manufacturing process.

Similarly to some PBF techniques such as EBM, DED refers to an AM category that applies focused energy sources such as laser or electron beam as its energy source to fuse or melt the materials. The main difference between PBF and DED is the material feeding method, where materials are fused or melted simultaneously as the feeding materials are deposited from nozzles in the DED process. Powder feeding and wire feeding are two major feeding forms, where the latter one is less accurate but more material-efficient [26]. Although polymer-based and ceramic-based materials have been applied in DED systems, DED is mostly used for manufacturing metal-based materials and is also usually known as the "metal deposition" technique. DED shows good capability in building dense objects or parts, while the obtained resolution (accuracy) and surface finish are limiting factors. In addition, the manufacturing speed is a limitation for most DED systems. To overcome these limitations, tuning of important manufacturing parameters such as sizes and power of focused energy sources is required.

Sheet Lamination refers to a typical AM process that deposits material in a sheet form (such as plastic, paper, and metal foil) followed by key operations including stacking, bonding, and cutting. Laminated object manufacturing (LOM) is the first commercialized SL technique using paper-based materials. To date, a wide range of materials, including paper, polymer, ceramic, metal, and composite materials, can be manufactured through the SL system, where ultrasonic welding, thermal bonding, adhesive bonding, and clamping are commonly used methods. The manufacturing

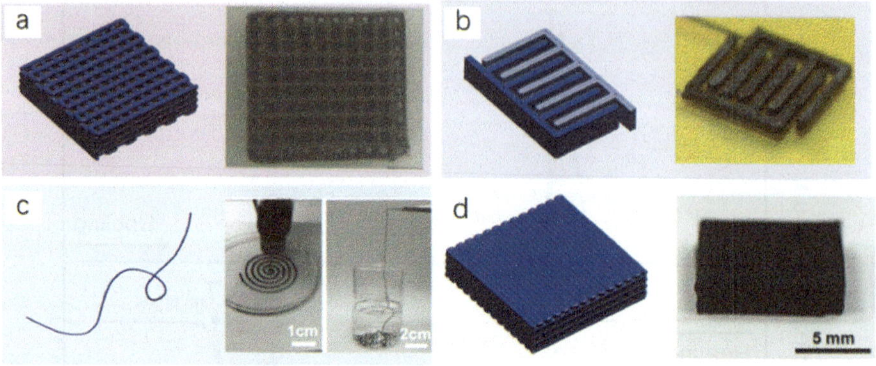

Fig. 2.5 Several designs of additively manufactured energy devices [27]: **a** woodpile [28], **b** interdigitated [29], **c** fibred [30], and **d** compact-structured [31] designs

speed is relatively fast, while the accuracy of the objects obtained is related to features like the thickness of the feedstock and printing parameters.

For energy storage applications, various AM techniques possess advantages and disadvantages with their unique inherent manufacturing properties. To select a suitable AM technique, two major perspectives, namely the use of additively manufactured materials and the type of energy-storing mechanisms, should be primarily considered for the fabrication of desirable energy storage devices with rational 3D designs. Some common 3D designs are displayed in Fig. 2.5, where the woodpile [28], interdigitated [29], fibred [30], and compact-structured [31] designs are included for various application scenarios [27]. For example, the interdigitated design is commonly used for on-chip and miniaturized energy storage devices where two interdigitated electrode fingers are placed at an in-plane dimension and act as cathode and anode. The interspace is normally filled with electrolytes for the full-scale energy storage devices. Another example, the fibered design, is usually employed for the flexible and wearable energy storage applications, where the robust mechanical properties should be also considered in addition to desirable electrochemical performance. In the following sections, diverse additively manufactured materials for energy storage applications and recent advancements in AM of rechargeable batteries and electrochemical capacitors are described and discussed.

2.3 Additively Manufactured Materials for Energy Storage

2.3.1 Polymer-Based Materials

Among various AM classifications, majority of them have demonstrated viability in the fabrication of polymer-based 3D objects, parts, and devices. To access the polymer-based AM processing, the feeding materials are key elements, of which powder, filament, and sheet forms have been used [32]. With specific AM processing methods, these feeding polymer materials are commonly processed by UV light, laser, or visible light in a layer-by-layer manner to obtain structures with pre-designed 3D structures. To realize the photocuring process, photosensitive feeding materials are essential while different mechanisms are used to initiate polymerization. For example, in SLA systems, the most widely used mechanism is via cationic photopolymerization or hybrid photopolymerization with photosensitive resin-based feeding materials [33]. Powder-formed polymer materials are adopted in the SLS technique. To fabricate polymer composite powders from pure polymer, two major routes are summarized in Fig. 2.6 [25]. Pure polymer pellets and pure polymer powder are used as typical starting materials, and some composite powder preparation methods can be employed in AM techniques. For instance, dry mixing, wet mixing, melt blending and mechanical grinding, solution mixing, and in situ polymerization are applied for SLS.

According to the formulation, additively manufactured polymer materials can be classified into two major categories, pure polymer materials and polymer-based composite materials. Since most polymers are electrically insulative, the manufactured pure polymer materials are normally not suitable to work as the current collector or electrode. However, it is worth noting that some conductive polymers have been developed with the AM technique. Yuk et al. reported a 3D printable conductive polymer ink based on poly(3,4-ethylenedioxythiophene):polystyrene sulfonate (PEDOT:PSS) and fabricated electrically conductive pure polymers with high-aspect-ratio and high-resolution microstructure by DIW [34]. The conductive polymer ink enabled superior 3D printability and can be readily integrated with

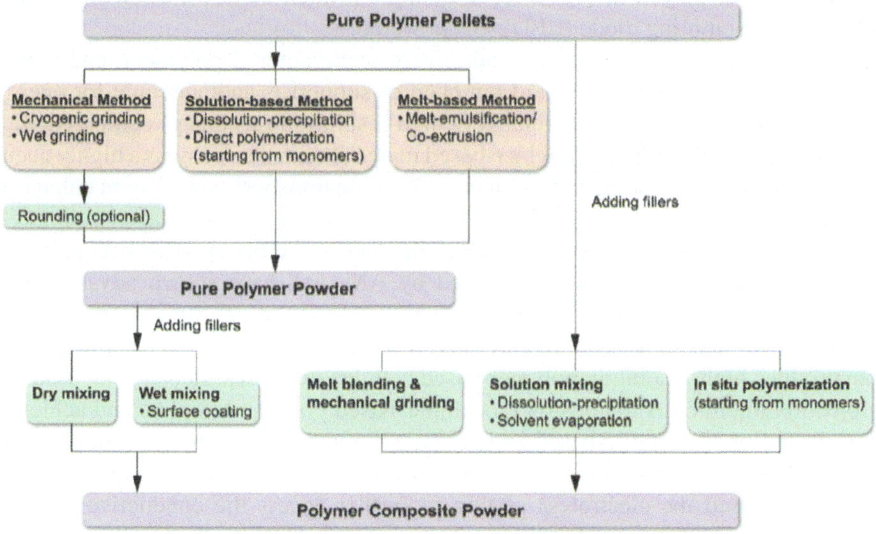

Fig. 2.6 Summary of the composite powder preparation methods [25]

other materials. Such a 3D-printed conductive polymer is a typical element for flexible electronic circuits and the soft neural probe. The development of AM conductive polymers is promising and possesses enormous potential for next-generation advanced energy storage applications.

Instead of being the electrode or current collector, 3D-printed polymers are ideal electrolyte materials, and research work on polymer-based AM has been reported. Solid-state polymer-based electrolytes can address some safety issues of liquid electrolytes such as overheating, leakage, and explosion. Using an SLA-type 3D gel printer, Rahman et al. fabricated gel polymer electrolytes containing N,N-dimethylacrylamide and polyvinylidene fluoride [36]. A high ionic conductivity of 6.5×10^{-4} S cm^{-1} was achieved at room temperature. AM of polymer-based templates is an alternative approach for fabricating electrolytes with controllable microstructure. LAGP-polymer hybrid electrolytes with a 3D bicontinuous microstructure were induced from 3D-printed cellular polymer templates, where various structures including cubic, gyroidal, diamond, and spinodal (bijel) structures were designed and obtained readily [10].

Additively manufactured polymer-based composite materials are also promising candidates for electrode components in energy storage applications. Reyes et al. applied FDM to achieve the AM of poly(lactic acid) (PLA)-based cathode, anode, and separator materials using a low-cost fused filament fabrication (FFF) 3D printer and obtained the 3D-printed Li-ion battery full-cells as shown in Fig. 2.7a, b [35]. The diameter and thickness of additively manufactured cathodes, anodes, and separators were 16 mm and 150 μm, respectively, and an infusion with 1 M $LiClO_4$ in ethyl methyl carbonate/PLA was conducted before the battery assembly. It is worth noting that the AM of the current collector by FDM using the electric filament was also reported. As displayed in Fig. 2.7c, d, a single printed battery was also reported, where a light-emitting diode (LED) was powered.

In addition to the direct use of additively manufactured polymers, employing polymer-based materials as a hard template or support is another route for energy storage applications due to its relatively high resolution (accuracy) compared with other materials. The derived polymer-based materials can thus possess a highly accurate microstructure, leading to fast kinetics for ionic transport as well as an enhanced electrochemical behavior.

Bio-inspired cellular structures such as the lattice and honeycomb structures are representative ones that can be obtained by AM, and have shown advantageous properties such as high surface area and light weight for energy storage applications compared to their block counterparts [5]. SLA is one of the good candidates that can produce high-resolution polymer-based cellular structures. Zheng et al. used the projection micro-SLA to construct a cellular-based polymer template and achieved the conversion to metallic-based and ceramic-based ones after post-treatment [37]. Sun et al. utilized projection SLA to fabricate the polymer-based octet-truss lattice (OTL) and used the electroless plating to further access the conductive Ni/OTL support [38]. For the supercapacitor application, the $CoNi_2S_4$/Ni/OTL and carbon materials/Ni/OTL electrodes were obtained, resulting in a high-performance lattice

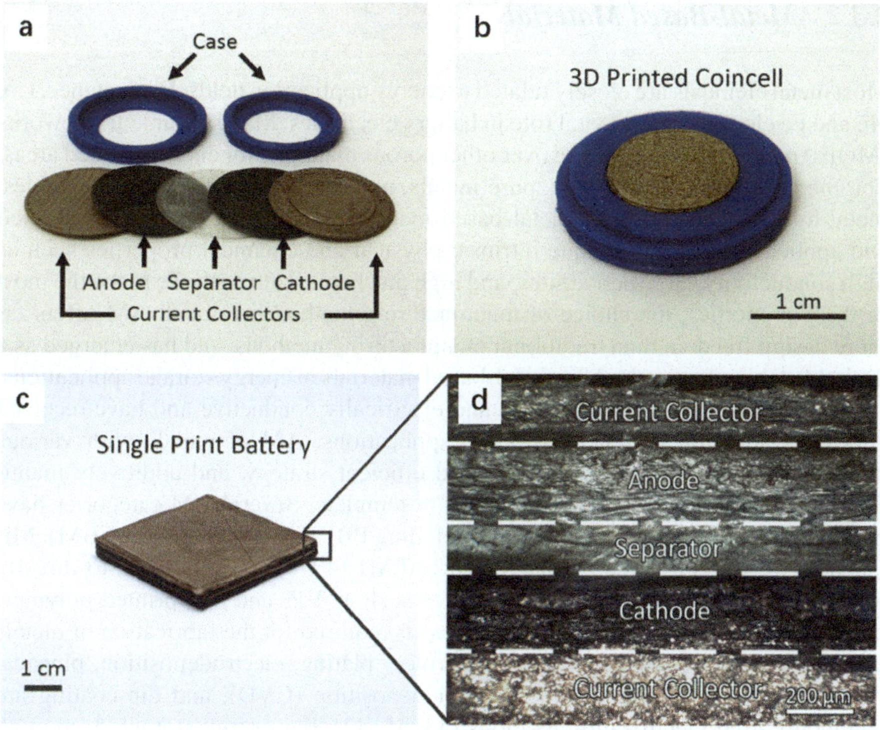

Fig. 2.7 Photographs of **a** individual components and **b** the assembled 3D-printed coin cell. **c** Photograph and **d** corresponding cross-sectional microscope images of a 3D-printed battery [35]

asymmetric all-solid-state supercapacitor, representing a typical AM example of polymer-based materials for energy storage.

The rapid development of wearable bioelectronic devices and information technologies has significantly driven the need for implantable energy storage devices. AM offers a platform for building implantable polymer-based materials, which hold good promise for implantable energy storage applications. Electrochemical performance, biocompatibility, and biodegradability of additively manufactured materials or devices are key determinant factors for implantable energy storage. Krishnadoss et al. presented ionic liquid functionalized biopolymers as electrolytes using the DIW technique [39]. The obtained polymer-based electrolytes could be interdigitated into soft micro-supercapacitor prototype structures, which exhibited superior electrochemical performance, good biocompatibility, and good electromechanical properties, demonstrating that additively manufactured polymers can act as a suitable material for implantable energy storage device components.

2.3.2 *Metal-Based Materials*

Most metal elements are closely related to energy application fields. For instance, Co, Ni, and Fe elements play a vital role in battery electrodes. Metal organic frameworks (MOFs) possess higher promise over other porous materials for energy-related areas. For energy storage applications, pure metals, metal oxides, metal dichalcogenides, metal hydroxides, and other metal-based materials have been extensively studied and applied due to their unique intrinsic physical and chemical properties such as high conductivity, large active sites, and high mechanical strength. To make the most of these properties, the choice of manufacturing method is critical. AM offers far more design freedom than traditional manufacturing methods, and has emerged as a desirable fabication method for metal-based materials in energy storage applications.

Pure metals are highly thermally and electrically conductive and have received enormous attention for energy-storage applications. AM of metals with various microstructure provides an emerging and efficient strategy, and additively manufactured metals can act as robust support or template. Several AM categories have been employed to manufacture metals including PBF (such as SLM and EBM), ME (such as FDM), VP (such as SLA), BJ, and MJ [40–44]. PBF is able to directly manufacture metals, while other categories such as VP- and ME-printed polymer-based templates and a metallization process is required for the fabrication of metals or metal-polymer composites [45]. Electroless plating, electrodeposition, physical vapor deposition (PVD), chemical vapor deposition (CVD), and dip-coating are commonly used metallization methods [43, 45–47]. For energy storage, Lim et al. reported a robust pure Cu porous framework by extrusion 3D printing for advanced lithium metal anodes [48]. With uniformly-plated Li, the 3D-printed Cu framework achieved a high areal capacity of 20 mAh cm^{-2}, demonstrating a promising application of additively manufactured pure metals as a support for energy storage.

Metal anodes are considered as one of the most promising anodes for next-generation energy storage. Among various metal anodes, Li metal is regarded as the ultimate anode owing to its ultrahigh theoretical capacity and low redox potential [49]. Currently, it is challenging to directly fabricate metal anodes with AM, thus the fabrication of additively manufactured supports is an alternative choice. Cao et al. used a cellulose nanofiber (CNF) ink to fabricate the carbonized CNF/Li metal anodes that enabled the formation of high-performance Li metal microbatteries [50].

With rationally designed porosity, pure metals offer more beneficial microstructural features including larger surface areas and more active sites, making them extremely suitable for energy storage applications. The fabrication of porous metallic structures is normally conducted through dealloying such as selective leaching and hydrothermal synthesis [51, 52]. In contrast, AM is a more advanced technology. According to the manufacturing principle, the precise structural control of metals at varying length scales and highly complex geometries can be readily achieved [53]. Saleh et al. compared porous solid truss Ag lattices with simple Ag block electrodes with the same overall sizes for lithium storage [40]. Porous solid truss Ag lattice fabricated via aerosol jet 3D printing showed an unprecedented improvement in electrochemical performance compared with the block one, which was highly related to the internal microstructure of electrodes. As displayed in Fig. 2.8a, b, as the unit cell increases, porous solid truss lattices exhibit a growing surface area to volume ratio than simple blocks, implying an increased porosity. Such an increase led to the enhancement of energy-storage mechanical and electrochemical properties as well

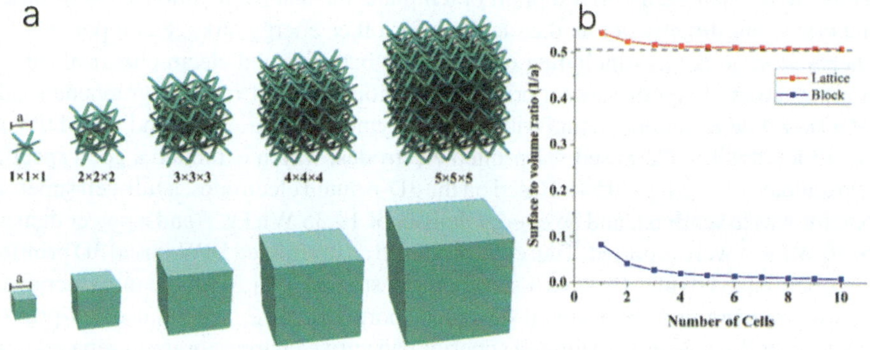

Number of Cells	1×1×1	2×2×2	3×3×3	4×4×4	5×5×5	6×6×6	7×7×7	8×8×8	9×9×9	10×10×10
Surface area per unit volume ratio (Micro-lattice/Block)	0.5	1.2	2.0	2.7	3.4	4.1	4.8	5.6	6.3	7.0
Surface area per unit mass ratio (Micro-lattice/Block)	6.8	15.8	24.9	34.1	43.2	52.4	61.6	70.8	80.0	89.2

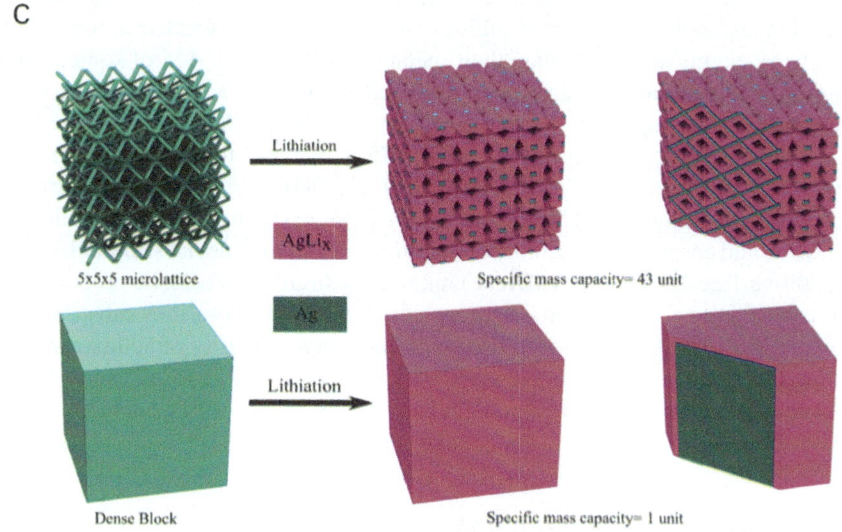

Fig. 2.8 **a** Schematic of porous solid truss lattices and simple blocks with the same overall sizes. **b** A comparison of the surface area to volume ratio as the number of cells increases. **c** Schematic of different lithiation processes for porous solid truss lattices and simple blocks with the same overall sizes [40]

(Fig. 2.8c). During the charge and discharge process, the 400% increase in specific capacity and 100% increase in areal capacity was found for porous latticed electrodes, demonstrating a full use of electrode from the volumetric perspective. The increased mechanical robustness was further confirmed with rationally designed pores.

Other metal-based materials such as metal oxides are excellent candidates for AM as energy storage electrode components. For both batteries and electrochemical capacitors, abundant active sites and desirable mass transport are necessary. Thus, rather than the use as a support or template, most abovementioned metal-based materials can directly act as the electrode or other energy storage components for energy storage devices including rechargeable batteries and electrochemical capacitors. Zhao et al. synthesized a printable ink consisting of reduced graphene oxide (rGO), α-MnO_2 nanorods, and triblock copolymer Pluronic F127 and used DIW to construct metal oxide-based 3D-printed electrodes, which exhibited a good specific capacitance of 422 F g^{-1} [54]. Based on the 3D-printed electrodes, a full-cell supercapacitor was assembled, and an energy density of 19.35 Wh kg^{-1} and a power density of 50 W kg^{-1} were reported. Tian et al. proposed an optimized DIW-based 3D printing of cobalt-based oxides for micro-sized energy storage [55]. With the introduction of holey graphene and the rational design of pore structure, as-obtained 3D-printed micro-pseudocapacitors exhibited superior capacitive charge storage compared with the non-optimized micro-pseudocapacitors, illustrating the importance of material and microstructural engineering for AM of metal oxides for high-performance energy storage.

Besides metal oxides, other metal-based materials have also demonstrated better electrochemical potential compared with the corresponding metal oxide counterparts. For example, some metal sulfides can exhibit higher electrical conductivity and electrochemical activity than their metal oxides [56, 57]. AM of some of these metal-based materials has also been explored as well to access desirable energy storage devices. Tang et al. fabricated leaf-inspired $Ni_{0.33}Co_{0.66}S_2$/graphene (3DP-NCS/G) aerogels via DIW-based 3D printing [58]. Notably, the Ni-Co-based sulfides were converted from $Ni_{0.33}Co_{0.66}(OH)_2{\cdot}xH_2O$ (NCOH) precursors via a sulfurization procedure following DIW. The obtained 3DP-NCS/G exhibited good electrochemical behaviors and ensured the assembly of hybrid energy storage devices. Orangi reported an additive-free 2D $Ti_3C_2T_x$ (MXene) ink with desirable viscoelastic properties and fabricated MXene-based micro-supercapacitors via DIW on various substrates such as paper and polymer films [59]. The maximum energy density of additively manufactured micro-supercapacitors reached 51.7 μW h cm^{-2}. The results show that the MXene-based ink is a promising choice for flexible energy storage devices.

2.3.3 Carbon-Based Materials

Carbon-based materials are the most widely investigated materials due to their unique properties such as ease of processability, low costs, tunable porosity, controllable electrochemical activity, and good chemical stability [60]. They are thus regarded as "star materials" for energy storage applications and have been extensively applied. Carbon materials can be classified into four major categories, being 0D carbon, 1D carbon, 2D carbon, and 3D carbon. Carbon-based materials are one of the first studied materials manufactured via AM processes. Most AM categories including VP, ME, MJ, BJ, PBF, and SL have demonstrated efficacy for the fabrication of carbon-based materials [61–65]. Through these AM processes, pure carbon and carbon-based materials have been both employed to fabricate geometrically controlled objects, which is a clear indication that 3D carbon-based materials can be readily manufactured with low-dimensional carbon components. Such manufacturing offers a powerful platform for diverse functional applications including advanced energy storage.

Activated carbon, carbon onions, and carbon quantum dots are normally regarded as representative 0D carbon that is also known as carbon nanoparticles. Activated carbon is the most common one which can be economically produced from carbon-based precursors through chemical or physical activation [66]. Areir et al. developed an activated carbon-polyvinyl alcohol (PVA)-based ink and fabricated activated carbon electrodes based on the obtained ink using FDM/paste deposition for supercapacitor application [67].

Besides AM of 0D carbon, AM of 1D carbon such as carbon nanotubes (CNTs) and carbon fibers has also attracted great attention in the field of energy storage [66]. 1D carbon offers a short diffusion length for ions as well as continuous transport channels for electrons, which are highly beneficial for improving the kinetics of charge carriers. AM provides an ideal platform for building 1D carbon-based materials with customized geometries. Yu et al. prepared a CNTs-based 3D printable ink by milling CNTs in a solution composed of isopropyl alcohol (IPA), ethylene glycol, and a dispersion agent [68]. CNTs-based micro-supercapacitors were fabricated via DIW and a high areal capacitance along with good cycling stability was measured. The same DIW method was used by Chen et al. to construct single-wall CNTs-based electrodes and fully packaged supercapacitors on flexible substrates [69]. As-printed supercapacitors array with three cells could be readily connected in series, clearly implying the scalability of additively manufactured carbon material-based electrochemical energy storage devices.

2D carbon becomes increasingly popular for a wide range of applications. Graphene is a representative single-atom-thick 2D honeycomb-structured carbon that was first isolated by Novoselov et al. and caused the "graphene gold rush" [70]. Mechanical exfoliation, CVD growth method, and liquid-phase exfoliation are commonly used methods to synthesize graphene [71]. Among these methods, although the liquid-phase exfoliation from graphite suffers from low quality, it is capable of yielding relatively high production. During the exfoliation process, an excellent dispersion state of graphene oxide (GO) nanosheets can be achieved as well, leading to an ease of GO-based 3D printable ink [72]. Thus, a wide range of previous research has been conducted to realize the AM of graphene-based objects for various applications such as gas sensing, tissue engineering, and energy [28, 73–75].

For energy storage, graphene and graphene-based AM attract growing attention, where the appealing intrinsic properties of graphene have been excavated to a certain extent. Zhu et al. used the high-concentration GO suspension to fabricate a 3D-printed periodic microlatticed graphene aerogel via DIW [76]. An enhanced Young's modulus was observed for as-manufactured 3D-printed graphene compared with those of bulk graphene materials, illustrating the advantageous mechanical property of additively manufactured graphene materials. Tang et al. prepared a hybrid-dimensional ink composed of 0D Fe_2O_3 nanoparticles, 1D Ag nanowires, and 2D graphene nanosheets, and fabricated hybrid-dimensional electrode-based micro-supercapacitors [77]. A maximum device areal capacitance of 412.3 mF cm^{-2} and good flexibility were both achieved, demonstrating the importance of ink formulation for desirable energy storage devices.

Similarly to metal-based materials, additively manufactured graphene is also suitable as the electrical support. Wang et al. employed DIW to construct interdigitated graphene framework as a superior support of metal oxide nanostructures [78]. With further filled NiO nanorods, a device areal capacity of 197.5 mC cm^{-2} was achieved. Besides NiO nanorods, MnO_2 nanosheets were also employed for high-performance symmetric micro-pseudocapacitors.

Graphene-CNTs hybrids are also promising candidates for advanced energy storage, where the intrinsic advantages of graphene and CNTs can be both retained. Tang et al. proposed a generalized DIW-based 3D printing process for mixed-dimensional hybrid aerogels including graphene/multiwalled CNTs hybrids [79]. As shown in scanning electron microscopy (SEM), transmission electron microscopy (TEM), and optical images (Fig. 2.9a–d), hybrid microstructural morphology was observed for as-printed objects. Compared with the conventional thick-film electrode

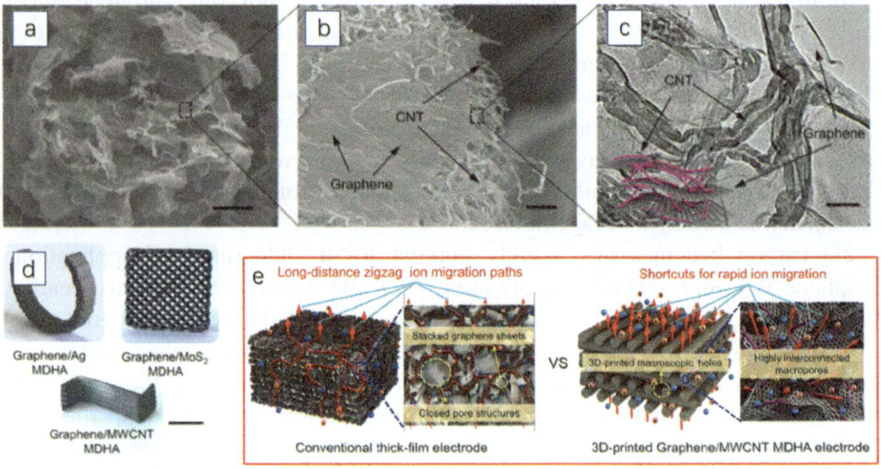

Fig. 2.9 **a**, **b** SEM images and **c** TEM image (the inset is the schematic) of graphene/multiwalled CNTs mixed-dimensional hybrid aerogels. **d** Optical image of various 3D-printed mixed-dimensional hybrid aerogels. **e** Schematics of comparison of the ion migration between conventional thick-film electrode and the 3D-printed graphene/multiwalled CNTs mixed-dimensional hybrid aerogel electrode [79]

that induced insufficient ion migrations, 3D-printed hybrid microstructure could ensure sufficient ion migrations (Fig. 2.9e) and result in enhanced electrochemical behaviors.

Carbon-based materials with various dimensions deliver their own advantages and disadvantages in energy storage. AM of these carbon-based materials for energy storage applications represents a huge advancement in the field of AM, whose scope is significantly expanded. However, although some progress has been made, there are still many limitations for AM of carbon-based materials, such as the precise control of material microstructure as well as the further improvement of electrochemical energy storage behaviors.

2.3.4 *Ceramic-Based Materials*

Ceramic materials are electrical insulators, which are inactive when working as electrodes for electrochemical energy storage. Instead of acting as electrodes, ceramic-based materials can be used as key components for electrolytes or separators. For solid-state energy storage, ceramics and polymers are two mainly investigated solid-state electrolyte material types that possess high ionic conductivity and intimate interface contacts, respectively [80]. AM of ceramic-based materials can be processed via PBF, ME, BJ, VP, and SL. For energy storage applications, most AM of ceramics was achieved using ceramic-based printable ink. Blake et al. proposed a N-methyl-2-pyrrolidone/glycerol solvent system and obtained the ink with a high ceramic loading for printing [81]. The as-printed ceramic-polymer electrolytes enabled good electrochemical performance (i.e., 5 C).

Cheng et al. engineered an elevated-temperature DIW 3D-printer and prepared a poly(vinylidene fluoride-hexafluoropropylene)-based electrolyte ink containing TiO_2 ceramic nanoparticles [82]. Based on the ceramic-polymer electrolyte, a 3D-printed interdigitated full cell battery was fabricated with six-layer-printed $LiFePO_4$ (LFP) and $Li_4Ti_5O_{12}$ (LTO) electrodes.

Geometrical customization is a key feature of additively manufactured materials. Liu et al. prepared various 3D printable $Li_{1.3}Al_{0.3}Ti_{1.7}(PO_4)_3$ (LATP)-based inks and obtained ceramic and hybrid solid-state electrolytes with customized geometries including "L", "T", and " + " shapes as shown in Fig. 2.10 [83]. It is worth noting that electrochemical behaviors of additively manufactured LATP ceramic-based customized electrolyte films were still retained at a high level, which are thus promising in AM fabrication for solid-state energy storage.

2.3.5 *Other Materials*

Other materials such as some multi-element materials and novel nanomaterials have also been demonstrated as important components for advanced energy

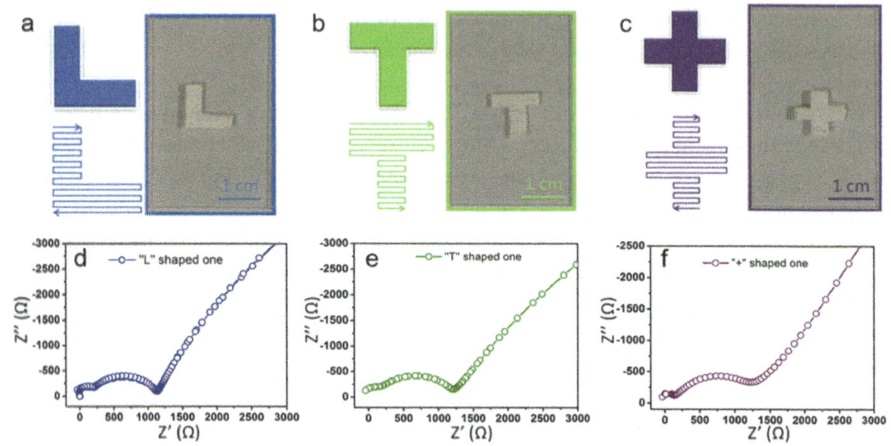

Fig. 2.10 **a–c** Schematics, printing routines, and optical images of various LATP ceramic solid-state electrolytes with customized geometries. **d–f** Corresponding impedance plots of LATP ceramic solid-state electrolytes [83]

storage. Lithium (or sodium) transition metal oxides including $LiCoO_2$ (LCO), LFP, $LiNi_xCo_yMn_{1-x-y}O_2$ (NCM), $LiNi_xCo_yAl_{1-x-y}O_2$ (NCA), $LiTi_2(PO_4)_3$ (LTP), $Li_3V_2(PO_4)_3$ (LVP), $NaTi_2(PO_4)_3$ (NTP), $Na_3V_2(PO_4)_3$ (NVP), and $Na_2MnFe(CN)_6$ have been extensively explored as promising cathodes for lithium(-ion)- and sodium(-ion)-based rechargeable batteries. Many multi-element material anodes (e.g., LTO, titanium niobium oxide (TNO), and $Na_2Ti_3O_7$ (NTO)) and non-metallic elements (e.g., Si, S, and Se) have also attracted extensive attention for rechargeable batteries. To satisfy the practical usage requirements, high energy, high power, high rate, and long cycling stability are some of the key considerations.

Conventional electrode fabrication based on cathode and anode materials is mainly conducted via the roller coating method, where the mass loading is relatively low for practical applications. Moreover, higher mass loading with bulk geometries usually leads to lower charge storage kinetics and ultimately to decreased energy storage behaviors. To this end, AM offers an emerging and efficient way to fabricate high-mass-loading electrodes with controllable geometries such as the above-mentioned cellular structures. For the Li–S battery application, Cai et al. reported a free-standing sulfur cathode containing a 3D scaffold and hierarchical pores via DIW [85]. With optimized electron/ion transport, the electrochemical kinetics of as-printed sulfur cathodes were expedited under the high mass loading condition. Accordingly, a high capacity of 693 mAh g^{-1} was measured with a good cycling stability. Wang and co-workers studied the geometric effects on electrochemical performance for 3D-printed LFP cathodes with high mass loading [84]. They fabricated three types of cathodes including the circle-grid pattern, circle-ring pattern, and circle-line pattern as shown in Fig. 2.11a–f, and compared their electrochemical performance (Fig. 2.11g). The measured results showed that the discharge capacities of the three types of as-printed cathodes were approximately the same during initial cycles, while the circle-line

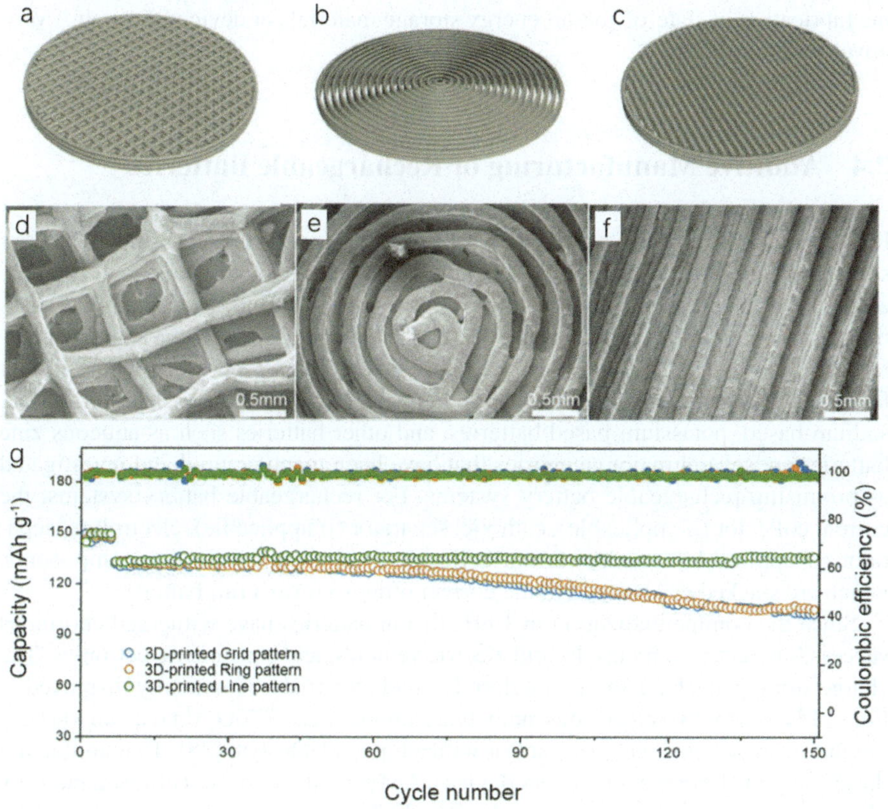

Fig. 2.11 **a–c** Schematics and **d–f** SEM images of 3D-printed LFP electrodes with three pattern types: **a, d** the circle-grid pattern, **b, e** the circle-ring pattern, and **c, f** the circle-line pattern. **g** Cycling performance of 3D-printed LFP electrodes with three pattern types [84]

patterned one delivered better cycling stability over other two types for over 150 cycles, demonstrating a clear advantage of circle-line patterned microstructure for energy storage. This indicates that microstructure play a vital role for desirable electrochemical behaviors for additively manufactured materials.

To date, numerous novel nanomaterials (i.e., black phosphorus) have drawn much research interest for energy applications. For example, black phosphorus provides a broad spacing channel (5.3 Å) that is larger than graphite (3.4 Å) [86]. This indicates more insertion and extraction space for ions, implying the great potential in black phosphorus materials. However, AM of black phosphorus materials is not readily conducted as the preparation of corresponding feeding materials is difficult. To solve the poor dispersibility of high-concentration black phosphorus-based solution and retain the good flow behavior of the ink, Xing et al. used Pluronic F127 to prepare a black phosphorous nanosheets/polypyrrole-based ink and fabricated high-performance electrodes via the DIW method [87]. It represents a possible route for the fabrication of state-of-the-art energy storage materials or devices through AM of novel nanomaterials.

2.4 Additive Manufacturing of Rechargeable Batteries

Rechargeable batteries have been widely employed as one of the most suitable energy source suppliers for large-scale electrical grids, electric vehicles, and small portable electronics. Rechargeable batteries store charges by reversible electrochemical reactions that occur mainly at two sides of cathodes and anodes, while the existing interface (such as the electrode–electrolyte interface) chemistry represents an important factor that guides the evolution of electrochemical cycling behaviors. Lithium-based, sodium-based, potassium-based batteries, and other batteries such as aqueous zinc batteries are several major categories that have been manufactured and investigated as promising rechargeable battery systems. For rechargeable battery systems, the current collector (if applicable), cathode, separator (if applicable), electrolyte, separator (if applicable), anode, current collector (if applicable) are key components, which are stacked or arranged in the correct order to form a full battery.

Since the commercialization in 1991, Li-ion batteries have witnessed enormous success for electric vehicles, hybrid electric vehicles, and consumer electronics. The charge storage mechanism of a modern Li-ion battery is schematically displayed in Fig. 2.12a, where two electrodes, namely a cathode (i.e., $LiCoO_2$) and an anode (i.e., graphite), are separated by a separator with a liquid electrolyte [88]. To manufacture these components or the full battery devices, AM provides a new but efficient building platform compared with conventional methods. The structure–property relationships can be explored easily with the structural RP capability in AM systems. As described

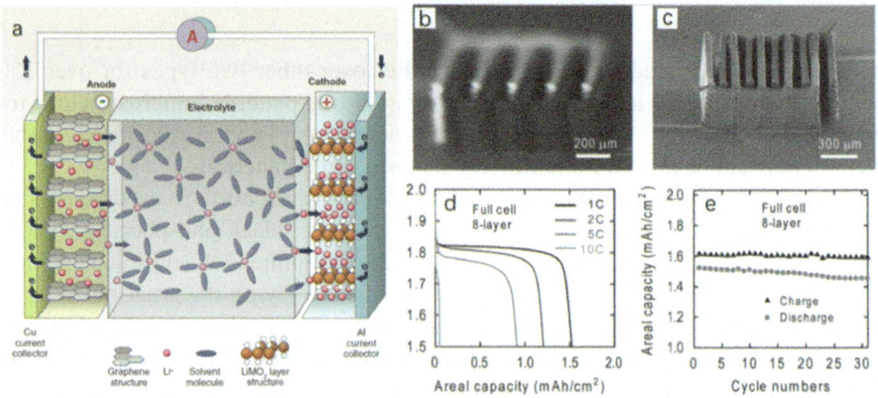

Fig. 2.12 **a** Schematic of a typical Li-ion battery [89]. **b** Optical and **c** SEM images of a 3D-printed interdigitated Li-ion microbattery with LFP//LTO electrode architectures. **d** Discharge profile and **e** cycling stability of the 3D-printed interdigitated Li-ion microbattery [90]

in the above section, through AM, various materials including polymer-based, metal-based, ceramic-based, and other materials have been demonstrated as candidates of various components for energy storage devices, and a huge range of advantages were found from robust mechanical properties to desirable electrochemical behaviors.

The rapid development of portable devices has induced the rising need for miniaturized Li-ion batteries. Interdigitated electrodes are the commonly utilized configuration, where the patterning of electrode materials is necessary for the full miniaturized Li-ion battery devices. In this regard, AM is very suitable for fabricating miniaturized Li-ion batteries. The first 3D interdigitated Li-ion microbattery was fabricated by Sun and co-workers via a facile DIW technique in 2013 [90]. In the microbattery, two in-plane interdigitated electrodes composed of LFP and LTO, respectively, were precisely arranged onto the glass substrate as shown in Fig. 2.12b, c. A good discharge profile was exhibited with a stable cycling performance for the as-printed interdigitated Li-ion microbattery, illustrating a significant progress for AM in the field of energy storage devices as well as functional materials and devices.

Besides interdigitated Li-ion microbattery, fiber-shaped Li-battery has also attracted extensive attention for wearable energy storage devices for electronic textiles and epidermal electronics. Wang et al. designed and reported a 3D-printed all-fiber Li-ion battery by DIW-based 3D printing with high-viscosity polymer inks based on LFP (the active material of the cathode) and LTO (the active material of the anode), respectively [91]. CNTs were introduced into the inks to act as the conductive additive, and the specific capacity of ~110 mAh g^{-1} with a good flexible property was realized for the resultant fiber-shaped Li-ion battery. The authors demonstrated that the DIW-based 3D printing approach is a low-cost, fast, and scalable method to manufacture wearable energy storage devices with desirable electrochemical performance and robust mechanic properties.

Other lithium-based batteries such as Li–S batteries and Li–Se batteries are among the most potential energy storage devices with several times higher energy density compared with conventional Li-ion batteries. Using AM methods, some key battery components such as cathode materials have been obtained with high mass loading and high electrochemical performance. Shen et al. first utilized the DIW-based printing technique to achieve the AM of sulfur copolymer-graphene cathode architecture with an external cellular geometry and internal well-designed macropores [92]. Sulfur particles and 1, 3-diisopropenylbenzene were added into the condensed GO dispersion for 3D printable ink. The as-printed cathodes delivered a specific capacity of 812.8 mAh g^{-1} and good cycle stability for Li–S batteries. Besides Li–S batteries, Shen and co-workers employed the same DIW-based AM fabrication approach to obtain the cellular cathode electrodes by preparing 3D printable KB/SeS_2-based composite inks [93]. With optimized hierarchical pores, a high areal discharge capacity of 9.5 mAh cm^{-2} and good cycling stability were obtained for the 3D-printed KB/SeS_2 composite cathodes. These previous works are good indications that AM parameters and material chemistry are two major optimization ways to access desirable energy storage materials and devices.

Li–O_2 batteries (or Li–air batteries) are also one of the desirable candidates for next-generation energy storage systems due to their ultrahigh theoretical energy density of 3458 Wh kg^{-1} [95]. Their charge storage mechanism is schematically shown in Fig. 2.13a, where porous structures are favorable for the high energy densities [89]. AM shows high capability in directly producing complex material architectures including porous structures. Lyu et al. provided a typical example of the AM fabrication of a porous conducting matrix derived from Co-MOF-based precursors [94]. The 3D printing process of the Co-MOF-based precursors of the porous conducting matrix is displayed in Fig. 2.13b, where the Pluronic F127 powder and raw Co-MOF crystals were utilized as the main content of the printable ink. The appearance and microstructure of the resultant 3D-printed Co-MOF-derived framework are shown in Fig. 2.13c–f, where the self-standing and hierarchically porous (including micrometer-sized pores, mesopores and micropores) structures were exhibited. For Li-O_2 batteries, this designed cathode possessed unique characteristics including

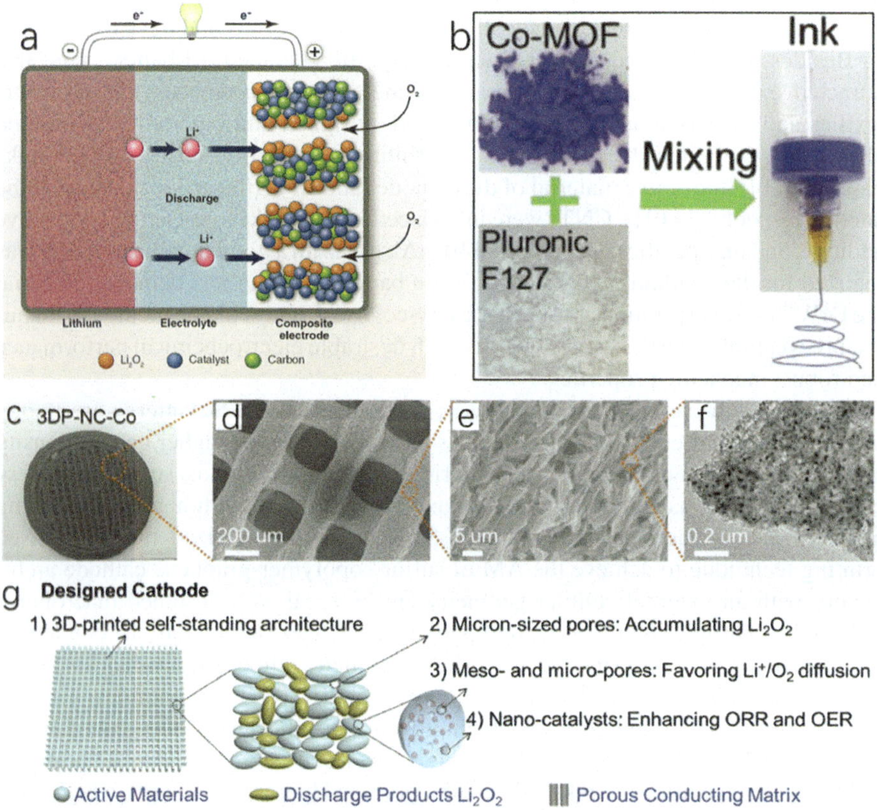

Fig. 2.13 **a** Schematic of a typical Li–O_2 battery [89]. **b** Schematic of a 3D printing process of Co-MOF-derived framework. **c** Photograph, **d, e** SEM images, and **f** TEM image of 3D-printed Co-MOF-derived framework. **g** Schematic of the designed cathode with 3D-printed self-standing and hierarchically porous catalyst framework [94]

accumulating Li_2O_2, favoring Li^+/O_2 diffusion, and enhancing the oxygen reduction reaction (ORR) and oxygen evaluation reaction (OER) (Fig. 2.13g). This work demonstrates that AM of Li–O_2 (or Li–air) battery materials and devices is promising.

In addition to lithium-based rechargeable batteries, other rechargeable batteries such as sodium-based, potassium-based, manganese-based batteries, and zinc-based batteries are also promising as next-generation rechargeable battery systems. Due to the abundance of sodium sources, sodium-based batteries such as sodium-ion batteries become lower-cost energy storage systems compared with lithium-based batteries, while similar intercalation chemistry is shared. Similarly, potassium-based batteries, manganese-based batteries, and zinc-based batteries also belong to low-cost battery systems. Through AM approaches, important components of these batteries have been fabricated with beneficial electrochemical behaviors. Microstructural control and materials chemistry are two major addressed points. However, although some works have been reported for AM of rechargeable batteries, the in-depth understanding and practical application are still in an infancy stage. Thus, more research for AM of advanced rechargeable batteries is highly expected in the future.

2.5 Additive Manufacturing of Electrochemical Capacitors

Electrochemical capacitors, which are also referred to as supercapacitors, hybrid capacitors, or ultracapacitors, are promising alternative energy storage systems besides rechargeable batteries. Similarly to rechargeable batteries, two electrodes (attached with current collectors, if applicable) are separated by the electrolyte and separators (if applicable) in electrochemical capacitors, and these contained parts are regarded as key components. However, an electrochemical capacitor mainly relies on the electrostatic absorption–desorption process of ions (i.e., electrolyte ions) or faradaic reactions occurred at the surface or near-surface sites of electrode materials. According to the charge storage mechanism classification, the electrical double-layer capacitor (EDLC) type and pseudocapacitor type can be categorized with the absorption–desorption-based (non-faradaic) process and the faradaic reaction-based process, respectively. The different charge storage mechanisms for EDLC electrode materials and pseudocapacitor electrode materials are schematically shown in Fig. 2.14 [96]. In electrochemical capacitors, two electrodes can be either different or the same to satisfy the charge storage process for the symmetric EDLC-type electrochemical capacitors.

To fabricate the above key components of electrochemical capacitors, AM has been regarded as a cost-efficient and powerful tool. Highly complex structures can be easily achieved with the assistance of CAD model files. Moreover, the full electrochemical capacitor device can be directly constructed as well via AM, without a further device assembly operation in conventional methods. To date, AM of electrochemical capacitors has garnered widespread interest and a large number of research

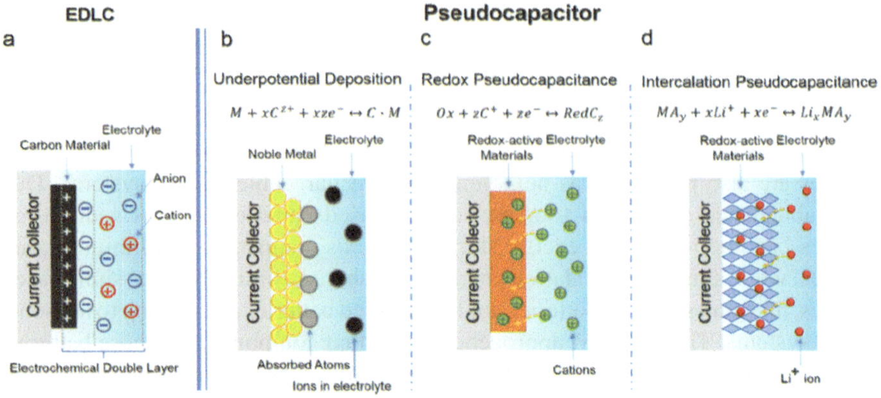

Fig. 2.14 Schematics of charge storage mechanisms of **a** EDLC materials, and **b–d** different types of pseudocapacitor materials [96]

has been conducted to demonstrate the advantageous electrochemical and mechanical features of additively manufactured electrochemical capacitor components or devices.

EDLCs are first developed electrochemical capacitors, where highly reversible charge–discharge processes occurred [97]. Carbon-based materials are EDLC materials manufactured from various methods including AM. As described in Sect. 2.3, the additively manufactured 3D carbon-based materials possess advantageous energy storage features compared with 2D ones fabricated by conventional methods. Moreover, AM-controlled approaches offer a digital tuning way for desirable microstructure, resulting in preferable electrochemical behaviors. Jiang et al. employed DIW-based 3D printing to fabricate programmable graphene aerogel microlattices from a GO based ink [28]. The 3D-printed graphene microlattices exhibited high capacitive performance, showing typical AM fabrication of highly controllable carbon-based materials for EDLCs.

Besides DIW, Xue et al. utilized DLP to realize the fabrication and tuning of hierarchically cellular polymer lattice templates and further used electroless plating to achieve the conductive and mechanically robust graphene lattices for quasi-solid EDLC supercapacitors [98]. The DLP-processed polymer lattice templates ensured good porous microstructure and the further coating of graphene led to the desirable capacitive charge storage process during the EDLC charge and discharge processes. The areal capacitance of 57.75 mF cm^{-2} was obtained along with a good rate capability and long lifespan for quasi-solid EDLC supercapacitors.

In addition to EDLCs, pseudocapacitors normally deliver higher specific capacitances and higher energy densities with the presence of faradic reactions. As described above, Tian et al. used an optimized DIW-based 3D printing to fabricate pseudocapacitive cobalt-based oxide electrodes for high-performance energy storage [55]. Apart from the direct AM of pseudocapacitive electrodes, AM of support for pseudocapacitive materials is also an efficient strategy for desirable energy storage. Liu et al. utilized direct metal laser sintering to obtain 3D hierarchical porous metallic scaffolds as the support of pseudocapacitive materials [100]. With a further deposition of a conducting polymer, MnO_2, and Mn_2O_3, 3D pseudocapacitor electrode materials were fabricated with enhanced charge storage performance.

Fibred-designed and interdigitated-designed electrochemical capacitors are becoming increasingly important as wearable and miniaturized electrochemical energy storage devices, respectively. With the inherent capability of controlling the material geometries digitally, AM has been demonstrated to achieve diverse desirable microstructure, leading to a high electrochemical performance output. Zhao et al. reported a typical work, where two 3D-printed fiber electrodes were manufactured via DIW and then used for the assembly of the fiber-shaped asymmetric supercapacitors [99]. The fabrication process is schematically shown in Fig. 2.15a, and the optical images of the extruded wet fiber illustrate a versatile patterning advantage (Fig. 2.15b, c). For the resultant fiber-shaped asymmetric supercapacitors, a closely stacked structure was observed (Fig. 2.15d), showing the mechanical robustness of 3D-printed fibered-designed materials. It is worth noting that AM is also an appropriate tool for fabricating interdigitated designed electrochemical capacitors. To date, numerous research efforts have been devoted to developing various interdigitated electrodes for high-performance miniaturized electrochemical capacitor devices, and significant progress has been made. However, some key performance parameters such as the energy density are still far from the requirements for practical energy applications. Thus, more efforts are still needed to address the electrochemical behavior issues.

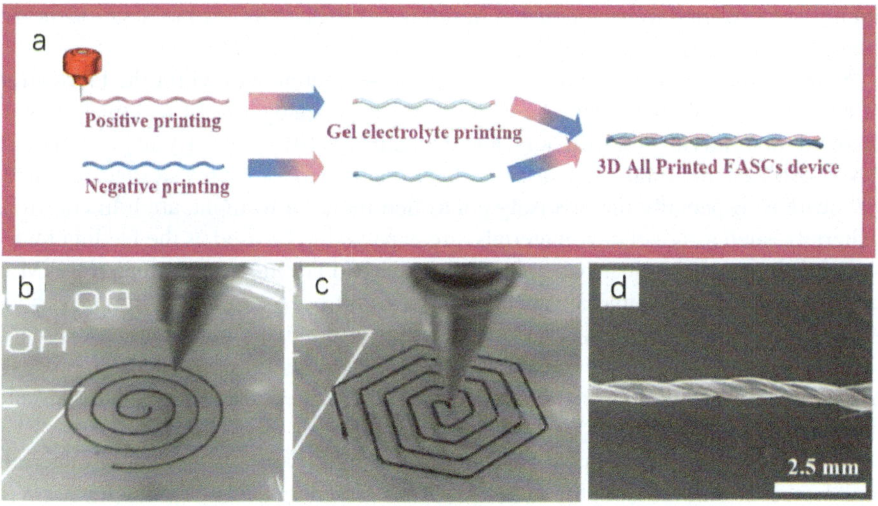

Fig. 2.15 **a** Schematic showing the fabrication process of fiber-shaped asymmetric supercapacitors. **b, c** Optical images of the extruded wet fiber during the printing process. **d** SEM image of the assembled fiber-shaped asymmetric supercapacitors [99]

2.6 Summary and Outlook

Increasing demand for customized designs and higher energy output has driven a growing research trend for advanced manufacturing technologies and novel materials. AM has emerged as an advanced manufacturing technology that relies on a digitally controlled way to produce objects or parts with customized geometries. For various application fields, the manufacturing principle of AM is highly advantageous due to the ease of tuning of external structures as well as internal microstructure. So far, seven basic AM categories have been classified, and an increasing number of AM techniques are being created, implying high sustainability and potential for continuous improvement. The application fields of AM also expanded from structural areas to functional materials or devices such as sensors, drug delivery systems, and energy storage devices.

Electrochemical energy storage devices, which are classified into rechargeable batteries and electrochemical capacitors, have witnessed great success in both large-scale and small-scale energy consumption devices including electric vehicles and portable devices. To realize desirable electrochemical performances, the manufacturing process and materials are two key factors. Using desirable materials for energy storage devices, AM provides an ideal platform for building high-performance energy storage devices or components. To date, numerous research has been conducted to investigate the pros and cons of AM for energy storage, and a wide range of additively manufactured materials have been reported with good electrochemical behaviors for advanced energy storage devices. While there still exist some limitations, the reported data demonstrated the huge potential in AM for energy storage applications. More studies are expected to be conducted in future to advance the AM of state-of-the-art energy storage materials and devices.

Some of the challenges and opportunities have been identified for the fabrication of next-generation energy storage devices. First, the categories of additively manufactured materials are not sufficient, and their study is still at an early stage. Although AM of different materials categories has been conducted for energy storage, the material designs, especially the precisely controlled material formula, are limited. More materials, such as novel nanomaterials, are expected to be used as the feeding materials for their unique electrochemical properties. In addition, multi-material AM is usually ignored, and some advancements have not been explored. Among various AM techniques, DIW offers a versatile feature for the materials category, while the rheological behaviors of the ink are a challenging element for the extrusion process. More novel printable inks are thus expected for AM of energy storage materials and devices in the future.

Second, the electrochemical performance of additively manufactured materials and devices is still far from the increasing need for energy consumption devices. To date, not all promising energy storage devices have been demonstrated via AM. The energy densities and power densities of reported AM- processes are expected to be further improved by controlling key printing parameters and material formulation.

Device hybridization is one of the appropriate approaches for producing high-energy and high-power energy storage devices via AM.

Third, an in-depth understanding of the relationship between energy materials formulation, AM-processed microstructure, and delivered electrochemical performance is lacking. During battery or electrochemical capacitor cycling, the decay of capacity (or capacitance) has been widely studied for materials fabricated via conventional methods. By providing an ideal platform for digital tuning of microstructure, AM presents a new approach to the in situ or ex situ investigation of the intrinsic charge storage mechanisms during the charge and discharge processes.

Acknowledgements The authors would like to acknowledge the financial support from the National Natural Science Foundation of China (21975230) and Zhejiang Provincial Natural Science Foundation of China (LY20B010001).

References

1. Ligon SC, Liska R, Stampfl J, Gurr M, Mülhaupt R (2017) Polymers for 3D printing and customized additive manufacturing. Chem Rev 117:10212–10290
2. Yang S, Zhao YF (2015) Additive manufacturing-enabled design theory and methodology: a critical review. Int J Adv Manuf 80:327–342
3. Cheng M, Deivanayagam R, Shahbazian-Yassar R (2020) 3D printing of electrochemical energy storage devices: a review of printing techniques and electrode/electrolyte architectures. Batteries Supercaps 3:130–146
4. Tian X, Jin J, Yuan S, Chua CK, Tor SB, Zhou K (2017) Emerging 3D-printed electrochemical energy storage devices: a critical review. Adv Energy Mater 7:1700127
5. Tian X, Zhou K (2020) 3D printing of cellular materials for advanced electrochemical energy storage and conversion. Nanoscale 12:7416–7432
6. Gibson I, Rosen D, Stucker B (2015) Vat photopolymerization processes. In: Gibson I, Rosen D, Stucker B (eds) Additive manufacturing technologies: 3D printing, rapid prototyping, and direct digital manufacturing. Springer, New York, New York, NY, pp 63–106
7. Appuhamillage GA, Chartrain N, Meenakshisundaram V, Feller KD, Williams CB, Long TE (2019) 110th anniversary: vat photopolymerization-based additive manufacturing: current trends and future directions in materials design. Ind Eng Chem Res 58:15109–15118
8. Gibson I, Rosen DW, Stucker B (2010) Design for additive manufacturing. Springer, Additive manufacturing technologies, pp 299–332
9. He Y, Chen S, Nie L, Sun Z, Wu X, Liu W (2020) Stereolithography three-dimensional printing solid polymer electrolytes for all-solid-state lithium metal batteries. Nano Lett 20:7136–7143
10. Zekoll S, Marriner-Edwards C, Hekselman AKO, Kasemchainan J, Kuss C, Armstrong DEJ, Cai D, Wallace RJ, Richter FH, Thijssen JHJ, Bruce PG (2018) Hybrid electrolytes with 3D bicontinuous ordered ceramic and polymer microchannels for all-solid-state batteries. Energy Environ Sci 11:185–201
11. Jacobs PF (1992) Fundamentals of stereolithography. In: 1992 international solid freeform fabrication symposium
12. Hinczewski C, Corbel S, Chartier T (1998) Stereolithography for the fabrication of ceramic three-dimensional parts. In: Rapid Prototyping J
13. Manapat JZ, Chen Q, Ye P, Advincula RC (2017) 3D printing of polymer nanocomposites via stereolithography. Macromol Mater Eng 302:1600553
14. Mu Q, Wang L, Dunn CK, Kuang X, Duan F, Zhang Z, Qi HJ, Wang TJ (2017) Digital light processing 3D printing of conductive complex structures. Addit Manuf 18:74–83

15. Tee YL, Tran P, Leary M, Pille P, Brandt M (2020) 3D Printing of polymer composites with material jetting: mechanical and fractographic analysis. Addit Manuf 36:101558
16. Qiu T, Palagi S, Fischer P (2015) 3D-printed soft microrobot for swimming in biological fluids. In: 2015 37th annual international conference of the IEEE engineering in medicine and biology society (EMBC). IEEE, pp. 4922–4925
17. Acquaticci F, Yommi MM, Gwirc SN, Lew SE (2017) Rapid prototyping of pyramidal structured absorbers for ultrasound. Open J Acoust 7:83
18. Furbank RJ, Morris JF (2004) An experimental study of particle effects on drop formation. Phys Fluids 16:1777–1790
19. Sweet RG (1964) High-frequency oscillography with electrostatically deflected ink jets. Stanford Univ Ca Stanford Electronics Labs
20. Sachs E, Cima M, Williams P, Brancazio D, Cornie J (1992) Three dimensional printing: rapid tooling and prototypes directly from a CAD model. J Eng Ind 114:481–488
21. Bai Y, Williams CB (2015) An exploration of binder jetting of copper. Rapid Prototyping J
22. Malda J, Visser J, Melchels FP, Jüngst T, Hennink WE, Dhert WJ, Groll J, Hutmacher DW (2013) 25th anniversary article: engineering hydrogels for biofabrication. Adv Mater 25:5011–5028
23. Skylar-Scott MA, Mueller J, Visser CW, Lewis JA (2019) Voxelated soft matter via multimaterial multinozzle 3D printing. Nature 575:330–335
24. Gulzar U, Glynn C, O'Dwyer C (2020) Additive manufacturing for energy storage: methods, designs and material selection for customizable 3D printed batteries and supercapacitors. Curr Opin Electrochem 20:46–53
25. Tan LJ, Zhu W, Zhou K (2020) Recent Prog Polymer Mater Addit Manuf 30:2003062
26. Gibson I, Rosen D, Stucker B (2015) Directed energy deposition processes. In: Gibson I, Rosen D, Stucker B (eds) Additive manufacturing technologies: 3D printing, rapid prototyping, and direct digital manufacturing. Springer, New York, New York, NY, pp 245–268
27. Tagliaferri S, Panagiotopoulos A, Mattevi CJMA (2021) Direct Ink Writ Energy Mater 2:540–563
28. Jiang Y, Xu Z, Huang T, Liu Y, Guo F, Xi J, Gao W, Gao C (2018) Direct 3D printing of ultralight graphene oxide aerogel microlattices. Adv Func Mater 28:1707024
29. Li W, Li Y, Su M, An B, Liu J, Su D, Li L, Li F, Song YJ (2017) Printing assembly and structural regulation of graphene towards three-dimensional flexible micro-supercapacitors. J Mater Chem A 5:16281–16288
30. Wang Y, Chen C, Xie H, Gao T, Yao Y, Pastel G, Han X, Li Y, Zhao J, Fu K (2017) 3D-printed all-fiber li-ion battery toward wearable energy storage. Adv Func Mater 27:1703140
31. Wu H, Xu Y, Ren X, Liu B, Engelhard MH, Ding MS, El-Khoury PZ, Zhang L, Li Q, Xu K (2019) Polymer-in-"Quasi-Ionic Liquid" electrolytes for high-voltage lithium metal batteries. Adv Energy Mater 9:1902108
32. Stansbury JW, Idacavage MJ (2016) 3D printing with polymers: Challenges among expanding options and opportunities. Dent Mater 32:54–64
33. Quan H, Zhang T, Xu H, Luo S, Nie J, Zhu XJBM (2020) Photo-curing 3D printing technique and its challenges 5:110–115
34. Yuk H, Lu B, Lin S, Qu K, Xu J, Luo J, Zhao X (2020) 3D printing of conducting polymers. Nat Commun 11:1604
35. Reyes C, Somogyi R, Niu S, Cruz MA, Yang F, Catenacci MJ, Rhodes CP, Wiley BJJAAEM (2018) Three-dimensional printing of a complete lithium ion battery with fused filament fabrication. ACS Appl Energy Mater 1:5268–5279
36. Rahman MS, Shiblee MNI, Ahmed K, Khosla A, Ogawa J, Kawakami M, Furukawa H (2020) Flexible and conductive 3D printable polyvinylidene fluoride and poly(N, N-dimethylacrylamide) based gel polymer electrolytes. Macromol Mater Eng 305:2000262
37. Zheng X, Lee H, Weisgraber TH, Shusteff M, DeOtte J, Duoss EB, Kuntz JD, Biener MM, Ge Q, Jackson JA, Kucheyev SO, Fang NX, Spadaccini CM (2014) Ultralight, ultrastiff mechanical metamaterials. Science 344:1373–1377

38. Song J, Chen Y, Cao K, Lu Y, Xin JH, Tao X (2018) Fully controllable design and fabrication of three-dimensional lattice supercapacitors. ACS Appl Mater Interfaces 10:39839–39850
39. Krishnadoss V, Kanjilal B, Hesketh A, Miller C, Mugweru A, Akbard M, Khademhosseini A, leijten J, Noshadi I (2021) In situ 3D printing of implantable energy storage devices. Chem Eng J 409:128213
40. Saleh MS, Li J, Park J, Panat R (2018) 3D printed hierarchically-porous microlattice electrode materials for exceptionally high specific capacity and areal capacity lithium ion batteries. Addit Manuf 23:70–78
41. Vyatskikh A, Delalande S, Kudo A, Zhang X, Portela CM, Greer JR (2018) Additive manufacturing of 3D nano-architected metals. Nat Commun 9:593
42. Rashed MG, Ashraf M, Mines RAW, Hazell PJ (2016) Metallic microlattice materials: A current state of the art on manufacturing, mechanical properties and applications. Mater Des 95:518–533
43. Su X, Li X, Ong CYA, Herng TS, Wang Y, Peng E, Ding J (2019) Metallization of 3D printed polymers and their application as a fully functional water-splitting system. Adv Sci 6:1801670
44. Yadav P, Fu Z, Knorr M, Travitzky N (2020) Binder jetting 3D printing of titanium aluminides based materials: a feasibility study. Adv Energy Mater 22:2000408
45. Egorov V, O'Dwyer C (2020) Architected porous metals in electrochemical energy storage. Curr Opin Electrochem 21:201–208
46. Schaedler TA, Jacobsen AJ, Torrents A, Sorensen AE, Lian J, Greer JR, Valdevit L, Carter WB (2011) Ultralight Metallic Microlattices 334:962–965
47. Juarez T, Schroer A, Schwaiger R, Hodge AM (2018) Evaluating sputter deposited metal coatings on 3D printed polymer micro-truss structures. Mater Des 140:442–450
48. Lim GJH, Lyu Z, Zhang X, Koh JJ, Zhang Y, He C, Adams S, Wang J, Ding J (2020) Robust pure copper framework by extrusion 3D printing for advanced lithium metal anodes. J Mater Chem A 8:9058–9067
49. Zhang Y, Zuo T-T, Popovic J, Lim K, Yin Y-X, Maier J, Guo Y-G (2020) Towards better Li metal anodes: challenges and strategies. Mater Today 33:56–74
50. Cao D, Xing Y, Tantratian K, Wang X, Ma Y, Mukhopadhyay A, Cheng Z, Zhang Q, Jiao Y, Chen L, Zhu H (2019) 3D printed high-performance lithium metal microbatteries enabled by nanocellulose. Adv Mater 31:1807313
51. Yao R-Q, Lang X-Y, Jiang QJJOI (2019) S.R. international, recent advances of nanoporous metal-based catalyst: synthesis, application and perspectives. J Iron Steel Res Int 26:779–795
52. Tompsett GA, Conner WC, KSJCaEjocp, Yngvesson P (2006) Chemistry, microwave synthesis of nanoporous materials 7:296–319
53. Mooraj S, Qi Z, Zhu C, Ren J, Peng S, Liu L, Zhang S, Feng S, Kong F, Liu YR (2020) 3D printing of metal-based materials for renewable energy applications. Nanoscale 1–28
54. Zhao X, Liu B, Pan P, Yang Z, He J, Li H, Wei J, Cao Z, Zhang H, Chang J, Bao Q, Yang X (2021) Fabrication of reduced graphene oxide/manganese oxide ink for 3D-printing technology on the application of high-performance supercapacitors. J Mater Sci 56:8102–8114
55. Tian X, Tang K, Jin H, Wang T, Liu X, Yang W, Zou Z, Hou S, Zhou K (2019) Boosting capacitive charge storage of 3D-printed micro-pseudocapacitors via rational holey graphene engineering. Carbon 155:562–569
56. Yu XY, Lou XW (2018) Mixed metal sulfides for electrochemical energy storage and conversion. Adv Energy Mater 8:1701592
57. Rui X, Tan H, Yan Q (2014) Nanostructured metal sulfides for energy storage. Nanoscale 6:9889–9924
58. Tang X, Zhu C, Cheng D, Zhou H, Liu X, Xie P, Zhao Q, Zhang D, Fan T (2018) Architectured leaf-inspired $Ni_{0.33}Co_{0.66}S_2$/graphene aerogels via 3D printing for high-performance energy storage. Adv Funct Mater 28:1805057
59. Orangi J, Hamade F, Davis VA, Beidaghi M (2019) 3D printing of additive-free 2D $Ti_3C_2T_x$ (MXene) ink for fabrication of micro-supercapacitors with ultra-high energy densities. ACS Nano 14:640–650

60. Zhang LL, Zhao XS (2009) Carbon-based materials as supercapacitor electrodes. Chem Soc Rev 38:2520–2531
61. Jabari E, Liravi F, Davoodi E, Lin L, Toyserkani E (2020) High speed 3D material-jetting additive manufacturing of viscous graphene-based ink with high electrical conductivity. Adv Mater 35:101330
62. Popov V, Fleisher A, Muller-Kamskii G, Shishkin A, Katz-Demyanetz A, Travitzky N, Goel S (2021) Novel hybrid method to additively manufacture denser graphite structures using binder jetting. Sci Rep 11:2438
63. Guo H, Lv R, Bai S (2019) Recent advances on 3D printing graphene-based composites. Nano Mater Sci 1:101–115
64. Luong DX, Subramanian AK, Silva GAL, Yoon J, Cofer S, Yang K, Owuor PS, Wang T, Wang Z, Lou J, Ajayan PM, Tour JM (2018) Laminated object manufacturing of 3D-printed laser-induced graphene foams. Adv Mater 30:1707416
65. Yi X, Tan Z-J, Yu W-J, Li J, Li B-J, Huang B-Y, Liao J (2016) Three dimensional printing of carbon/carbon composites by selective laser sintering. Carbon 96:603–607
66. Yu Z, Tetard L, Zhai L, Thomas J (2015) Supercapacitor electrode materials: nanostructures from 0 to 3 dimensions. Energy Environ Sci 8:702–730
67. Areir M, Xu Y, Zhang R, Harrison D, Fyson J, Pei E (2017) A study of 3D printed active carbon electrode for the manufacture of electric double-layer capacitors. J Manuf Process 25:351–356
68. Yu W, Zhou H, Li BQ, Ding S (2017) 3D printing of carbon nanotubes-based microsupercapacitors. ACS Appl Mater Interfaces 9:4597–4604
69. Chen B, Jiang Y, Tang X, Pan Y, Hu S (2017) Fully packaged carbon nanotube supercapacitors by direct ink writing on flexible substrates. ACS Appl Mater Interfaces 9:28433–28440
70. Novoselov KS, Geim AK, Morozov SV, Jiang D, Zhang Y, Dubonos SV, Grigorieva IV, Firsov AA (2004) Electric field effect in atomically thin carbon films. Science 306:666–669
71. Yi M, Shen Z (2015) A review on mechanical exfoliation for the scalable production of graphene. J Mater Chem A 3:11700–11715
72. Wang J, Liu Y, Fan Z, Wang W, Wang B, Guo ZJAC, Materials H (2019) Ink-based 3D printing technologies for graphene-based materials: a review. Adv Compos Hybrid Mater 2:1–33
73. Loh HA, Graves AR, Stinespring CD, Sierros KA (2019) Direct ink writing of graphene-based solutions for gas sensing. ACS Appl Nano Mater 2:4104–4112
74. Gnanasekaran K, Heijmans T, van Bennekom S, Woldhuis H, Wijnia S, de With G, Friedrich H (2017) 3D printing of CNT- and graphene-based conductive polymer nanocomposites by fused deposition modeling. Appl Mater Today 9:21–28
75. Jakus AE, Shah RN (2017) Multi and mixed 3D-printing of graphene-hydroxyapatite hybrid materials for complex tissue engineering. J Biomed Mater Res, Part A 105:274–283
76. Zhu C, Han TY-J, Duoss EB, Golobic AM, Kuntz JD, Spadaccini CM, Worsley MA (2015) Highly compressible 3D periodic graphene aerogel microlattices. Nat Commun 6:6962
77. Tang K, Ma H, Tian Y, Liu Z, Jin H, Hou S, Zhou K, Tian X (2020) 3D printed hybrid-dimensional electrodes for flexible micro-supercapacitors with superior electrochemical behaviours. Virtual Phys Prototyping 15:511–519
78. Wang T, Li L, Tian X, Jin H, Tang K, Hou S, Zhou H, Yu X (2019) 3D-printed interdigitated graphene framework as superior support of metal oxide nanostructures for remarkable micro-pseudocapacitors. Electrochim Acta 319:245–252
79. Tang X, Zhou H, Cai Z, Cheng D, He P, Xie P, Zhang D, Fan T (2018) Generalized 3D printing of graphene-based mixed-dimensional hybrid aerogels. ACS Nano 12:3502–3511
80. Qian S, Chen H, Wu Z, Li D, Liu X, Tang Y, Zhang S (2021) Designing ceramic/polymer composite as highly ionic conductive solid-state electrolytes. Batter Supercaps 4:39–59
81. Blake AJ, Kohlmeyer RR, Hardin JO, Carmona EA, Maruyama B, Berrigan JD, Huang H, Durstock MF (2017) 3D printable ceramic-polymer electrolytes for flexible high-performance Li-Ion batteries with enhanced thermal stability. Adv Energy Mater 7:1602920
82. Cheng M, Jiang Y, Yao W, Yuan Y, Deivanayagam R, Foroozan T, Huang Z, Song B, Rojaee R, Shokuhfar T, Pan Y, Lu J, Shahbazian-Yassar R (2018) Elevated-temperature 3D printing of hybrid solid-state electrolyte for Li-Ion batteries. Adv Mater 30:1800615

83. Liu Z, Tian X, Liu M, Duan S, Ren Y, Ma H, Tang K, Shi J, Hou S, Jin H, Cao G (2021) Direct ink writing of $Li_{1.3}Al_{0.3}Ti_{1.7}(PO_4)_3$-based solid-state electrolytes with customized shapes and remarkable electrochemical behaviors. Small 17:2002866
84. Wang J, Sun Q, Gao X, Wang C, Li W, Holness FB, Zheng M, Li R, Price AD, Sun X, Sham T-K, Sun X (2018) Toward high areal energy and power density electrode for Li-Ion batteries via optimized 3D printing approach. ACS Appl Mater Interfaces 10:39794–39801
85. Cai J, Fan Z, Jin J, Shi Z, Dou S, Sun J, Liu Z (2020) Expediting the electrochemical kinetics of 3D-printed sulfur cathodes for Li–S batteries with high rate capability and areal capacity. Nano Energy 75:104970
86. Liu H, Hu K, Yan D, Chen R, Zou Y, Liu H, Wang S (2018) Recent advances on black phosphorus for energy storage, catalysis, and sensor applications. Adv Mater 30:1800295
87. Xing R, Xia Y, Huang R, Qi W, Su R, He ZJCC (2020) Three-dimensional printing of black phosphorous/polypyrrole electrode for energy storage using thermoresponsive ink. Chem Commun 56:3115–3118
88. Xie J, Lu Y-C (2020) A retrospective on lithium-ion batteries. Nat Commun 11:2499
89. Dunn B, Kamath H, Tarascon J-M (2011) Electrical energy storage for the grid: a battery of choices. Science 334:928–935
90. Sun K, Wei TS, Ahn BY, Seo JY, Dillon SJ, Lewis JA (2013) 3D printing of interdigitated Li-Ion microbattery architectures. Adv Mater 25:4539–4543
91. Wang Y, Chen C, Xie H, Gao T, Yao Y, Pastel G, Han X, Li Y, Zhao J, Fu K, Hu L (2017) 3D-printed all-fiber Li-Ion battery toward wearable energy storage. Adv Funct Mater 27:1703140
92. Shen K, Mei H, Li B, Ding J, Yang S (2018) 3D printing sulfur copolymer-graphene architectures for Li-S batteries. Adv Energy Mater 8:1701527
93. Shen C, Wang T, Xu X, Tian X (2020) 3D printed cellular cathodes with hierarchical pores and high mass loading for Li–SeS_2 battery. Electrochim Acta 349:136331
94. Lyu Z, Lim GJ, Guo R, Kou Z, Wang T, Guan C, Ding J, Chen W, Wang J (2019) 3D-printed MOF-derived hierarchically porous frameworks for practical high-energy density Li–O_2 batteries. Adv Func Mater 29:1806658
95. Lee HC, Park JO, Kim M, Kwon HJ, Kim J-H, Choi KH, Kim K, Im DJ (2019) High-energy-density Li-O_2 battery at cell scale with folded cell structure. Joule 3:542–556
96. Shao Y, El-Kady MF, Sun J, Li Y, Zhang Q, Zhu M, Wang H, Dunn B, Kaner RB (2018) Design and mechanisms of asymmetric supercapacitors. Chem Rev 118:9233–9280
97. Najib S, Erdem EA (2019) Current progress achieved in novel materials for supercapacitor electrodes: mini review. Nanoscale 1:2817–2827
98. Xue J, Gao L, Hu X, Cao K, Zhou W, Wang W, Lu Y (2019) Stereolithographic 3D printing-based hierarchically cellular lattices for high-performance quasi-solid supercapacitor. Nano-Micro Lett 11:46
99. Zhao J, Zhang Y, Huang Y, Xie J, Zhao X, Li C, Qu J, Zhang Q, Sun J, He B (2018) 3D printing fiber electrodes for an all-fiber integrated electronic device via hybridization of an asymmetric supercapacitor and a temperature sensor. Adv Sci 5:1801114
100. Liu X, Jervis R, Maher RC, Villar-Garcia IJ, Naylor-Marlow M, Shearing PR, Ouyang M, Cohen L, Brandon NP, Wu B (2016) 3D-printed structural pseudocapacitors. Adv Mater Technol 1:1600167

Chapter 3
4D Printing of Stimuli-Responsive Materials

Chunze Yan, Xiao Yang, and Hongzhi Wu

3.1 Introduction

Additive manufacturing technology not only exhibits an outstanding capability of printing complex geometries but also can precisely program and control the geometry, materials, structures, and properties of printed objects, offering the products tremendous design flexibility, high customizability, and significant material savings. However, conventional additive manufacturing technology, also known as three-dimensional (3D) printing, suffers a big limitation that its printed components are rigid, static, and inanimate. In other words, the 3D-printed structures cannot alter/transform with changes in the surrounding dynamic environment. This nature makes the 3D-printed components unable to meet the increasing demands for intelligent performances of devices, such as self-adaption, self-assembly, self-repair, self-learning, self-sensing, and decision making. In this context, four-dimensional (4D) printing, an innovative branch of additive manufacturing, emerges based on the multidisciplinary integration of materials, machinery, mechanics, information, and so on [6–10].

C. Yan (✉) · X. Yang · H. Wu
State Key Laboratory of Materials Processing and Die and Mould Technology, School of Materials Science and Engineering, Huazhong University of Science and Technology, Wuhan 430074, China
e-mail: c_yan@hust.edu.cn

X. Yang
e-mail: yangxiao520@hust.edu.cn

H. Wu
e-mail: whz@hust.edu.cn

K. Zhou (ed.), *Additive Manufacturing*,
https://doi.org/10.1007/978-3-031-04721-3_3

3.1.1 Definition of 4D Printing

At the Technology Entertainment and Design (TED) conference in 2013, the director of the Self-assembly Lab at the Massachusetts Institute of Technology (MIT), Skylar Tibbits, showed an additively manufactured soft strand that could transform into the letters "MIT" automatically when it was immersed into water [11, 12]. Through this demonstration, he first proposed the term "4D printing". Here, the fourth dimension is described as the shape transformation over time. This marked the start of 4D printing technology. In October 2014, the bimonthly American magazine, Foreign Affairs, published an article entitled "Prepare for the 4D printing revolution", which aroused worldwide attention to 4D printing from all walks of life. Since then, a research upsurge focusing on 4D printing has been raised in academia and industry.

4D printing was initially defined as "3D printing plus the fourth dimension of time". And the transformation of 4D-printed components over time mainly refers to the shape change [13–16]. But with the deepening advances of 4D printing, the concept of 4D printing has been continuously evolved and enriched. In 2017, Yusheng Shi, professor from Huazhong University of Science and Technology, organized relevant experts in China to hold the first academic seminar on 4D printing technology in Wuhan, China. He proposed a more comprehensive definition of 4D printing [17]. That is, 4D printing is an additive manufacturing technology of dynamic components, which can automatically and controllably change their shapes, properties, and/or functionalities with time and/or space under predetermined stimuli, such as heat, moisture, light, pH, magnetism, and electricity. This definition highlights that the stimuli-responsive behaviors of 4D-printed structures are not only limited to shape changes but also include property and functionality changes. Furthermore, the predictability and controllability of the stimuli-responsive changes are also emphasized. The new definition enriches the connotation of 4D printing, conducive to accelerating the development of 4D printing from simple phenomenon demonstration to practical application. It should be noted that 4D printing is still at a preliminary stage currently. In the future, the definition of 4D printing will be further enriched and expanded as the development of 4D printing goes further.

3.1.2 Prospect of 4D Printing

In traditional manufacturing technologies, the preparation of materials, design of structures, and realization of functionalities are commonly three independent links. Differently, 4D printing integrates the design of structures' intelligent behaviors into the fabrication processes, realizes the integrated manufacturing of material, structure, and functionality, as well as simplifies the process from concept design to final products [18–21]. It is promising to change traditional design concepts and manners, bringing a great revolution to the traditional manufacturing technology. 4D printing is also expected to bring an innovative development to high-end and microscale intelligent devices because it enables the direct implantation of intelligent performances into the printed objects without external electromechanical systems [22, 23]. It not only has the potential to eradicate the inherent weakness of electromechanical systems in traditional intelligent devices but also provides an alternative solution to the complex intelligent devices which are difficult to be created through conventional manufacturing technologies. In addition, utilizing the stimuli-responsive shape-morphing ability of 4D-printed components, the products can be first printed into a structure easy to be stored compactly and then activated to the required one with full volume when used [9]. This helps to reduce the space for storage and transportation, thus saving the logistic cost. These great advantages make 4D printing attractive in numerous fields including aerospace, biomedicine, automobile, robotics, and architecture. Therefore, 4D printing has become the forefront and hot spot of research at home and abroad. The market research predicts that the market share of 4D printing in the global industry will reach 555.6 million in 2025.

3.2 Materials for 4D Printing

4D printing is an additive manufacturing technology for fabricating intelligent structures with controllable stimuli-responsive characteristics. At present, most forming techniques used in 4D printing are traditional 3D printing techniques because the fourth dimension of 4D-printed components refers to the stimuli-responsive characteristics rather than the forming techniques. The stimuli-responsive abilities of 4D-printed components are commonly achieved by rational design, employment, combination, and arrangement of stimuli-responsive materials.

As the prime criterion, the stimuli-responsive materials for 4D printing should have good printability. That is, the materials can be well fabricated through additive manufacturing technologies. This is the prerequisite for achieving 4D printing. Currently, the evaluating criteria of the printability of 4D printing materials are almost the same as that of traditional 3D printing materials, where the criteria mainly depend on the forming techniques during printing. For instance, for the 4D printing based on stereolithography apparatus (SLA), the materials used should be photosensitive liquid materials with fast curing speeds, small curing shrinkage, low viscosity, and appropriate projection depths. For the 4D printing based on selective laser melting (SLM), the materials used should be powder materials with fine and uniform particle sizes, high sphericity, good fluidity, and high apparent density.

In addition, the materials for 4D printing can achieve controllable stimuli-responsive behaviors after additive manufacturing. This is the foundation and key of realizing 4D printing because the stimuli-responsive characteristics of 4D-printed structures are mainly dominated by the properties of the materials used as well as their combination and arrangement in 3D space. Intelligent materials are a group of materials, which can sense the alteration in the surrounding environment and respond to it via changing the shapes and/or physical and/or chemical properties [24–26]. Therefore, intelligent materials are naturally the primary candidate stimuli-responsive materials for 4D printing. However, not all intelligent materials can be used in 4D printing because some of them may be unable to be fabricated through additive manufacturing technologies, and some of them may lose their stimuli-responsive abilities after additive manufacturing. On the other hand, although the majority of current reports of 4D printing are based on intelligent materials, 4D printing can also be achieved through the use of some non-intelligent materials (traditional materials), which themselves do not have any controllable stimuli-responsive capacity but can achieve one through the appropriate programming of material composition and structure, presetting of stimulus signal, or combination of multiple materials. [17]. All in all, both intelligent materials and non-intelligent materials can be candidate materials for 4D printing, for which the key is whether they can realize controllable stimuli-responsive behaviors after additive manufacturing.

Although dozens of materials for 4D printing have been developed in the past decade, most of them can only meet the performance requirements of 4D-printed structures in a proof-of-concept or phenomenon demonstration stage. The lack of high-performance stimuli-responsive materials for 4D printing has become the greatest challenge of the development and application of 4D printing. Thus, it is urgent to conduct research and develop new stimuli-responsive materials for 4D printing to improve the performance of 4D-printed components and promote their practical applications. Thus, it is important to timely review the research advances in the stimuli-responsive materials for 4D printing. Currently, the stimuli-responsive materials for 4D printing can be classified as polymers and their composite materials [27–34], metals and their composite materials [35], and ceramics and their composite materials [36] according to their chemical components. In this section, the research and developing status of the above three categories of stimuli-responsive materials for 4D printing will be introduced separately.

3.2.1 4D Printing of Polymers and Their Composite Materials

Stimuli-responsive polymers and their composite materials are the most widely used category of materials for 4D printing because they are not only rich in material systems but also easy to be fabricated by additive manufacturing technologies. In addition, stimuli-responsive polymers and their composite materials also have the advantages of low price, low density, and simple design as compared with metals and ceramics. The additive manufacturing techniques most popularly used in their 4D printing mainly include direct ink writing (DIW), fused deposition modeling (FDM), selective laser sintering (SLS), PolyJet, stereolithography apparatus (SLA), and digital light processing (DLP). Currently, according to the stimuli types, the available stimuli-responsive polymers and their composite materials for 4D printing are mainly classified into thermo-responsive, moisture-responsive, magnetism-responsive, electricity-responsive, and light-responsive polymeric materials.

3.2.1.1 Thermo-responsive Polymers and Their Composite Materials

Thermo-responsive polymers and their composite materials can sense the temperature variation in surroundings and respond to it by changing the shapes, properties, or functionalities. At present, the thermo-responsive polymers and their composite materials mainly include thermo-responsive shape-memory polymers (SMPs), liquid crystal elastomers (LCEs), and thermo-responsive gel materials.

(1) Thermo-responsive shape-memory polymers

Thermo-responsive SMPs are one group of smart polymer materials featured by the ability to recover their original shape from a temporary shape upon heating [37–39]. Thermo-responsive SMPs have advantages of good printability, light weight, low price, abundant raw materials, large recovery strain, and simple shape-programming process. They have been the most popular thermo-responsive materials for 4D printing. Thermo-responsive SMPs respond to the ambient temperature variation through physical and chemical cross-links related to the transition temperature T_t (commonly the melting temperature T_m for a semicrystalline polymer and the glass transition temperature T_g for an amorphous polymer). They are usually composed of a stationary phase for memorizing the initial shape and a reversible phase that can process a reversible transformation between two states under external stimuli. The stationary phases for the thermo-responsive SMPs are commonly physical or chemical cross-linked structures with high T_t values. They have sufficient strength to ensure that only elastic deformation of molecular chains occurs rather than permanent plastic deformation when the SMPs are deformed under external loads. The reversible phases are commonly physical or chemical cross-linked structures with low T_t values. They can transform between the soft and hardening states reversibly with the variation of temperature. The shape-memory process of thermo-responsive SMPs contains three steps:

(a) First, the thermo-responsive SMP is heated to above the T_t of the reversible phase but below the T_t of the stationary phase to soften the reversible phase. The external stress is applied at this temperature to deform the SMP into a required temporary shape. Here, the stationary phase is in an elastic stretched state.
(b) Then, the deformed SMP is cooled down to below the T_t of the reversible phase under the applied stress. The reversible phase goes harden. And the molecular chains of the stationary phase are frozen in a stretched state, resulting in the entropy elasticity being stored in the stationary phase. When the external load is removed, the SMP is fixed into a temporary shape.
(c) Finally, when the deformed SMP is heated to above the T_t of the reversible phase, the reversible phase goes soft. And the molecular chains of the stationary phase are unfrozen and recover under the entropy elasticity. Molecular chains

of the reversible phase deform driven by the connection between the stationary and reversible phases. As a result, the deformed SMP recovers to its original shape.

Due to the large recovery strain and adjustable T_t values, thermo-responsive SMPs and their composite materials have been widely used to create 4D-printed structures with thermo-responsive shape-changing characteristics [40–50]. Yang et al. printed a series of models including aircraft, rocket, flower, and gripper using the polyurethane SMP and FDM technology. The printed flower can automatically fold from a two-dimensional (2D) plane structure to a three-dimensional structure upon heating. The printed gripper could perform a grasping motion when it was heated to above the T_t. Zarek et al. [50] fabricated the cardiovascular stent, Eiffel Tower, and bird models by printing the polycaprolactone SMP via SLA technology. These models could automatically recover from temporary shapes to the original shapes upon heating. Figure 3.1a demonstrated the shape recovery process of a 4D-printed bird model. Ge et al. [49] succeeded in the 4D printing of multi-material thermo-responsive structures through printing the photo-curable methacrylate-based SMPs with different T_t values in single construction via a high-resolution projection micro-stereolithography (PμSL) technology. The printed multi-material flowers could sequentially open their

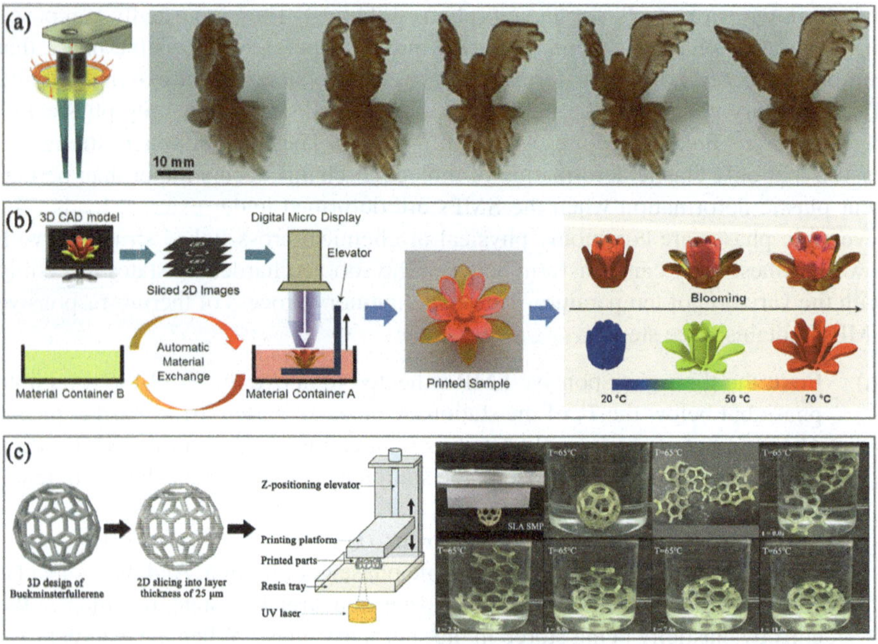

Fig. 3.1 **a** Illustration of SLA printer and shape recovery process of a printed bird at 70 °C; **b** Schematics of multi-material 4D printing based on PμSL technology and sequential recovery process of a 4D printing multi-material flower [49]. **c** SLA process of Buckminsterfullerene and its shape recovery process when immersed into a 65 °C water [47]

outer and inner petals at 50 °C and 70 °C, respectively (Fig. 3.1b). Choong et al. [47] developed a tert-butyl acrylate-co-di(ethylene glycol) diacrylate (tBA-co-DEGDA) network with thermo-responsive shape-memory characteristic based on the dual-component phase switching mechanism. They fabricated a Buckminsterfullerene using this SMP via SLM SLA technology. The printed Buckminsterfullerene can recover from an unfolded shape to the original bucky-ball shape when soaked into the 65 °C water (Fig. 3.1c). Wu et al. [43] developed a novel acrylate-based photosensitive thermo-responsive SMP for DLP. The structures printed using this SMP could be fixed into a temporary shape through deformation under external stress and subsequent freeze–thaw treatment. When placing the deformed structures into a 73 °C water bath, they could recover the original shape automatically. The above studies demonstrated the possibility of 4D printing of thermo-responsive SMPs. However, these 4D-printed structures of thermo-responsive SMPs suffer from a disadvantage that they cannot automatically change from the original shape in as-printed state, and a shape programming step under external force is required. To achieve reversible shape-changing behaviors in the 4D-printed structures of thermo-responsive SMPs, special structural design, multi-material combination, and composite printing are usually needed.

As widely reported, the uniform internal strain will be stored in the SMP structures prepared by FDM technology because the melted polymer filaments are extruded and stretched during printing. This internal strain can be released when the FDM-printed SMP structures are heated to above their T_t, leading the SMP filaments to contract in the extrusion direction but expand in the thickness direction. Utilizing this anisotropy of FDM-printed SMPs, 4D-printed structures, which can achieve controllable thermo-responsive shape-changing behaviors with no need for the shape programming step, can be created. For instance, Zhang et al. fabricated 2D lattice materials using polylactic acid (PLA) SMP and FDM technology. The circular rings in the printed lattice materials could change to hexagons or quadrangles upon heating because of the heat shrinkage of PLA filament along the extrusion direction. Manen et al. [46] programmed the anisotropic heat-shrinkable/expandable properties in planar PLA structures by adjusting the printing parameters during the FDM process as well as designing the patterns of PLA filaments on stacked layers (Fig. 3.2a). These planar PLA structures could sequentially self-bend, self-twist, or self-fold into the desired complex 3D shapes upon heating. Wang et al. [41, 44] fabricated thermo-responsive shape-changing bilayer structures through printing the polyester (PE) SMP on a paper substrate via FDM technology. Here, the polymer layer served as a stimuli-responsive active layer, while the paper layer served as a passive strain-limiting layer (Fig. 3.2b). When the bilayer structures were heated above the T_t of PE, the PE layer would shrink along the filament orientation, resulting in a deformation of the bilayer structures due to the resistance of the paper layer against contracting deformation of the PE layer. By controlling the angle between the contracting direction of the PE layer and the length direction of the paper sheet, the bilayer structures could achieve various deformations including bending, twisting, and spiral.

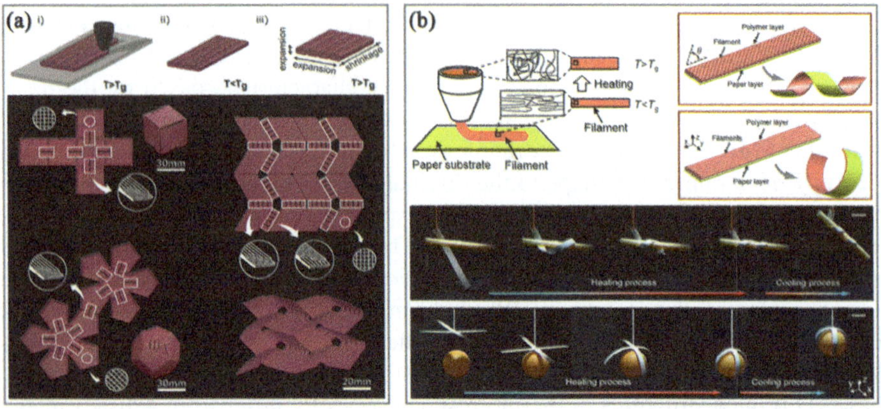

Fig. 3.2 **a** Fabrication schematic, structure design, and sequential shape-changing process of planar structures with various patterns of PLA filaments. Active regions in all structures are highlighted by white rectangles [46]. **b** FDM fabrication schematic, design, and shape-changing behaviors of PE-paper-based structures [44]

Multi-material SMP systems consisting of materials with different thermal expansion coefficients are also widely used to create thermo-responsive 4D printing structures. Zhang et al. [48] developed a direct 4D printing approach for smart lightweight structures through printing PLA strips on a fixed paper sheet. These composite sheets remained in a flat state due to the uniform release of internal stress upon heating but transformed into the designed 3D configuration due to the thermal elastic deformation caused by the mismatching coefficient of thermal expansion when cooled down to room temperature.

(2) Liquid crystal elastomers

LCEs are one kind of flexible polymeric material with low cross-linked density, which is composed of cross-linked main-chain or side-chain liquid crystal molecules [51]. When the LCEs are heated, they will transform from the liquid crystal phase with lengthened molecular chains to the isotropic phase with contracted molecular chains, resulting in a large anisotropic contraction deformation on a macroscopic scale. This unique heat-shrinkage performance of LCEs has been widely used to create 4D printing structures with thermo-responsive shape-changing characteristics [52–60].

Ambulo et al. [59] exploited the shear forces imposed on the LCEs inks during the DIW process to align their filaments along the print path. The printed structures can process locally controlled and reversible temperature responses by controlling the printing path. Kotikian et al. [57] developed a high-operating temperature DIW technology for LCEs. Programing the anisotropic heat contraction of LCEs through controlling the print path of DIW, they created a series of LCE architectures that could achieve 2D-to-3D and 3D-to-3D′ shape-morphing behaviors upon heating (Fig. 3.2a). Similarly, Ren et al. proposed a 4D printing programming approach for fabricating single-material structures of LCEs, which could exhibit various shape-changing behavior through locally regulating the printing speed and printing path in the DIW process to program the alignment degree of molecules. López-Valdeolivas et al. realized digital control of the local anisotropy of LCEs using DIW technology and created thermo-response LCE structures. Zhang et al. [56] reported that there was a temperature gradient between two sides of the LCE layer printed by DIW technology, resulting in an orientation gradient of mesogen units across the thickness in the LCE layer. Through post-photo-cross-linking treatment, this orientation gradient could be further fixed (Fig. 3.2b). Patterning this orientation gradient through liquid-assistant printing and programmed structure design to integrate bending and contraction deformation into the single printed structure, the LCE structure could perform controllable complex shape change upon heating.

Saed et al. [55] succeeded in adjusting the phase transition temperatures of the polymerizable LCE by controlling the crosslinking strategy. Through printing multiple LCE inks with different phase transition temperatures into a single structure via DIW technology and patterning their arrangement, they fabricated 4D-printed structures with sequential, reversible, ad multiple thermo-responsive shape-changing characteristics (Fig. 3.2c).

(3) Thermo-responsive gel materials

Thermo-responsive gel materials are self-adoptable macromolecular physically or chemically interconnected networks, which can demonstrate expansion and contraction to change the structure in response to the temperature variation. Thermo-responsive gel materials and their composites are another kind of attractive polymeric materials in 4D printing [61–66].

Arslan et al. [62] fabricated bilayer gel actuators by printing orthogonally continuous thermo-unresponsive poly(ethylene glycol) (PEG) elements in thermo-responsive poly(N-isopropyl acrylamide) (PNIPAm) matrix. The heat-swelling and shrinkage of the active isotropic PNIPAm matrix along the direction of continuous PEG filaments were restricted by the passive anisotropic PEG pattern. By controlling the geometry and orientation of the PEG elements, the printed multi-material structures could perform various motions from bending to twisting (Fig. 3.3a). Chen et al. [64] used the electrospinning technique to prepare porous substrates of PNIPAm with randomly arranged fibers. Subsequently, different well-designed rigid PNIPAm/clay patterns were printed on the porous substrates through 3D printing technology to achieve programmed anisotropic swelling performances. In this way, they created a series of thermo-responsive gel actuators with both fast deformation and enhanced designability in 3D shapes. In another work, Chen et al. [63] used DIW technology to print a bilayer leaf consisting of two different gel materials, namely the PNIPAm gel and the polyacrylamide (PAAm) gel. When the gel leaf was immersed in 50 °C water, the PNIPAm layer shrank, while the PAAm layer swelled, resulting in a curling deformation of the gel leaf. By designing the printing path to program the anisotropy of the printed structures, various shape-changing behavior could be achieved in an identical structure. Liu et al. [61] designed and fabricated the 4D-printed gel tubes composed of a thermo-responsive swelling PNIPAm gel and a thermo-nonresponsive PAAm gel. By controlling the symmetric and segmented arrangements of the active

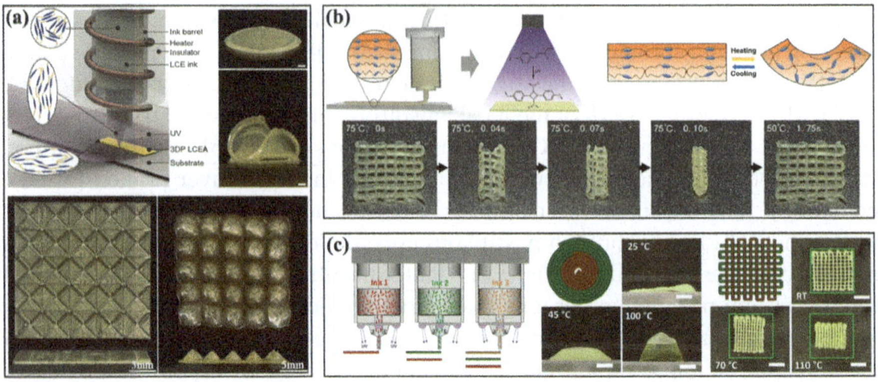

Fig. 3.3 **a** High operating temperature DIW process of LCE ink and shape-changing behaviors of printed LCE structures [57]. **b** Forming mechanism of orientation gradient of mesogen units during DIW process and shape change process of the printed bilayer structures upon heating [56]. **c** Printing schematic of DIW process to fabricate multi-material structures with three LCE inks and shape change processes of the printed multi-material structures with different print paths [55]

PNIPAm gel and passive PAAm gel in the tubes, these gel tubes could achieve various thermo-responsive shape changes such as axial elongation, radial expansion, and bending (Fig. 3.3b). Jin et al. developed a composite gel with enhanced self-supporting printability through adding Laponite nanoclay into the PNIPAm gel and a composite gel with enhanced temperature responsivity through adding graphene oxide (GO) into the PNIPAm gel. Due to the different shrinking ratios and mechanical properties between the PNIPAm-Laponite and PNIPAm-GO composite gels, the multi-material structures could deform in a predictable way upon heating.

Guo et al. [66] developed a shape-memory gel material by in-situ polymerization of acrylamide in the agarose matrix, which could be directly printed in the air. This gel material could process a reversible transition between soft and hard states upon heating and cooling based on the sol–gel transformation behavior of agarose, enabling a thermo-responsive shape change from 3D structures into different patterns (Fig. 3.3c).

3.2.1.2 Moisture-responsive Polymers and Their Composite Materials

Hydrogels are one kind of extraordinary hydrophilic polymers, which have a 3D network structure formed through physical and chemical crosslinking of monomers with a hydrophilic group. They can swell up to 200% of their original volume after absorbing water yet maintain their original structures without being dissolved. However, the swelling capacity of conventional single-component hydrogels is commonly isotropic. That is, only a linear expansion occurs in the single-component

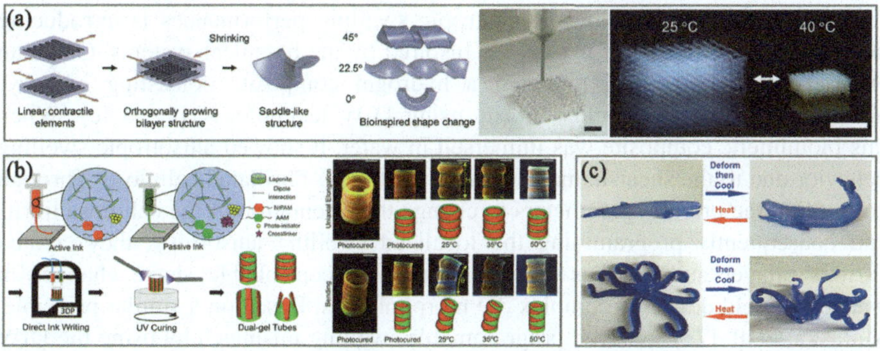

Fig. 3.4 **a** 4D printing process schematic of creating bilayer structures with orthogonally contractile elements and shape-changing behavior of printed representative bilayer gel structures [62]. **b** Shape-changing processes of 4D-printing gel tubes with symmetric and segmented arrangements of active PNIPAm (red) and passive PAAm (green) segments [61]. **c** Softening and hardening cycles of whale-like and octopus-like 4D printing gel structures [66]

hydrogel structure. To achieve complex and controllable moisture-responsive shape-changing behaviors including bending, twisting, folding, et al., programmable anisotropic swelling performances should be introduced in the hydrogel structures.

One strategy to create asymmetrical swelling is creating regions with different crosslinking densities in the structure of single hydrogel chemical nature through controlling the monomer conversion and crosslinking degrees. Jamal et al. [67] fabricated a photopatterned bilayer consisting of two PEG hydrogels with different molecular weights. When this PEG-based bilayer was immersed in aqueous solutions, it could demonstrate a self-folding behavior due to the differential swelling of the two PEG bilayers. Zhao et al. [68–70] developed a grayscale photo-printing method based on DLP to create programmable continuous variations of crosslinking density in the single-hydrogel structures (Fig. 3.4a). The uncured oligomers inside the regions with low crosslinking density diffused out of the structure when the printed structure was put into water, leading to a contraction in these regions. Consequently, a folding deformation occurs in the structure. Based on this principle, they designed and fabricated a series of complex origami structures, which could self-fold from planar configurations into 3D shapes.

In addition, Ji et al. [71] proposed a novel strategy to introduce controllable asymmetrical swelling in the hydrogel architectures through creating programmable secondary structure grooves on one side of the featured architectures (Fig. 3.4b). This asymmetrical swelling caused by the structural asymmetry would result in local curvatures, leading to a bending, twisting, or even combinational deformation in the hydrogel architectures depending on the patterns of grooves. Through coding the patterns of grooves, the printed structures could process controllable and reversible moisture-responsive shape changes from one-dimensional (1D) strips, 2D sheets, and even 3D structures to 3D morphologies. This approach had the advantages of flexible design, easy manufacturing, and simple material compositions.

Another strategy to achieve anisotropic swelling performances is introducing anisotropic particles in the hydrogels. Inspired by the botanical systems, Gladman et al. [15] designed and prepared a hydrogel composite consisting of acrylamide hydrogel and cellulose fibrils using DIW technology (Fig. 3.4c). When this biomimetic composite was immersed in water, it showed anisotropic swelling behavior due to the shear-induced alignment of nano-fibrillated cellulose. Through controlling the printing parameters to change the orientation of the cellulose fibrils and consequently programming the localized swelling anisotropy, they created various bilayer architectures that could process controllable shape changes in water and yield desirable complex 3D morphologies. Based on a similar principle, Mulakkal et al. [72] achieved programmable swelling mismatch by using the DIW technology to precisely align the cotton-derived pulp fibers in a carboxymethyl cellulose hydrogel. Utilizing this method, they fabricated a petal architecture, which could be deployed to a planar shape upon hydration and recovered to the original petal structure upon dehydration.

Integrating multiple materials with different swelling behaviors in single structures is also a popular strategy to achieve anisotropic swelling. Raviv et al. [12] printed a hydrophilic acrylate hydrogel and a rigid polymer in a single structure and achieved various moisture-responsive shape-changing behaviors in the printed structure through coding the proportion, shape, dimension, and position of the two materials. Baker et al. [73] demonstrated origami-inspired trilayer structures composed of a hydrophilic core and hydrophobic skins (Fig. 3.4d). Here, the core was polyurethane hydrogel, and the skin was polyurethane elastomer. They combined localized bilayer regions (active hinges) and a global trilayer composition (passive) within a single additive manufacturing structure, and finally achieved predictable shape changes through designing the spatial location of bilayer regions within the structures. Zhao et al. [74] fabricated a bilayer structure using DLP technology, which was constituted by an (ethylene glycol) diacrylate hydrophilic layer and a poly (propylene glycol) dimethacrylate hydrophobic layer. When the bilayer structure was put into water, a deformation would take place due to the strain mismatch between two layers during swelling.

3.2.1.3 Light-responsive Polymers and Their Composite Materials

Light is a bewitching stimulus. It has the advantages of non-contact control, precise focusing, clean/no pollution, easy trigger, and fast switching. Light-responsive polymers and their composite materials are one kind of polymeric material that can process physical and/or chemical reactions in response to light variation. These reactions usually result in shape, property, or functionality changes of the polymers and their composites. Light-responsive polymers and their composite materials have attracted increasing interest in 4D printing. Wales et al. [75] developed a novel photochromic nanostructured polyoxometalate, which could change from colorless to blue when exposed in the visible light and then return to colorless through the redox reaction when exposed to oxygen. This photochromic molecule was added into a bespoke polymeric ionic-liquid matrix, preparing a photochromic polymeric material for additive manufacturing and yielding macroscopic functional devices with reversible photochromism.

Gelebart et al. [76] reported an actuator with rewritable and reprogrammable light-responsive characteristics using liquid crystal polymers containing a pH- and light-responsive purple azomerocyanine (1-Am) dye. Different dyes were responsive to light irradiation of different wavelengths, enabling programmable folding deformation of the actuators by using light irradiation with different colors. Liu et al. [77] patterned inks with different colors on the polymer sheets. These inks exhibited different light absorbability depending on the wavelength of the light. When exposed to external light, the printed color inks discriminately generated heat due to differential light absorption, causing sequentially relief of local strain to induce sequential folding deformation in the ink-covered areas. Based on this principle, these areas covered by the color inks can act as a programmable light-repsonisve hinge. Utilizing this phenomenon, Liu et al. created 2D polymer sheets that were capable to self-fold into 3D architectures sequentially when activated by the external light.

Incorporating the materials showing photo-thermal effects (such as graphene, carbon nanotubes, carbon black, and gold nanorods) with the thermo-responsive polymeric materials is also a popular strategy to develop light-responsive materials for 4D printing. These photo-thermal particles can produce heat under the irradiation of light to trigger the shape, property, and/or functionality changes of the thermo-responsive polymeric matrix. Yang et al. [78] developed a light-responsive SMP by adding the photo-thermal carbon black (CB) into a thermo-responsive polyurethane (PU). Utilizing this composite material and FDM technology, they created light-responsive sunflower, which could bloom under a light source of 87 mW/cm^2 (Fig. 3.5a). Hua et al. [79] developed a light-responsive composite material containing PLA thermal SMP and multi-walled carbon nanotubes (MWCNTs) and created flexible light-responsive shape-changing actuators through printing it on paper substrates via FDM technology (Fig. 3.5b). Zhang et al. [80] developed a light-responsive polymeric composite by introducing the agent aniline trimer (AT) into a thermo-reversible shape-memory PU. Printing this photothermal DA-reactive PU through DIW technology, they fabricated butterfly and hand-like objects with programmable light-responsive shape-memory characteristics. Cheng et al. [81] used DIW technology to print the PNIPAm ink containing photothermal multiwalled MWCNTs on a polydimethylsiloxane (PDMS) substrate. This bilayer structure would bend upon heating due to the asymmetric thermal responsiveness of the PNIPAm and PDMS layers. Through designing the shape/position of PNIPAm and PDMS parts in a single structure, an artificial tendril was created, which could make a rotational bending motion under rotating light irradiation.

Moreover, Luo et al. [83] presented a simple approach for printing near-infrared-triggered cell-laden scaffolds with programmable shape change using the alginate/polydopamine (PDA) and alginate/gelatin methacryloyl (GelMA) bio-inks. The deformation of these orthogonal bilayer scaffolds could be adjusted by the light intensity, irradiation time, and design patterns of alginate/PDA struts in the scaffolds. Zhao

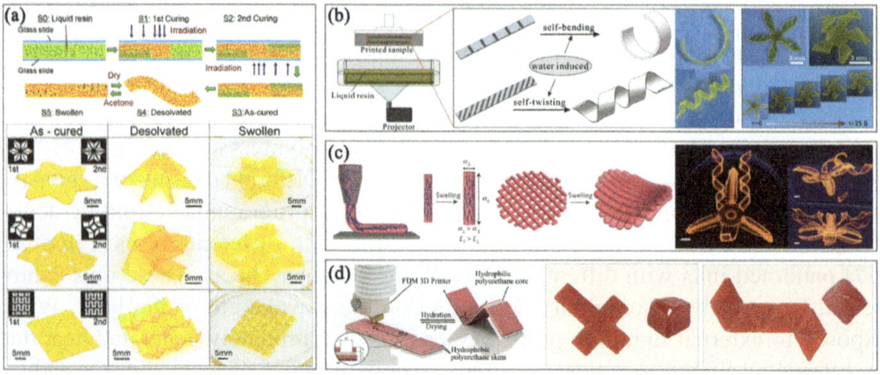

Fig. 3.5 **a** Schematic figure showing the grayscale photo-printing process and the shape-changing process induced by desolvation and swelling [69]. **b** Self-deform mechanism of 4D printing strips with programmed secondary microstructures and shape-changing behaviors of printed structures triggered by water [71]. **c** Schematic of preparing hydrogel-based composite containing aligned cellulose fibrils during DIW process and shape-changing behaviors of printed architecture [15]. **d** Fabrication schematic of hydrogel-elastomer trilayers and shape-changing behaviors of printed polygons from dehydrated to hydrated [73]

et al. [82] exploited the volume shrinkage of acrylate resin during photopolymerization to develop an easy method for fabricating light-responsive origami structures (Fig. 3.5c). An attenuated light field was produced after adding photo-absorbers into the polymer resin, causing non-uniform curing along the thickness direction. As a result, a nonuniform volume shrinkage was induced, further leading to a nonuniform stress field and driving the sheet bend to a 3D structure.

3.2.1.4 Magnetism-responsive Polymers and Their Composite Materials

The shape, property, and/or functionality change behaviors of magnetism-responsive polymers and their composite materials are triggered by the external magnetism field. Currently, some of the magnetism-responsive polymeric materials for 4D printing are developed based on thermo-responsive polymers. They are essentially innovative extensions of thermo-responsive polymers. This kind of magnetism-responsive polymers are prepared by adding magnetic particles (such as Fe_3O_4, NdFeB, Fe, and Ni nanoparticles) into the thermo-responsive polymers. These magnetic particles would reciprocate under the external alternating magnetic field. Their friction and collision with molecules will generate heat, causing an increase in the temperature and consequently triggering the shape, property, and/or functionality changes of the thermo-responsive polymeric matrix. Based on above on the above principle, Wei et al. [84] developed a magnetic-field-response SMP by introducing Fe_3O_4 nanoparticles into the PLA SMP. Utilizing this SMP composite, they created a series of magnetism-responsive biological scaffolds (Fig. 3.6a). Similarly, Lin et al. fabri-

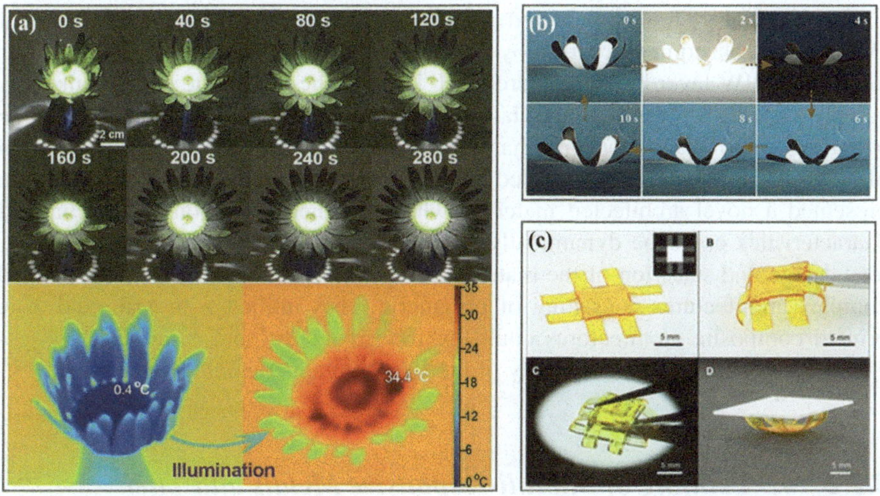

Fig. 3.6 **a** Shape variation of a printed sunflower under light illumination and its infrared image before and after illumination [78]. **b** Light-triggered shape-changing behavior of semicircular paper-based flower that was fabricated by PLA-MWCNTs composite [79]. **c** Shape changes of origami structures under illuminations [82]

cated the 4D-printed shape-memory occluders using the thermo-responsive PLA materials incorporated with Fe_3O_4 nanoparticles. These occluders could revert from a deformed shape to their original shape after implantation under the control of an external magnetic field. Zhu et al. developed a composite ink composed of PDMS and iron nanoparticles, using which magnetism-responsive 4D-printed structures with a fast response were fabricated.

Kim et al. [85] developed a DIW printer with a magnetic field around dispensing nozzle as well as a special composite ink containing NdFeB microparticles and fumed silica nanoparticles. During printing, the nanoparticles were reoriented along the applied field, and the patterned magnetic polarity is imparted to printed filaments (Fig. 3.6b). Utilizing this method, they realized the programming of ferromagnetic domains in complex soft material systems and created various structures demonstrating complex shape changes under applied magnetic fields. Shinoda et al. [86] incorporated a programed magnetic anisotropy into UV-curable gel material dispersed with two kinds of magnetic particles (strontium-ferrite particles and carbonyl iron powders) using an external magnetic field. Aligned magnetic particles in different portions of the printed structures generated different rotational moments under an external magnetic field, resulting in a prescribed movement. Utilizing this method, they created various magnetic-field-response biomimetic actuators. Wu et. al fabricated 4D-printed magnetism-responsive grippers by SLS technology using a composite powder composed of magnetic NdFeB and thermoplastic polyurethane. The printed grippers showed remote-controllable deformation in an external magnetic field.

Bastola et al. [87, 88] fabricated a new class of 4D-printed magnetorheological (MR) hybrid elastomer structure by encapsulating an MR fluid into an elastomer matrix via DIW layer by layer. Here, each layer of the composite architecture was composed of the MR fluid and the elastomer. These printed MR-elastomer composite structures could change their mechanical properties controllably with the variation of the external magnetic field. Based on a similar principle, Jackson et al. [89] also presented a novel architected material, whose magnetism-responsive mechanical characteristics could be dynamically controlled and timely tuned by the meticulous design and selection of the material composition and structures. They printed complex architectures consisting of polymeric tubes infilled with MR fluid. The printed composite architectures could exhibit great, fast, and reversible stiffness changes by controlling the intensity of the applied magnetic field (Fig. 3.6c).

3.2.2 4D Printing of Stimuli-responsive Metals and Their Composite Materials

Although the stimuli-responsive metals and their composites are higher in weight and material cost as compared with the stimuli-responsive polymers and their composite materials, they are attracting increasing interest in 4D printing because of their superior carrying and actuation capabilities, better high/low-temperature performances, and faster response rate. At present, stimuli-responsive metals and their composite materials for 4D printing are mainly the shape-memory alloys (SMAs) and their composite materials. SMAs are one group of functional metal materials characterized by shape-memory effect (SME) and super-elasticity (SE). The SME is a phenomenon that the alloy fixed into a temporary shape under specific conditions (such as the action of applied stress or a magnetic field) can recover the original shape under certain stimuli (such as the variation of temperature or magnetic field) [90]. The SE is a phenomenon that the alloy can be deformed into a temporary shape under the external load and then recovers its original shape upon unloading, with no need for any other stimulus. Both the SME and SE originate from a reversible transformation between the austenite (stable at high temperature) and martensite (stable at low temperature) phases. Due to the unique SME and SE, SMAs can realize the combinational functionalities of sensing, driving, controlling, carrying, energy conversation, self-assembly, and self-adaption. Thus, SMAs show great application potential in 4D printing. At present, the most popular additive manufacturing technologies used in the 4D printing of SMAs mainly include selective laser melting (SLM), selective electron beam melting (SEBM), laser engineered net shaping (LENS), wire and arc additive manufacturing (WAAM), and binder jetting. Multiple SMA systems have been used in 4D printing. According to the stimulus types, they can be classified into two categories, namely the thermo-responsive and the magnetism-responsive SMAs.

3.2.2.1 Thermo-responsive Shape-Memory Alloys and Their Composite Materials

(1) Mechanisms of thermal shape-memory effect and super-elasticity

The SME of thermo-responsive SMAs is triggered by heat. There are two mechanisms for the thermo-responsive SME: (a) stress-induced reorientation of thermal martensite; (b) stress-induced martensitic transformation. In mechanism (a), the SMAs are in a martensite state at the deformation temperature. When the self-accommodating thermal martensite (twin martensite) variants are subjected to external stress, they will be reoriented to the martensite variant favorable for the stress direction, producing the detwinned martensite. As a result, a macroscopic deformation will take place in the SMAs. And the deformed shape is kept after the applied stress is removed because the detwinned martensite is retained. When heating above the finish temperature of reverse martensitic transformation A_f, the detwinned martensite reverses to the original austenite, causing recovery from the deformed shape to the initial one. In mechanism (b): The SMAs are at an austenite state at the deformation temperature. The deformation of the SMAs under external stress is assumed by the stress-induced martensitic transformation. After unloading, the stress-induced martensite retains at this deformation temperature, and thus the deformed shape is kept. When the SMAs are heated above A_f, the stress-induced martensite reverses to the original austenite. Consequently, the deformed alloy recovers the original shape.

Unlike the SME in which the deformed alloy recovers its original shape with a need for heating or varying the magnetic field, the shape recovery in SE takes place after the removal of applied stress with no need for any other stimuli. The mechanism for SE is as follows: when the external stress is applied to SMAs above A_f but below M, the stress-induced martensitic transformation occurs and leads to a macroscopic deformation. Because the stress-induced martensite is unstable at this deformation temperature, it will retransform to austenite when the external stress is removed. As a result, the macroscopic deformation recovers upon unloading.

(2) Ni–Ti-based shape-memory alloys

Ni–Ti-based, Cu-based, and Fe-based SMAs are currently the three categories of the most widely studied and best-developed thermo-responsive SMAs. More specifically, current studies on 4D printing of SMAs mainly focus on Ni–Ti alloys due to their best SME and SE as well as good biocompatibility, high recovery stress, excellent corrosion resistance, superb wear resistance, good functional stability, and outstanding mechanical properties [91–94].

The high-quality additive manufacturing of Ni–Ti SMAs with high performances is the foundation of 4D printing of Ni–Ti SMAs. Thus, many research works have been performed focusing on the process optimization and performance control of the additively manufactured Ni–Ti SMAs. Saedi et al. [94] systematically investigated the effect of laser power and scanning velocity on the microstructures, martensitic transformation behavior, and SE of Ni–Ti SMAs fabricated by SLM technology. It was revealed that to obtain fully dense Ni–Ti parts, the combinations of high laser power with a high scanning velocity or low laser power with a low scanning velocity were suggested. The Ni–Ti parts fabricated using low laser parameters process better

SE than that fabricated using high laser parameters. Adopting an optimized combination of 100 W with 125 mm/s, they succeeded in creating dense Ni–Ti with a high relative density of about 99% and a high SE recovery strain of 5.77%. Lu et al. [95] studied the effect of laser volumetric energy density on martensitic transformation behaviors, mechanical properties, and SME of Ni–Ti SMAs fabricated by SLM technology. The results showed that the martensitic transformation temperatures decrease as lowering the laser volumetric energy density. The Ni–Ti SMAs printed by using the laser volumetric energy density of 222 J/mm^3 exhibited a high tensile strength of 776 MPa and a large elongation of 7.2%, while the Ni–Ti SMAs printed by using the laser volumetric energy density of 155 J/mm^3 demonstrated the largest recovery strain of 4.99%. Zhao et al. [96] used a bending test to measure the SME in the SLM-printed Ni–Ti SMAs fabricated using mechanically mixed powders and high laser volumetric energy densities of 375–675 J/mm^3. The recovery strain in the Ni–Ti samples fabricated with 675 J/mm^3 reached 5.17% when bent at room temperature by 6%.

Ma et al. [97] created a U-shaped Ni–Ti piece with multi-stage shape recovery behaviors via SLM (Fig. 3.7a, b). It was achieved by changing the hatch spacing at different sections of the U-shaped piece to control the martensitic transformation temperatures at corresponding sections, creating a location-dependent active response. Hao et al. [98, 99] created a Ni–Ti lath and a Ni–Ti scaffold using SLM. Both of them could almost completely recover from deformed shapes to their original shape upon heating (Fig. 3.7c).

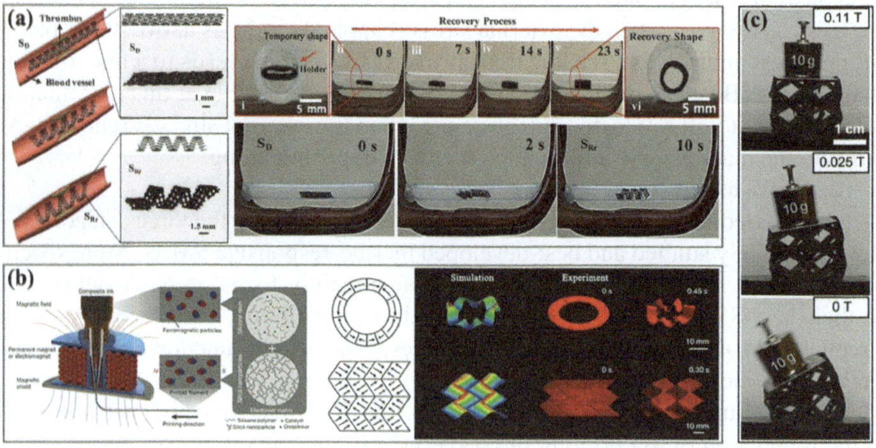

Fig. 3.7 **a** Shape-morphing process of composite cylinder and scaffold in an alternating magnetic field of 30 kHz [84]. **b** Schematic of programming ferromagnetic domains in complex soft material systems by using a novel DIW printer with a magnetic field around dispensing nozzle and shape change of printed structures under a uniform magnetic field of 200 mT [85]. **c** Shape changing process as reducing the applied magnetic field of MR fluid-filled lattice supporting a static 10 g mass [89]

Zhou et al. [100] created Ni–Ti SMAs via SEBM. A tensile strength of 1411.0 ± 59.3 MPa and an elongation of 11.8 ± 0.9% were obtained in the SEBM Ni–Ti SMAs. Wan et al. [101] prepared Ni–Ti SMAs by laser solid forming (LSF). The SE recovery strain of the as-deposited Ni–Ti SMAs samples reached 9.2% when it was compressively deformed by 10% at room temperature. Hou et al. [102] used LENS technology to locally melt the near-eutectic mixed Ni and Ti powders to create an Ni–Ti nanocomposite, which contained a substantial fraction of Ni_3Ti intermetallic compound in a NiTi binary matrix. The printed Ni–Ti nanocomposite exhibited small hysteresis, quasi-linear SE behavior, and stability of over 1 million compression cycles in the mechanical behavior and electrocaloric response.

(3) Cu-based shape-memory alloys

Cu-based SMAs have good SME and SE (recovery strain up to 8%), excellent thermal and electrical conductivity, high damping performance, low costs (about the one-tenth cost of Ni–Ti-based alloys), as well as an adjustable martensitic transformation temperature range (−90–120 °C). Thus, Cu-based SMAs have also attracted increasing interest in 4D printing. According to the chemical composition, Cu-based SMAs can be mainly classified into Cu-Zn–Al-based, Cu–Al–Mn-based, and Cu–Al–Ni-based SMAs. Amongst, Cu-Al-Ni-based SMAs have aroused the widest attention due to their good stabilization against aging phenomena.

Mazzer et al. [103] first prepared a Cu–Al–Ni–Mn alloy using the SLM technology. Highly dense Cu–Al–Ni–Mn samples with fine grains (10–100 μm), no cracks, and single monoclinic martensite microstructures were obtained. Afterward, Gustmann et al. [104–106] further studied the effect of processing parameters, scanning strategies, and additional re-melting steps on the microstructures and mechanical properties of the Cu–Al–Ni–Mn alloy fabricated by SLM technology. The results showed that the optimum laser volumetric energy density is between 30 and 40 J/mm^3. When a laser power of 300 W and a scanning velocity of 700 mm/s were used, the printed samples could exhibit a high relative density of 99% and a full martensite microstructure. The fracture strength and fracture strain of the SLM-printed samples were 1515 ± 50 MPa and 18 ± 1.7%, respectively. These values were comparable to the fracture strength (1500 ± 30 MPa) and fracture strain (16 ± 0.2%) of dense suction-cast samples. Tian et al. [107] also fabricated a Cu–Al–Ni–Ti alloy through SLM technology successfully. The average grain size and relative density of the SLM-printed Cu–Al–Ni–Ti samples were 43 μm and 99.5%, respectively. However, these samples only have an elongation of 7.6% and ultimate tensile strength of 541 MPa, which were still poorer than that of cast samples.

All the above works prove the feasibility of producing dense and fine Cu-based SMAs through additive manufacturing. But the limited reports mainly focus on the optimization of process parameters as well as the resultant evolutions of microstructures and mechanical properties in additively manufactured Cu-based SMAs. Yet, the investigations about the shape-memory performances of additively manufactured Cu-based SMAs have been hardly reported. Further studies are still urgent to unlock the potential of Cu-based SMAs in 4D printing.

(4) Fe-based shape-memory alloys

Thermo-responsive Fe-based SMAs can be mainly classified into Fe–Mn–Si-based, Fe–Mn–Al-based, and Fe–Ni–Co–Al-based SMAs. Although the Fe-based SMAs are slightly poorer in SME and SE as compared with Ni–Ti-based and Cu-based SMAs, they have advantages of the lowest cost, excellent formability, and good mechanical properties. They have broad application prospection in the 4D printing of intelligent structural members, especially large-scale productions. In 2016, Niendorf et al. [109] fabricated a Fe–Mn–Al–Ni alloy through SLM technology for the first time. The printed sample subjected to simple solution treatment and subsequent aging treatment exhibited a high SE recovery strain of 7.5% when compressed by about 11% at −100 °C. This work demonstrates that the Fe-based SMAs with good shape-memory performance can be obtained through additive manufacturing technologies. However, to the authors' best knowledge, no relevant investigations have been further reported. The potential of Fe-based SMAs in 4D printing still needs to be further tapped.

3.2.2.2 Magnetic Shape-Memory Alloys and Their Composite Materials

Magnetic SMAs are also known as ferromagnetic SMAs. Their SME is triggered by the variation of the magnetic field. As compared with the thermo-responsive SMAs, magnetic SMAs has a much higher actuation frequency because they are activated by the magnetic fields and consequently are not hindered by the relatively slow heat transfer mechanism. Similar to the thermal SME, there are also two mechanisms for the magnetic SME: (1) magnetically induced reorientation of martensite; (2) magnetic-field-induced martensitic transformation. At present, magnetic SMAs mainly include the Ni-Mn-based and Co–Ni-based SMAs. Among them, the Ni–Mn–Ga SMAs are currently the most studied magnetic SMAs in the 4D printing field due to their high magnetic-field-induced strain.

Ternary ferromagnetic Ni–Mn–Ga Heusler structured alloys are one kind of distinct magnetism-responsive SMAs, which have attracted wide interest since a large magnetic field induced strain of up to 10% was reported in single Ni–Mn–Ga crystals. Caputo et al. [110–112] successfully fabricated complex Ni–Mn–Ga magnetism-responsive SMA parts with different porosities using binder jetting technology. After a thermo-magneto-mechanically training treatment, the printed Ni–Mn–Ga samples could exhibit a reversible magnetism-responsive strain of about 0.01%. This work proved the possibility of 4D printing of magnetism-responsive SMAs (Fig. 3.8). Afterward, Mostafaei et al. [113] systematically investigated the effect of sintering temperatures on the microstructures, martensitic transformation characteristics, and magnetism-responsive behaviors of Ni–Mn–Ga SMAs manufactured by binder jetting technology. It was found that their relative density increased from 45% to 99% with increasing the sintering temperature from 1273 K to 1373 K. X-ray diffraction patterns indicated that the 14 M martensite was the main phase in the Ni–Mn–Ga samples sintered at 1273–1363 K for 2 h. The Ni–Mn–Ga samples sintered

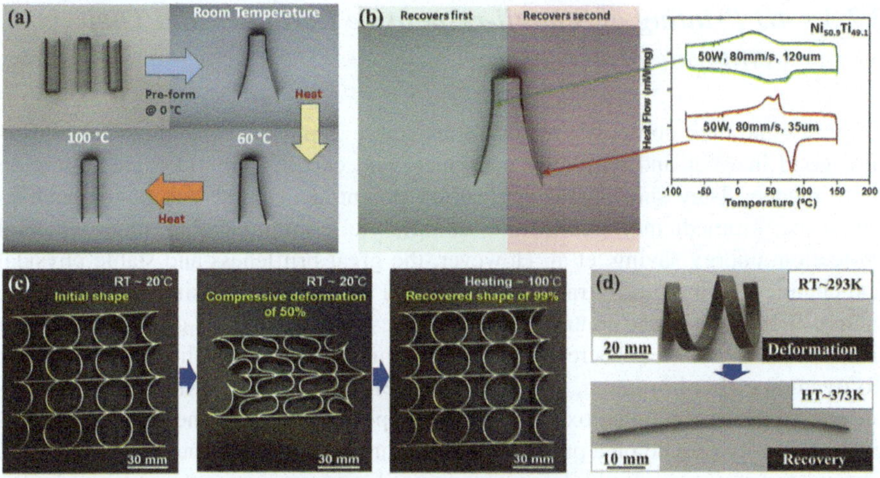

Fig. 3.8 **a** Shape-memory process of SLM-printed U-shaped sample with location-dependent active response. **b** martensitic transformation temperatures in different sections of the U-shaped sample (shown in the DSC curves on the right) [97]. **c** Shape-memory process of Ni–Ti scaffold fabricated by SLM technology [99]. **d** Shape-memory process of Ni–Ti lath fabricated by SLM technology [98].

at 1343 K for 2 h showed a maximum saturation magnetization of –56.5 Am^2/kg, martensitic transformation temperature of about 346 K, and Curie temperature of 363 K.

Laitinen et al. [114] fabricated the Ni–Mn–Ga SMAs through SLM technology and investigated the effect of process parameters on their microstructures. It was found the loss of Mn corresponding concentration variation of Ni was a big concern for the SLM fabrication of Ni–Mn–Ga SMAs. Thus, a minor over-alloying of Mn into the initial powder to counteract the observed loss was required. Increasing the volume energy density increased the loss of Mn. Additionally, the results revealed highly dissimilar effects of the applied process parameters on the composition and relative density of the as-fabricated material, with a clear optimum where high relative density material could be produced with minimal loss of Mn. Using the optimal process parameters, the SLM Ni–Mn–Ga magnetism-responsive SMAs with a high relative density of 98.3% and homogenous compositions were obtained.

Taylor et al. [115] proposed a new additive manufacturing strategy to fabricate porous polycrystalline Ni–Mn–Ga micro-architectures. This method mainly includes three steps: (1) creation of micro-structures via extrusion-based 3D printing, (2) removal of binder and residual solvents via annealing treatment, and (3) homogenization of composition via high-temperature sintering. The Ni–Mn–Ga micro-trusses created using this method showed a Curie temperature of 85–90 °C, a saturation magnetization up to 56 Am^2/kg, and reversible martensitic transformation temperatures between 45 °C and 90 °C.

3.2.3 4D Printing of Stimuli-responsive Ceramics and Their Composite Materials

Ceramics and their composite materials are stable in physical and chemical properties, good in wear and corrosion resistance, and excellent in electrical insulation performance. They show great application potential in numerous fields, such as aerospace, biomedicine, architecture, automobile, communication, environmental protection, energy-saving, et al. However, the great brittleness and stable physical and chemical properties of traditional ceramics and their composite materials make them difficult to achieve controllable stimuli-responsive changes of shapes, properties, and functionalities. Therefore, they are commonly hardly applied in 4D printing. To overcome this limitation, Lu et al. [36] developed a novel ZrO_2 nanoparticle-reinforced poly(dimethylsiloxane) matrix composite material. The nanocomposites could be easily printed on a pre-stretched precursor substrate through DIW process. The printed structures could deform into complex ceramic precursors driven by the elastic energy stored in the precursor substrate and then transform into elastomer-derived ceramics with robust mechanical properties after heat treatment (Fig. 3.9). The ceramic precursors could perform programmable self-shaping through designing their patterns. This work enables the fabrication of complex ceramic origami as well as the 4D printing of ceramic architectures (Fig. 3.10).

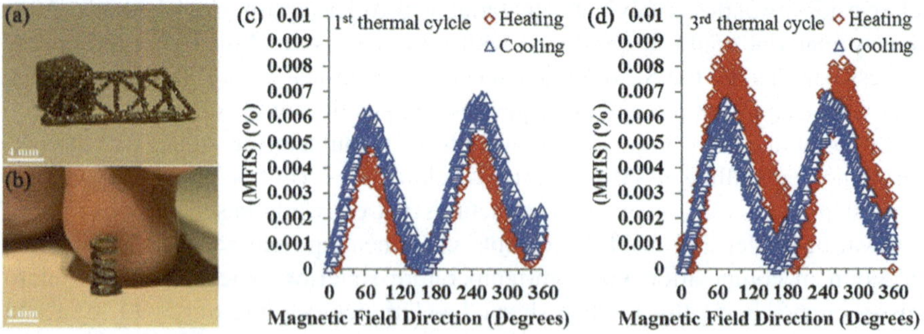

Fig. 3.9 **a** and **b** Ni–Mn–Ga magnetic SMA parts created via the binder jetting technology. **c** and **d** variation of magnetic-field-induced strain with magnetic field direction during the **c** first and **d** third heating–cooling cycles of Ni–Mn–Ga parts [111]

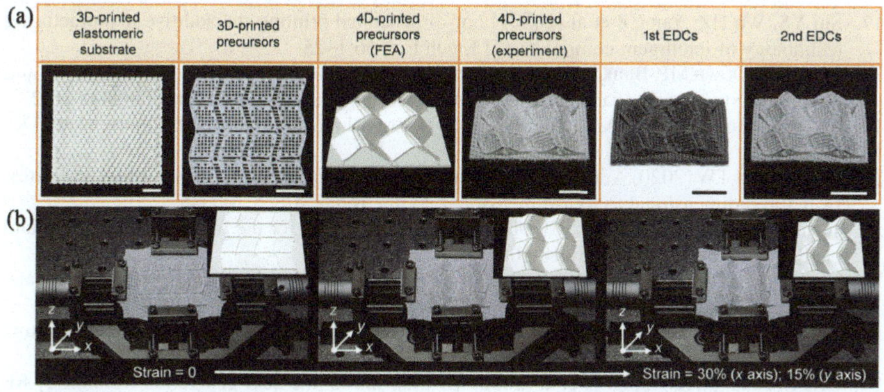

Fig. 3.10 **a** Representative design of 4D-printed elastomer-derived ceramics. **b** 4D printing of elastomer-derived ceramics [36]

References

1. Liu SY, Shin YC (2019) Additive manufacturing of Ti6Al4V alloy: a review. Mater Des 164:107552
2. Ngo TD, Kashani A, Imbalzano G et al (2018) Additive manufacturing (3D printing): a review of materials, methods, applications and challenges. Compos B 143:172–196
3. Bose S, Ke D, Sahasrabudhe H et al (2018) Additive manufacturing of biomaterials. Prog Mater Sci 93:45–111
4. Bandyopadhyay A, Heer B (2018) Additive manufacturing of multi-material structures. Mater Sci Eng R Rep 129:1–16
5. Sames WJ, List FA, Pannala S et al (2016) The metallurgy and processing science of metal additive manufacturing. Int Mater Rev 61:315–360
6. Wan X, Luo L, Liu Y et al (2020) Direct ink writing based 4D printing of materials and their applications. Adv Sci 7:2001000
7. Rafiee M, Farahani RD, Therriault D (2020) Multi-material 3D and 4D printing: a survey. Adv Sci 7:1902307
8. Ionov L (2018) 4D biofabrication: materials, methods, and applications. Adv Healthcare Mater 7:1800412
9. Kuang X, Roach DJ, Wu J et al (2019) Advances in 4D printing: materials and applications. Adv Func Mater 29:1805290
10. Wan Z, Zhang P, Liu Y et al (2020) Four-dimensional bioprinting: current developments and applications in bone tissue engineering. Acta Biomater 101:26–42
11. Tibbits S (2014) 4D printing: multi-material shape change. Archit Des 84:116–121
12. Raviv D, Zhao W, McKnelly C et al (2014) Active printed materials for complex self-evolving deformations. Sci Rep 4:7422
13. Khoo ZX, Teoh JEM, Liu Y et al (2015) 3D printing of smart materials: a review on recent progresses in 4D printing. Virtual Phys Prototyping 10:103–122
14. Ghi A, Rossetti F (2016) 4D printing: an emerging technology in manufacturing? Digitally Supported Innovat 171–178
15. Gladman AS, Matsumoto EA, Nuzzo RG et al (2016) Biomimetic 4D printing. Nat Mater 15:413–419
16. Leist SK, Zhou J (2016) Current status of 4D printing technology and the potential of light-reactive smart materials as 4D printable materials. Virtual Phys Prototyping 11:249–262

17. Shi YS, Wu HZ, Yan CZ et al (2020) Four-dimensional printing-the additive manufacturing technology of intelligent components. J Mech Eng 56:1–25
18. Ryan KR, Down MP, Banks CE (2021) Future of additive manufacturing: overview of 4D and 3D printed smart and advanced materials and their applications. Chem Eng J 403:126162
19. Zolfagharian A, Kaynak A, Kouzani A (2020) Closed-loop 4D-printed soft robots. Mater Des 188:108411
20. Liu Y, Chou TW (2020) Additive manufacturing of multidirectional preforms and composites: from three-dimensional to four-dimensional. Mater Today Adv 5:100045
21. Joshi S, Rawat K, et al KC (2020) 4D printing of materials for the future: opportunities and challenges. Appl Mater Today 18:100490
22. Hu Y, Wang Z, Jin D et al (2019) Botanical-inspired 4D printing of hydrogel at the microscale. Adv Func Mater 30:1907377
23. Spiegel CA, Hippler M, Münchinger A et al (2019) 4D printing at the microscale. Adv Func Mater 30:1907615
24. Li YC, Zhang YS, Akpek A et al (2016) 4D bioprinting: the next-generation technology for biofabrication enabled by stimuli-responsive materials. Biofabrication 9:012001
25. Lui YS, Sow WT, Tan LP et al (2019) 4D printing and stimuli-responsive materials in biomedical aspects. Acta Biomater 92:19–36
26. Rastogi P, Kandasubramanian B (2019) Breakthrough in the printing tactics for stimuli-responsive materials: 4D printing. Chem Eng J 366:264–304
27. Ke D, Chen Z, Momo ZY et al (2020) Recent advances of two-way shape memory polymers and four-dimensional printing under stress-free conditions. Smart Mater Struct 29:023001
28. Falahati M, Ahmadvand P, Safaee S et al (2020) Smart polymers and nanocomposites for 3D and 4D printing. Mater Today 40:215–245
29. Shie MY, Shen YF, Astuti SD et al (1864) Review of polymeric materials in 4D printing biomedical applications. Polymers 2019:11
30. Ma SQ, Zhang YP, Wang M et al (2019) Recent progress in 4D printing of stimuli-responsive polymeric materials. Sci China Technol Sci 63:532–544
31. Henríquez GCM, Vallejos SMA, Hernandez RJ (2019) Polymers for additive manufacturing and 4D-printing: materials, methodologies, and biomedical applications. Prog Polym Sci 94:57–116
32. Dong Y, Wang S, Ke Y et al (2020) 4D printed hydrogels: fabrication, materials, and applications. Adv Mater Technol 5:2000034
33. Champeau M, Heinze DA, Viana TN et al (2020) 4D printing of hydrogels: a review. Adv Func Mater 30:1910606
34. Ahmed K, Shiblee MDNI, Khosla A et al (2020) Review: recent progresses in 4D printing of gel materials. J Electrochem Soc 167:037563
35. Atli KC, Johnson L, Ranaiefar M et al (2019) 4D printing of metallic functional materials. Adv Mater Processes 177:16–21
36. Liu G, Zhao Y, Wu G et al (2018) Origami and 4D printing of elastomer-derived ceramic structures. Sci Adv 4: eaat0641
37. Xia Y, He Y, Zhang F et al (2020) A review of shape memory polymers and composites: mechanisms, materials, and applications. Adv Mater e2000713
38. Tan LJ, Zhu W, Zhou K (2020) Recent progress on polymer materials for additive manufacturing. Adv Func Mater 30:2003062
39. Scalet G (2020) Two-way and multiple-way shape memory polymers for soft robotics: an overview. Actuators 9
40. Keneth SE, Lieberman R, Rednor M et al (2020) Multi-material 3D printed shape memory polymer with tunable melting and glass transition temperature activated by heat or light. Polymers 12:710
41. Wang W, Yu CY, Serrano APA et al (2019) Soft grasping mechanisms composed of shape memory polymer based self-bending units. Compos B Eng 164:198–204
42. Zhang B, Zhang W, Zhang Z et al (2019) Self-healing four-dimensional printing with an ultraviolet curable double-network shape memory polymer system. ACS Appl Mater Interfaces 11:10328–10336

43. Wu HZ, Chen P, Yan CZ et al (2019) Four-dimensional printing of a novel acrylate-based shape memory polymer using digital light processing. Mater Des 171:107704
44. Wang W, Li C, Cho M et al (2018) Soft tendril-inspired grippers: shape morphing of programmable polymer-paper bilayer composites. ACS Appl Mater Interfaces 10:10419–10427
45. Ding Z, Yuan C, Peng X et al (2017) Direct 4D printing via active composite materials. Sci Adv 3:e1602890
46. Manen VT, Janbaz S, Zadpoor AA (2017) Programming 2D/3D shape-shifting with hobbyist 3D printers. Mater Horiz 4:1064–1069
47. Choong YYC, Maleksaeedi S, Eng H et al (2017) 4D printing of high performance shape memory polymer using stereolithography. Mater Des 126:219–225
48. Zhang Q, Zhang K, Hu G (2016) Smart three-dimensional lightweight structure triggered from a thin composite sheet via 3D printing technique. Sci Rep 6:22431
49. Ge Q, Sakhaei AH, Lee H et al (2016) Multimaterial 4D printing with tailorable shape memory polymers. Sci Rep 6:31110
50. Zarek M, Layani M, Cooperstein I et al (2016) 3D printing of shape memory polymers for flexible electronic devices. Adv Mater 28:4449–4454
51. Ge F, Zhao Y (2019) Microstructured actuation of liquid crystal polymer networks. Advan Funct Mater 30
52. Roach DJ, Yuan C, Kuang X et al (2019) Long liquid crystal elastomer fibers with large reversible actuation strains for smart textiles and artificial muscles. ACS Appl Mater Interfaces 11:19514–19521
53. Ambulo CP, Ford MJ, Searles K et al (2020) 4D-printable liquid metal-liquid crystal elastomer composites. ACS Appl Mater Interfaces
54. Ren L, Li B, He Y et al (2020) Programming shape-morphing behavior of liquid crystal elastomers via parameter-encoded 4D printing. ACS Appl Mater Interfaces 12:15562–15572
55. Saed MO, Ambulo CP, Kim H et al (2019) Molecularly-engineered, 4D-printed liquid crystal elastomer actuators. Adv Func Mater 29:1806412
56. Zhang C, Lu X, Fei G et al (2019) 4D printing of a liquid crystal elastomer with a controllable orientation gradient. ACS Appl Mater Interfaces 11:44774–44782
57. Kotikian A, Truby RL, Boley JW et al (2018) 3D printing of liquid crystal elastomeric actuators with spatially programed nematic order. Adv Mater 30:1706164
58. Lopez-Valdeolivas M, Liu D, Broer DJ et al (2018) 4D printed actuators with soft-robotic functions. Macromol Rapid Commun 39:1700710
59. Ambulo CP, Burroughs JJ, Boothby JM et al (2017) Four-dimensional printing of liquid crystal elastomers. ACS Appl Mater Interfaces 9:37332–37339
60. Yuan C, Roach DJ, Dunn CK et al (2017) 3D printed reversible shape changing soft actuators assisted by liquid crystal elastomers. Soft Matter 13:5558–5568
61. Liu J, Erol O, Pantula A et al (2019) Dual-gel 4D printing of bioinspired tubes. ACS Appl Mater Interfaces 11:8492–8498
62. Arslan H, Nojoomi A, Jeon J et al (2019) 3D printing of anisotropic hydrogels with bioinspired motion. Advanced Science 6:1800703
63. Chen Z, Zhao D, Liu B et al (2019) 3D printing of multifunctional hydrogels. Adv Func Mater 29:1900971
64. Chen TT, Bakhshi H, Liu L et al (2018) Combining 3D printing with electrospinning for rapid response and enhanced designability of hydrogel actuators. Adv Func Mater 28:1800514
65. Jin Y, Shen Y, Yin J et al (2018) Nanoclay-based self-supporting responsive nanocomposite hydrogels for printing applications. ACS Appl Mater Interfaces 10:10461–10470
66. Guo JH, Zhang RR, Zhang L et al (2018) 4D printing of robust hydrogels consisted of agarose nanofibers and polyacrylamide. ACS Macro Lett 7:442–446
67. Jamal M, Kadam SS, Xiao R et al (2013) Bio-origami hydrogel scaffolds composed of photocrosslinked PEG bilayers. Adv Healthcare Mater 2:1142–1150
68. Zhao Z, Qi HJ, Fang D (2019) A finite deformation theory of desolvation and swelling in partially photo-cross-linked polymer networks for 3D/4D printing applications. Soft Matter 15:1005–1016

69. Zhao Z, Wu J, Mu X et al (2017) Desolvation induced origami of photocurable polymers by digit light processing. Macromol Rapid Commun 38:1600625
70. Wu J, Zhao Z, Kuang X et al (2018) Reversible shape change structures by grayscale pattern 4D printing. Multifunctional Mater 1:015002
71. Ji Z, Yan C, Yu B et al (2019) 3D printing of hydrogel architectures with complex and controllable shape deformation. Adv Mater Technol 4:1800713
72. Mulakkal MC, Trask RS, Ting VP et al (2018) Responsive cellulose-hydrogel composite ink for 4D printing. Mater Des 160:108–118
73. Baker AB, Bates SRG, Llewellyn-Jones TM et al (2019) 4D printing with robust thermoplastic polyurethane hydrogel-elastomer trilayers. Mater Des 163:107544
74. Zhao Z, Kuang X, Yuan C et al (2018) Hydrophilic/hydrophobic composite shape-shifting structures. ACS Appl Mater Interfaces 10:19932–19939
75. Wales DJ, Cao Q, Kastner K et al (2018) 3D-printable photochromic molecular materials for reversible information storage. Adv Mater 30:1800159
76. Gelebart AH, Mulder DJ, Vantomme G et al (2017) A rewritable, reprogrammable, dual light-responsive polymer actuator. Angewandate Chemie-Int Edition 56:13436–13439
77. Liu Y, Shaw B, Dickey MD et al (2017) Sequential self-folding of polymer sheets. Sci Adv 3:e1602417
78. Yang H, Leow WR, Wang T et al (2017) 3D printed photoresponsive devices dased on shape memory composites. Adv Mater 29:1701627
79. Hua DC, Zhang XQ, Ji ZY et al (2018) 3D printing of shape changing composites for constructing flexible paper-based photothermal bilayer actuators. J Mater Chem C 6:2123–2131
80. Zhang Y, Yin X-Y, Zheng M et al (2019) 3D printing of thermoreversible polyurethanes with targeted shape memory and precise in situ self-healing properties. J Mater Chem A 7:6972–6984
81. Cheng Y, Chan KH, Wang XQ et al (2019) Direct-ink-write 3D printing of hydrogels into biomimetic soft robots. ACS Nano 13:13176–13184
82. Zhao Z, Wu J, Mu X et al (2017) Origami by frontal photopolymerization. Sci Adv 3:e1602326
83. Luo Y, Lin X, Chen B et al (2019) Cell-laden four-dimensional bioprinting using near-infrared-triggered shape-morphing alginate/polydopamine bioinks. Biofabrication 11:045019
84. Wei H, Zhang Q, Yao Y et al (2017) Direct-write fabrication of 4D ative shape-changing structures based on a shape memory polymer and its nanocomposite. ACS Appl Mater Interfaces 9:876–883
85. Kim Y, Yuk H, Zhao R et al (2018) Printing ferromagnetic domains for untethered fast-transforming soft materials. Nature 558:274–279
86. Shinoda H, Azukizawa S, Maeda K et al (2019) Bio-mimic motion of 3D-printed gel structures dispersed with magnetic particles. J Electrochem Soc 166:B3235–B3239
87. Bastola AK, Paudel M, Li L (2018) Development of hybrid magnetorheological elastomers by 3D printing. Polymer 149:213–228
88. Bastola AK, Hoang VT, Li L (2017) A novel hybrid magnetorheological elastomer developed by 3D printing. Mater Des 114:391–397
89. Jackson JA, Messner MC, Dudukovic NA et al (2018) Field responsive mechanical metamaterials. Sci Adv 4:eaau6419
90. Jani MJ, Leary M, Subic A et al (2014) A review of shape memory alloy research, applications and opportunities. Mater Des 56:1078–1113
91. Liu S, Zhu J, Lin XD et al (2021) Coupling effect of stretch-bending deformation and electric pulse treatment on phase transformation behavior and superelasticity of a Ti-50.8 at.% Ni alloy. Mater Sci Eng A, 799:140164
92. Song YN, Sun QD, Guo K et al (2020) Effect of scanning strategies on the microstructure and mechanical behavior of 316L stainless steel fabricated by selective laser melting. Mater Sci Eng, A 793:139879
93. Wang XB, Yu JY, Liu JW et al (2020) Effect of process parameters on the phase transformation behavior and tensile properties of NiTi shape memory alloys fabricated by selective laser melting. Addit Manuf 36:101545

94. Saedi S, Moghaddam SN, Amerinatanzi A et al (2018) On the effects of selective laser melting process parameters on microstructure and thermomechanical response of Ni-rich NiTi. Acta Mater 144:552–560
95. Lu HZ, Yang C, Luo X et al (2019) Ultrahigh-performance TiNi shape memory alloy by 4D printing. Mater Sci Eng, A 763:138166
96. Zhao CY, Liang HL, Luo SC et al (2020) The effect of energy input on reaction, phase transition and shape memory effect of NiTi alloy by selective laser melting. J Alloy Compd 817:153288
97. Ma J, Franco B, Tapia G et al (2017) Spatial control of functional response in 4D-printed active metallic structures. Sci Rep 7:46707
98. Zhang Q, Hao S, Liu Y et al (2020) The microstructure of a selective laser melting (SLM)-fabricated NiTi shape memory alloy with superior tensile property and shape memory recoverability. Appl Mater Today 19:100547
99. Xiong ZW, Li ZH, Sun Z et al (2019) Selective laser melting of NiTi alloy with superior tensile property and shape memory effect. J Mater Sci Technol 35:2238–2242
100. Zhou Q, Hayat MD, Chen G et al (2019) Selective electron beam melting of NiTi: microstructure, phase transformation and mechanical properties. Mater Sci Eng, A 744:290–298
101. Wan XM, Feng Y, Lin X et al (2019) Large superelastic recovery and elastocaloric effect in as-deposited additive manufactured Ni50.8Ti49.2 alloy. Appl Phys Lett 114:221903
102. Hou H, Simsek E, Ma T et al (2019) Fatigue-resistant high-performance elastocaloric materials made by additive manufacturing. Nature 366:1116–1121
103. Mazzer EM, Kiminami CS, Gargarella P et al (2014) Atomization and selective laser melting of a Cu-Al-Ni-Mn shape memory alloy. Mater Sci Forum 802:343–348
104. Gustmann T, Neves A, Kühn U et al (2016) Influence of processing parameters on the fabrication of a Cu-Al-Ni-Mn shape-memory alloy by selective laser melting. Addit Manuf 11:23–31
105. Gustmann T, Santos DJM, Gargarella P et al (2016) Properties of Cu-based shape-memory alloys prepared by selective laser melting. Shape Memory and Superelasticity 3:24–36
106. Gustmann T, Schwab H, Kühn U et al (2018) Selective laser remelting of an additively manufactured Cu-Al-Ni-Mn shape-memory alloy. Mater Des 153:129–138
107. Tian J, Zhu WZ, Wei QS et al (2019) Process optimization, microstructures and mechanical properties of a Cu-based shape memory alloy fabricated by selective laser melting. J Alloy Compd 785:754–764
108. Zhuo L, Song B, Li R et al (2020) Effect of element evaporation on the microstructure and properties of CuZnAl shape memory alloys prepared by selective laser melting. Opt Laser Technol 127:106164
109. Niendorf T, Brenne F, Krooß P et al (2016) Microstructural evolution and functional properties of Fe-Mn-Al-Ni shape memory alloy processed by selective laser melting. Metall and Mater Trans A 47:2569–2573
110. Caputo MP, Waryoba DR, Solomon CV (2020) Sintering effects on additive manufactured Ni-Mn-Ga shape memory alloys: a microstructure and thermal analysis. J Mater Sci 55:5311–5321
111. Caputo MP, Berkowitz AE, Armstrong A et al (2018) 4D printing of net shape parts made from Ni-Mn-Ga magnetic shape-memory alloys. Addit Manuf 21:579–588
112. Caputo MP, Solomon CV (2017) A facile method for producing porous parts with complex geometries from ferromagnetic Ni-Mn-Ga shape memory alloys. Mater Lett 200:87–89
113. Mostafaei A, Vecchis RDP, Stevens EL et al (2018) Sintering regimes and resulting microstructure and properties of binder jet 3D printed Ni-Mn-Ga magnetic shape memory alloys. Acta Mater 154:355–364
114. Laitinen V, Sozinov A, Saren A et al (2019) Laser powder bed fusion of Ni-Mn-Ga magnetic shape memory alloy. Addit Manuf 30:100891
115. Taylor SL, Shah RN, Dunand DC (2018) Ni-Mn-Ga micro-trusses via sintering of 3D-printed inks containing elemental powders. Acta Mater 143:20–29

116. Khalil-Allafi J, Dlouhy A, Eggeler G (2002) Ni4Ti3 a -precipitation during aging of NiTi shape memory alloys and its influence on martensitic phase transformations. Acta Mater 50:4255–4274
117. Frenzel J, George EP, Dlouhy A et al (2010) Influence of Ni on martensitic phase transformations in NiTi shape memory alloys. Acta Mater 58:3444–3458
118. Saedi S, Turabi AS, Andani MT et al (2017) Texture, aging, and superelasticity of selective laser melting fabricated Ni-rich NiTi alloys. Mater Sci Eng, A 686:1–10
119. Saedi S, Turabi AS, Andani TM et al (2016) The influence of heat treatment on the thermo-mechanical response of Ni-rich NiTi alloys manufactured by selective laser melting. J Alloy Compd 677:204–210
120. Wang XB, Speirs M, Kustov S et al (2018) Selective laser melting produced layer-structured NiTi shape memory alloys with high damping properties and Elinvar effect. Scripta Mater 146:246–250
121. Dadbakhsh S, Speirs M, Kruth JP et al (2015) Influence of SLM on shape memory and compression behaviour of NiTi scaffolds. CIRP Ann Manuf Technol 64:209–212
122. Li W, Yang Y, Liu J et al (2017) Enhanced nanohardness and new insights into texture evolution and phase transformation of $TiAl/TiB_2$ in-situ metal matrix composites prepared via selective laser melting. Acta Mater 136:90–104
123. Li W, Zhi W, Zhao J et al (2018) Cinnamaldehyde protects VSMCs against ox-LDL-induced proliferation and migration through S arrest and inhibition of p38 JNK/MAPKs and NF-kappaB. Vascular Pharmacol 108:57–66
124. Belyaev S, Resnina N, Iaparova E et al (2019) Influence of chemical composition of NiTi alloy on the martensite stabilization effect. J Alloy Compd 787:1365–1371
125. Ren LQ, Zhou XL, Xue JZ et al (2019) Thermal metamaterials with site-specific thermal properties fabricated by 3D magnetic printing. Adv Mater Technol 4:1900296
126. Lu L, Guo P, Pan YY (2017) Magnetic-field-assisted projection stereolithography for three-dimensional printing of smart structures. J Manuf Sci Eng 139:071008
127. Kawaguchi M (2016) Dispersion stabilities and rheological properties of fumed silica suspensions. J Dispersion Sci Technol 38:642–660
128. Xie JQ, Zhang HP, Shao YY et al (2020) Investigation of the performance of fumed silica as flow additive in polyester powder coatings. Coatings 10
129. Yu R, Yang X, Zhang Y et al (2017) Three-dimensional printing of shape memory composites with epoxy-acrylate hybrid photopolymer. ACS Appl Mater Interfaces 9:1820–1829
130. Zhao TT, Yu R, Li XP et al (2018) 4D printing of shape memory polyurethane via stereolithography. Eur Polymer J 101:120–126
131. Kuang X, Chen K, Dunn CK et al (2018) 3D printing of highly stretchable, shape-memory, and self-healing elastomer toward novel 4D printing. ACS Appl Mater Interfaces 10:7381–7388
132. Zhang X, Ai JW, Ma Z et al (2019) Binary cooperative flexible magnetoelectric materials working as self-powered tactile sensors. J Mater Chem C 7:8527–8536
133. Huang L, Jiang R, Wu J et al (2017) Ultrafast digital printing toward 4D shape changing materials. Adv Mater 29:1605390
134. Zhao Q, Qi HJ, Xie T (2015) Recent progress in shape memory polymer: New behavior, enabling materials, and mechanistic understanding. Prog Polym Sci 49–50:79–120
135. Zhang Y, Huang L, Song H et al (2019) 4D printing of a digital shape memory polymer with tunable high performance. ACS Appl Mater Interfaces 11:32408–32413
136. Chen Y, Molnárová O, Tyc O et al (2019) Recoverability of large strains and deformation twinning in martensite during tensile deformation of NiTi shape memory alloy polycrystals. Acta Mater 180:243–259

Chapter 4
Personalized Medicine: Manufacturing Oral Solid Dosage Forms Through Additive Manufacturing

Yinfeng He, Maria Inês Evangelista Barreiros, and Hatim Cader

4.1 Background

Accurately providing the correct therapeutic to a specific patient is not a simple task. The drug dosages may vary depending on the patient's age, weight, health conditions, race, etc. Currently, solid dosage forms are all made by mass production where the tablets/capsules of each batch will have the same shape, drug loading, and releasing profiles. Such a 'one size fit for all' principle will require the pharmacist to design different pill combinations depending on the course of treatment and the sequence of taking for each patient, which could be complicated when a patient has complex health conditions. However, the current mass production process can only provide standardized medicines, leaving the patient with limited choices of drug loading and releasing profiles. This not only brings troubles to the patient, but also significantly increases the risk of missing or taking the wrong pills which could be fatal for the patient. The ultimate goal will be to design and manufacture dosages based on each individual's actual needs while this is not realistic through the existing pharmaceutical manufacturing technique. As a destructive manufacturing technology that appeared in the late 20th century, additive manufacturing (AM) offered a new production concept that can achieve this goal.

The idea of using AM, also called three-dimensional (3D) printing technology, for pharmaceutical applications lies in its portability, easiness of use, and the possibility to have personalized medicines that can be patient-specific. There is an idea that 3D printing can allow the medicine to be manufactured anywhere and at any time, depending on the patient's demand. A patient who nowadays needs to take several

Y. He (✉) · M. I. E. Barreiros · H. Cader
Faculty of Engineering, University of Nottingham, University Park, Nottingham NG7 2RD, UK
e-mail: yinfeng.he@nottingham.ac.uk

Y. He
Nottingham Ningbo China Beacons of Excellence, Research and Innovation Institute, No.211 Xinguang Road, Ningbo, China

K. Zhou (ed.), *Additive Manufacturing*,
https://doi.org/10.1007/978-3-031-04721-3_4

pills a day will someday be able to take one pill and its therapeutic effect will be the same as taking several tablets. However, there is still a long way from that. Firstly, one should consider the market size and whether pharmaceutical companies would allow for the manufacturing process to be taken from their hands and given to the pharmacists, the ones who would produce the pills based on the doctor's prescriptions. The market of pharmaceutical industry is huge. Going from $1.12 trillion in 2018 to a prediction value of $1.5 trillion by 2023, the industry is showing no signs of recession or growth-stop. By considering the value of the market and how much the industry is willing to spend on the research and development of new and improved products, it is arguable that the industry would swiftly approve of this shift from big manufacturing sites to smaller printers in a hospital/pharmacy/ or at home.

For example, orphan drugs are something the industry has been spending more money on. The revenue is significant as the cost is extremely high, which will make it a promising area for 3D printing. A drug is considered as an orphan drug when the disease it is used to treat, prevent or diagnose is life threatening and has a prevalence of less than 5 in 10,000 people. Those are the criteria for status of orphan drug given by the European Medicines Agency. With a smaller number of people needing such treatment, this could be an area where 3D printing could be used. Having the possibility of personalisation and ability to produce smaller quantities than batch production would be the greatest advantage of using this technique. Using 3D printing, while still controlling by the industry, would allow for less waste of materials and consequently decrease in costs.

In addition, the pharmaceutical industry follows a set of rules called Good Manufacturing Practice (GMP). This ensures that all products are of similar and high quality, there is good quality risk management, and it covers everything from the starting materials, the location where the manufacturing takes place, the equipment that is used, and even the people who work in the process directly or indirectly. Not only that but the materials, equipment, processes, and products are all regulated and assessed according to the country/region agency, such as the Food and Drug Administration (FDA) in the US and European Medicines Agency (EMA) in the European Union. When starting materials arrive at the facilities, these are put in quarantine while their properties and quality are assessed. Several tests are also performed in between manufacturing processes. Even though the process is supposed to be simplified when using 3D printing, quality still needs to be assured. In every country and region, there are guidelines that need to be followed on the final product, tests that need to be done, and within limits to ensure its quality. In the UK, there is the British Pharmacopeia, which includes the monographs from the European Pharmacopeia. It contains a "comprehensive collection of authoritative official standards for UK pharmaceutical substances and medicinal products". With so many tests and analyses required by the Pharmacopeia for excipients, drugs, and finished products that rely on laboratory equipment, it is difficult to manufacture and follow the quality of the printed dosage in a hospital/pharmacy.

Contrary to traditional manufacturing techniques, the most important advance brought by AM is the possibility of it being used for the personalization of dosage, leading the dosage form to be more patient-tailored [1] rather than the 'one size fits all' approach that is used with traditional manufacturing. This personalization can englobe not only different drug loadings [2], the use of multiple active pharmaceutical ingredients (APIs) in one dosage form—the concept of a polypill—[3–5], but also the control of the releasing profile through shape [6–9] or material deposition [10] beyond a formulation approach. Besides these, AM could also allow the dosage form to be made as needed at the hospital/pharmacy, shortening the supply chain and reducing production cost/waste. Therefore, AM has been widely recognized as a revolutionary technique that can achieve the goal of providing affordable bespoke therapeutics for individuals and will significantly improve wellbeing and health for human beings. However, the concept of producing solid dosage forms through AM is still in the research stage and has not been applied in mass production yet. The limitations include the production rate of current manufacturing technology, choice of excipient/APIs that is compatible for additive manufacturing, quality control, and relevant regulations for additively manufactured solid dosage forms as well as the machine cost. This chapter will mainly cover different AM techniques that have been used to manufacture oral solid dosage forms, through which we will discuss the advantages and limitations of each additive manufacturing technique. Similar to existing manufacturing methods, the future of additive manufacturing of solid dosage forms is not likely to be only one technique, instead, different AM techniques might be needed based on specific APIs and excipients. Therefore, we hope this chapter will help the reader to understand and discover the suitable AM techniques for their target, and accelerate their research in AM of solid dosage forms.

4.2 Traditional Routes for Manufacturing Solid Oral Dosage Forms

A solid oral dosage form is a drug delivery device, such as a tablet or capsule, administered by ingestion [11]. Tablets are popular among patients as the route of delivery is the least invasive, easily understood by the patient, and administered by the patient themselves. For pharmaceutical manufacturers, tablets are a relatively inexpensive technology that allows the API to be deliveried in its most stable form [11].

Tablet manufacturing is a highly regulated process required to follow current Good Manufacturing Practices (cGMP). This can require the preparation of tablets in completely enclosed systems that minimize handling [12].

The first phase of producing tablets begins with decreasing the physical size of the API and its excipients through milling processes (e.g., compressing and cutting). This improves the performance of the tablet by maximizing the surface area and uniformity. Granulation of the ingredients is normally the next step where characteristics of flowability and compressibility are achieved. Sizing and blending of granules occur, commonly through wet or dry granulation (other processes exist too). This is a low-energy process so that negative-phase transformations can be detected (although chances of occurring are low). Further additives for excipients, such as lubricants, can be involved. Compression and encapsulation of the dosage forms occur to form stable tablets. Applying a coating is the final step for modified drug release or aesthetic and taste alterations [13].

As shown above, the standard practice of manufacturing tablets involves multiple steps. Within each step, except for granulation sizing and blending, the probability of phase transformations of the drug can be significant. Such transformations can refer to polymorphism, crystallization, solvate, and salt formations [14], all of which can affect the tablet's physical, chemical, and mechanical properties. Important properties, including the solubility, bioavailability, and stability of the tablet, can be negatively impacted as a result of these changes as well [15]. The potential routes of manufacturing are illustrated in Fig. 4.1.

At each stage, the risk of a phase transformation (potentially causing a change in API behaviour in the body) exists. Different polymorphisms can alter drug properties, negatively affecting behaviour such as dissolution, stability, and toxicity of the drug [16]. As the manufacturing route progresses, the danger of a phase transformation of the drug or an excipient remains and needs to be monitored. The complexity of the production routine also makes current manufacturing techniques hard to be used for small-scale production, considering the exceptional marginal cost behind.

Although the conventional tablet manufacturing route has drawbacks, it also has its strengths. Most commonly cited are that the system is well suited for mass manufacturing of tablets at a relatively low cost. Another advantage over the current state of 3D printing technology is quality control. Most 3D printers currently used in laboratory environments lack an effective quality control system, making it difficult for a manufacturer to get live information on the amount of material being deposited. Repeatability and consistency are the key for tablet manufacturing. There are however numerous examples from academic investigations illustrating how in-line quality control procedures can be incorporated into the manufacturing process.

4.3 Additively Manufactured Oral Solid Dosage Forms

4.3.1 Personalized Medicine

Personalized medicine is one of the key applications for AM over the conventional process. The possibility of producing the number of tablets needed, with a specific drug loading that is suitable to a specific patient and a desired drug-release profile without wasting materials, can be realized through 3D printing. To produce a particular dosage form for a more appropriate clinical procedure, medical professionals should take into account the patient's age, gender, or current ailments [17]. Currently, there are set dosages that are available for each market authorised medication. But when taking into account a patient's gender, age and weight, the dosage needed can differ. One that would be suitable and needed for a male adult is different than one

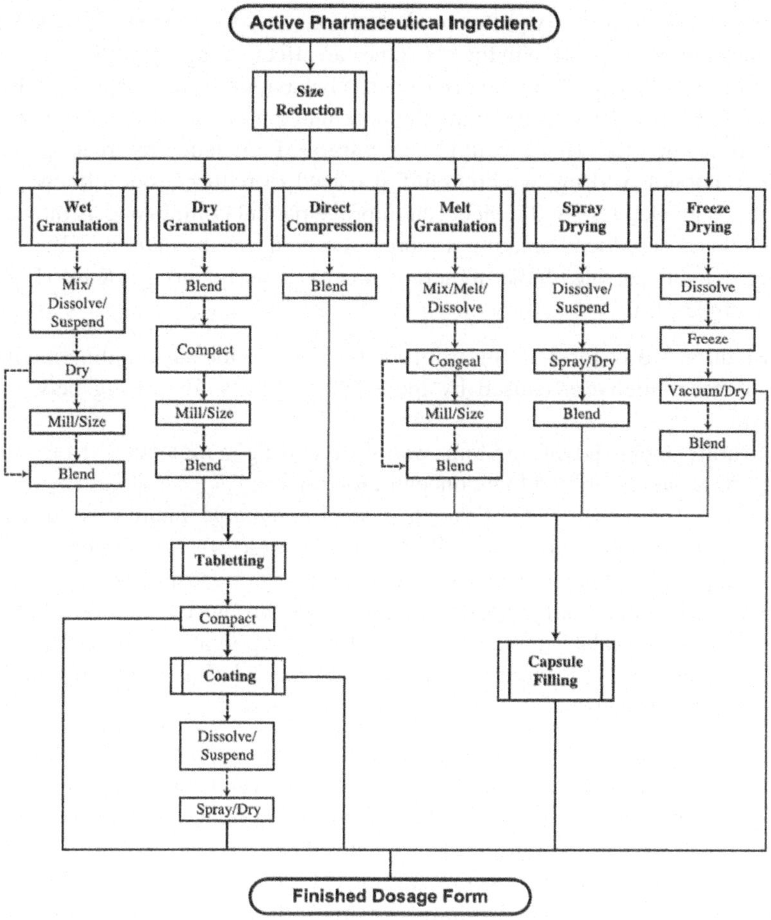

Fig. 4.1 A overview of different manufacturing routes for solid oral dosage form production (reproduced from [13])

needed by a child or an elderly person [18]. The geometric and material freedom provided by AM offers the opportunity to build complex tablet structures that can have controlled drug loading and releasing profiles [9] and therefore uniquely accomplish the development of a particular therapy for a patient. This can be achieved by changing the shape of a tablet [19] having a different dosage [20] or even combining drugs, depending on whether to treat or prevent certain diseases.

The main challenge for an effective controlled drug-release for oral systems is the variation in environment and transit time in the gastrointestinal tract. Challenges include the variation between patients in terms of gastric residence time and the conditions in which tablets have been ingested (e.g., if the patient is in a fasted or fed state) [21]. The pH value is another significant variable of the gastrointestinal tract. The pH value is usually low in the stomach (at least two in the fasted state but can increase to up to five after food) and increases to approximately 6.5 in the small intestine. The variation in the pH value can affect a drug's stability, absorption, and solubility [21]. Changing the drug-release behaviour such as onset and duration can address these problems, allowing the optimum amount of drug to be released at specific times or sites. There would be an improved efficiency by treating ailments with less amount of drug, which could also lead to reduced side effects [22]. In addition, controlled released tablets can improve patient compliance as the need for repeated doses is reduced, making it more convenient for the patient [22].

Drug delivery devices have numerous mechanisms for drug release (Fig. 4.2). These include [23]:

- Dissolution—transfer the drug from a solid phase to the surrounding medium.
- Partitioning—delivery caused by the relative affinity for a drug between two phases.
- Diffusion—independent random walks of a drug's molecules from a high-concentration region to a low-concentration region.
- Osmosis—a flow of water through a semi-permeable membrane to equalise concentrations of impermeable solutes on either side of a membrane.
- Swelling—the uptake of water by a delivery system causing an increase in the volume of the device, allowing the release of drug through dissolution or diffusion.
- Erosion and degradation—the release of drug as a polymer disintegrates
- Regional delivery and targeting—localised direct administration of a drug near or at the site or action.

Often, several of the above mechanisms can occur in sequence during drug release where the mass transport step is the rate-limiting step, which will control drug-release speed [24]. These mechanisms can be controlled by material selection or device design to achieve control of drug-releasing speed.

The geometric design freedom that AM provides gives the user the ability to create complex tablet structures that can influence drug-releasing behaviour [25]. Creating

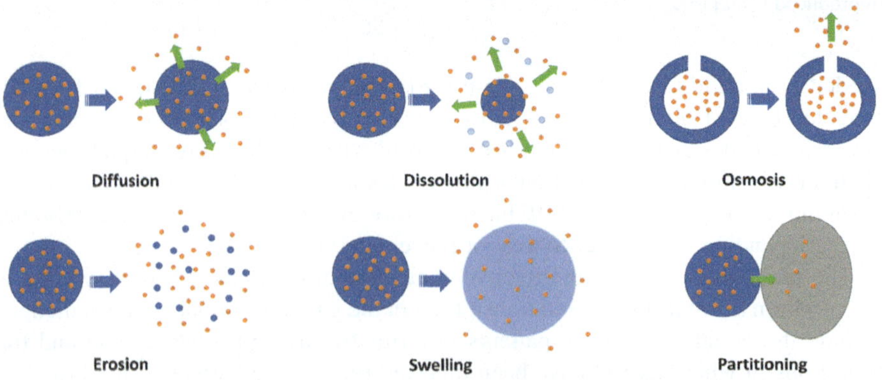

Fig. 4.2 Schematic of releasing mechanisms for encapsulated active ingredients

a specific release profile for one or multiple drugs in a single tablet can uniquely be accomplished by AM. As a general concept, a lower infill percentage increases the surface area to volume ratio of the object, therefore increasing the release rate [26, 27]. Take floating tablet as an example, where a longer floating time is desired, a lower infill percentage means the object is less dense and will stay afloat for a longer period, which in turn will delay the release of the drug [28–30]. When a polymer such as polyvinyl alcohol (PVA) is used, a higher infill percentage equals a higher amount of it, impairing the swelling and consequently decreasing the drug-release rate [25, 31]. But the change of infill percentage can also be coupled with a change in shape or design of the printed tablet [30, 32] and with a change in polymeric materials [27, 33–35].

Even though the infill percentage might have a relevant effect on the drug-release rates, polymer excipients can also be chosen depending on the drug-release profile that is required and modified by changing the infill percentage [27, 35]. Channeled shapes such as the one made by Arafat et al. are also another example of how to increase the rate and even achieve immediate release [1]. But different shapes independent of the infill percentage can also lead to a modified release rate. Goyanes et al. [9] and Khaled et al. [8] show how different shapes of the same formulation can affect the release rate. Clear examples can also be seen in Khaled et al.'s work [36], where two different Guaifenesin-based formulations were used to have an initial burst release of the drug, followed by a sustained release of the same drug. In further work by Khaled et al., different drugs' release rates were extended by using different amounts of hydroxypropyl methylcellulose (HPMC) in a gel form, which acted as a hydrophilic matrix [4, 5]. Pietrzak et al. demonstrated how to use different polymers, whether purposely made for immediate or sustained release, can alter the profile as well [37].

The ease with which multiple materials can be incorporated into a printer gives another possible route to design the drug-release behaviour. For example, porous barriers can be printed to contain the API to control the releasing speed, core–shell structures with single or multiple cores can enable a sequential release of one or multiple APIs. The ability to print drug-loaded inks together with excipient ink can manipulate the spatial distribution of the drug in the tablet. With the help of modelling, the user will be able to design the drug-release computationally as needed. This would add to the prospect of a programmable drug-release system.

4.3.2 Customizable Multi-active Therapeutics

The concept of a multi-active delivery system is not recent. On the contrary, combination therapy has been widely studied, used, and commercialized as a treatment method for several different conditions. An example of a widely known and used types of combination therapy is the combined oral contraceptive pill [38]. Other forms of treatments that already require a combination approach include tuberculosis [39] and HIV [39, 40], which might be administered through a different route. Depending on the route of administration and the type of condition that is being treated, a multi-active delivery approach will have different advantages and challenges. However, achieving such systems is not easy nor straightforward, as traditional manufacturing techniques tend to produce products in larger batches and different patients might need different doses of different APIs even when the condition being treated is the same. AM or 3D printing opens up a new market for supplying cost-effective bespoke products for individuals. This includes the idea of providing customized therapeutics. AM enables the production of specific drug delivery systems with different APIs that might be needed with defined the releasing profiles of each drug [5, 41].

4.3.3 Shorten Supply Chain

The application of AM for oral solid dosage production can also shorten the current long, slow, and expensive supply chains [42]. Storage time would be an exceptional challenge for unstable and quick-degrading drugs. With 3D printable formulations, liquid or paste filled cartridges could be delivered to, stored, and prepared in pharmacies or hospitals ready to be printed [43]. A major source of capital loss in this industry is in its inventory levels where long replenishment time requires greater stock productions with estimated losses of US $20–25 billion [42]. Additively manufactured tablets could be printed on demand by a pharmacist or physician when required by a patient [43]. A greater reduction in wasted of materials also decreases the overall cost of production.

4.4 Current AM Techniques for Solid Dosage Form Manufacturing

According to the American Society for Testing and Materials (ASTM), AM techniques can be put into seven main catalogues [50]: binder jetting, direct energy deposition, material extrusion, material jetting, powder bed fusion, sheet lamination, and vat photopolymerization. Techniques under each catalogue may have different machine set up but sharing a similar manufacturing principle. Six out of the seven techniques have been used for printing solid dosage forms, where the binder jetting technique from Aprecia has received FDA approval for one of their tablets [51]. This section will look into research in material extrusion, material jetting, binder jetting, vat polymerization, and powder bed fusion techniques for manufacturing oral solid dosage forms.

4.4.1 Material Extrusion

The most commonly explored technique in research is material extrusion, which can be divided into two different sub-catalogues: fused deposition modelling (FDM) and semisolid extrusion.

4.4.1.1 FDM

FDM is the most explored extrusion technique for additively manufactured tablets. During the FDM process, the thermoplastic filament is prepared and fed into a heated nozzle. The melted filament is then extruded and deposited onto the building platform which solidifies as the extrude cools down. The 3D object is built layer-wise by stacking up the extruded and solidified filament melt [25, 44]. For printing oral solid form dosages, the filaments used, have either been impregnated with the APIs [9, 10, 25], produced by hot-melt extrusion (HME) to have a solid dispersion as the feedstock (Fig. 4.3) [45–47], or used directly to solely print a capsular device that would then be loaded with the APIs [48]. Although the tooling cost of the FDM process is lower than most of the other AM techniques, the high processing

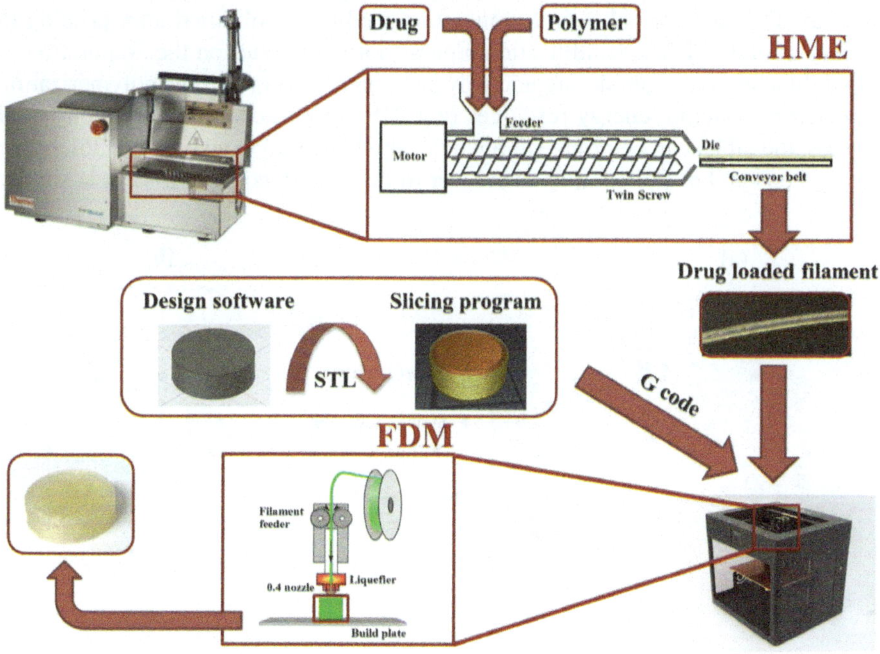

Fig. 4.3 FDM tablet printing coupled with HME for the production of drug-loaded filaments (reproduced from [47])

temperature during API loaded filament preparation and FDM printing might be a hurdle when dealing with thermally unstable drugs [9, 10, 25]. In addition, the available polymeric excipients can hardly achieve faster drug-release profiles [1, 33, 34, 49]. Another relevant disadvantage to FDM printing is that drug loading is normally restricted [31, 50].

4.4.1.2 Semisolid Extrusion

The semisolid extrusion process is another extrusion-based AM technique (Fig. 4.4), but without the need for high processing temperature and therefore more friendly to thermally unstable APIs and excipients. In a semisolid extrusion process, a paste or gel form formulation is normally used. During the printing process, the formulation is transferred to a syringe or cartridge and extruded through a nozzle by either an extrusion screw or compressed air [50, 51]. Contrary to an FDM process, the API loaded material (formulation) does not need to be made into filament form through the HME process. Therefore the whole printing can be done at room temperature, overcoming the issue of temperature-induced degradation and increasing the number of materials that can be used. For a semisolid extrusion-based process, as commonly used pharmaceutical excipients can be included in the formulations [4, 5, 8, 51–53], the material availability is much larger, and higher drug loadings can easily be obtained [32, 52]. The paste or gel form formulations need to be solidified after printing to achieve a solid pill. The solidification process normally relies on the evaporation of liquid diluents (e.g. volatile organic solvent, water, etc.) or further polymerization triggered by external energy resources (e.g. UV light). For the evaporation based process, the tablet can be printed as a whole and then dried in an oven [5, 36], or the drying step can be applied after each layer of material deposition [54]. It is similar

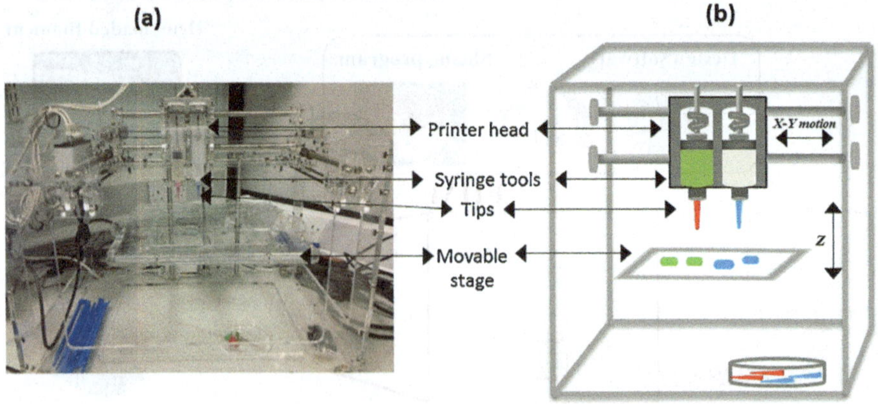

Fig. 4.4 Photo (**a**) and schematic (**b**) illustrating a dual heads semisolid extrusion printing process (reproduced from [36])

for external energy resources based solidification that can happen either during the printing process or at the end when the whole structure is finished [55, 56].

4.4.1.3 Oral Solid Dosage Forms: Tablet, Floating Devices, and Capsules

Material extrusion has been widely used to print the final dosage form in research. As mentioned before, the filaments required are either hot-melt extruded, which has combined the carrier polymer with the drug in this step, or a polymeric filament can be soaked in a drug-containing solution where the filament becomes impregnated with the drug molecules. Nonetheless, in both cases, FDM is the last step to obtain a tablet. Recently, Goyanes et al. [57] and Fanous et al. [33] have reported an alternative to the need for prior production of filaments used in FDM by having a one-step direct powder printing process (Fig. 4.5). While there is a previous step of mixing the formulation powders until homogenously distributed, the powder mixture is fed into the cartridge where it is heated until able to directly extrude from the nozzle, creating the desired 3D shape. Even though it does not eliminate the need for high temperatures which might be a hurdle for thermolabile materials, this does facilitate and quickens the process for obtaining a tablet. However, there are cases where FDM is not the final step into producing a solid oral dosage form. It can also be coupled with traditional manufacturing steps such as coating [10]. Goyanes et al. have demonstrated that a coating step after FDM printing a tablet, led to a modified release profile of the drug in question, budesonide. Even though this shows that a printing technique can be coupled with traditional manufacturing techniques as well, this is a further step in the production of a pharmaceutical product and the same might be obtained through the use of other materials or by a change in product design.

Material extrusion can also be used to produce floating devices [29, 30, 58–60] and special capsular [48, 61, 62] devices that would not be possible otherwise (Fig. 4.6). In those cases, the drug-containing part of the dosage form is produced by another

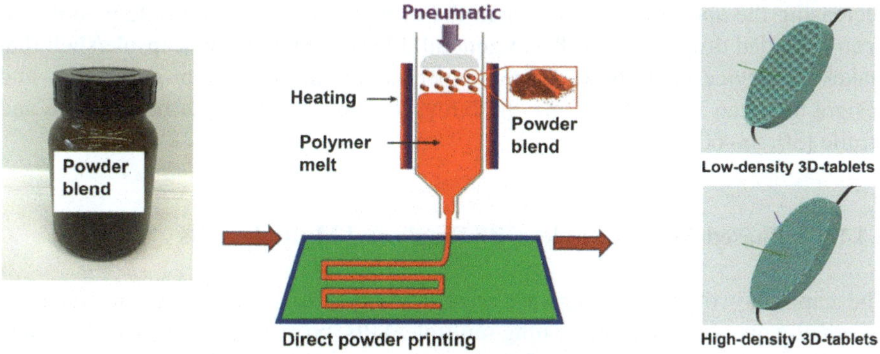

Fig. 4.5 Illustrative example of the direct powder printing process (reproduced from [33])

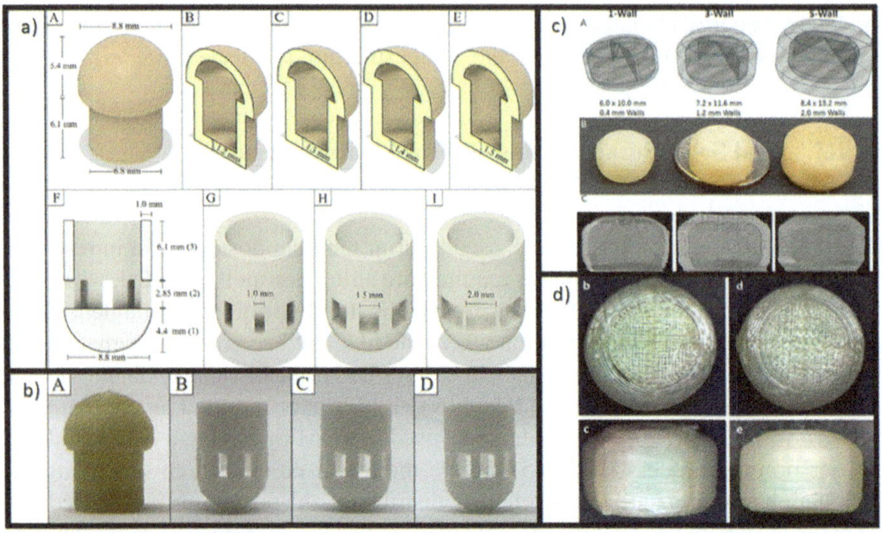

Fig. 4.6 Examples of printed floating devices and capsules **a–b** illustrate the design and printed floating capsular device, respectively (reproduced from [58]) **c** (reproduced from [62]) and **d** (reproduced from [61]) are examples of capsules printed by FDM

process such as direct compression [30, 59, 60] or HME [48]. Using FDM to print capsules that could later be filled with a solid or liquid formulation, Smith et al. for example, used FDM to print capsules that could later be filled with a solid or liquid formulation [61, 62]. By changing the thickness of the capsule, the induction time for the drug to release can be reduced or increased, which consequently impacts where in the digestive tract the release occurs [62]. To obtain a floating device, FDM has been used to print shells that while carrying a compressed tablet inside, allow the system to float [30, 58–60]. Charoenying et al. used two different polymer materials, PVA and PLA, to print the cap and body of the capsular device, respectively. The group concluded that a thicker cap would impact positively the floating time while decreasing the aperture of the holes in the printed body of the capsule would lead to an extended drug-release [58]. In general, FDM printing of the capsule/shell that makes the device float allows for the experimentation of different geometries and designs that can impact not only the buoyancy time but also the release rate and profile [30, 58–60].

4.4.1.4 Phased-Release Tablet with Single and Multiple APIs

The capability of manufacturing phased-release tablets, designing the releasing speed/strategy, as well as integrating multiple APIs, provide materials extrusion and the other AM techniques more flexibility in producing bespoke dosage forms. In an attempt to produce an oral solid dosage form to treat tuberculosis, Genina et al.

[48] used FDM to manufacture a dual-compartmental dosage unit with polylactic acid (PLA) that allowed the administration of two different drugs that required a phased-release in different parts of the digestive tract. The two drugs were in different compartments and the delayed release of drugs was achieved through putting a cap on the compartment. Using a different approach to obtain the concomitant but phased-release administration of the same two drugs from that Genina et al. used (rifampicin and isoniazid), Tabriz et al. produced a filament for each drug using different polymers [26]. To have isoniazid release in the stomach, it was hot-melt extruded with hydroxypropyl cellulose (HPC), while rifampicin was hot-melt extruded with hypromellose acetate succinate (HPMCAS) so that it would be released in the upper intestine. The two different filaments were then used to FDM print a single tablet [26]. The two examples showcase how extrusion printing can be used in different ways to achieve a similar result, which is especially important when working with several drugs that are required to be taken together, will interact differently and be absorbed in different parts of the digestive tract (Fig. 4.7).

As mentioned, one of the most significant advantages is the possibility of having more than one drug in a single tablet. However, it is imperative that the different drugs do not interact. With a layer-by-layer deposition process, 3D printing allows for a more defined and separated structure, if a compartment device is not used, as it is the case of the work by Genina et al. [62]. However, with the same issue to solve and the same drugs needed for the treatment of tuberculosis, Tabriz et al. obtained a tablet with no drug interactions [44]. Using the same approach of different polymer-drug combinations, Gioumouxouzis et al. printed a dual-layer tablet with no drug-drug or drug-polymer interaction and with the drugs in their amorphous form [63]. Pereira et al., also successfully obtained a multiple drug FDM printed tablet where some

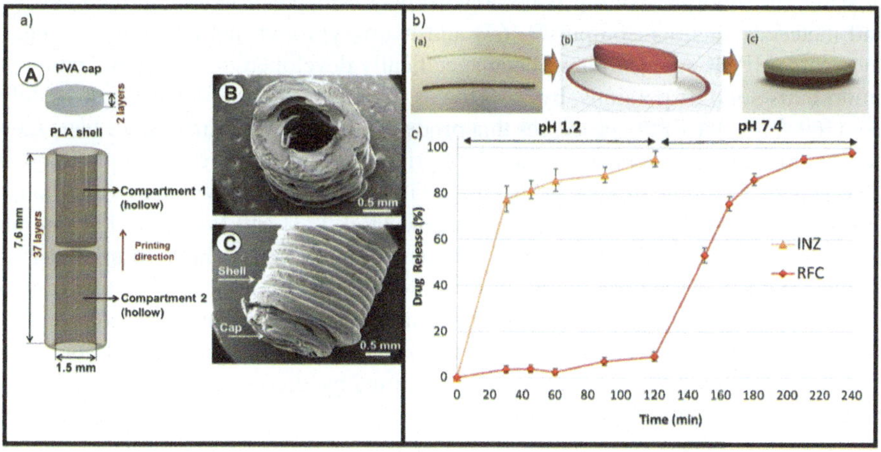

Fig. 4.7 FDM printed phased-release device: **a** shows an FDM printed capsular device that allows for the phase-release of two different drugs (reproduced from [48]), **b** shows a dual-drug FDM-printed tablet where different polymers allowed for different release times (reproduced from [26])

drugs were amorphous and some were crystalline [64]. Similar multiple drug tablets can also be manufactured by semisolid extrusion printing. Khaled et al. have reported three-in-one [4] and five-in-one [5] tablets where each drug was separated, and no interaction was observed. As room temperature and common pharmaceutical excipients were used in the paste formulations, all drugs were crystalline, though they were not poorly soluble which would require them to be amorphous. In HIV treatment, several drugs are usually administrated, and Siyawamaya et al. have printed a three-drug tablet that even achieved enhanced relative bioavailability when compared to a commercial product [63], which illustrates how 3D printing can aid pharmaceutical production.

4.4.1.5 Choice of APIs

A lot of different drugs have been used for the printing of oral solid dosage forms, both with FDM and semisolid extrusion. This has been the case for both single and multiple drug administration, as one of the advantages and promises of using 3D printing for manufacturing of pharmaceutical products is the fact that a single tablet with multiple drugs is possible. And this, for a patient with comorbidities or a disease that requires the concomitant administration of multiple drugs, can increase patient compliance [65]. While formulating for water-soluble drugs is a much easier task, poorly soluble drugs require further consideration on which technique or materials to use. One of the most used approaches in traditional manufacturing to aid solubilization of poorly soluble drugs is amorphization, in which amorphous solid dispersion (ASD) is made. This is used as the amorphous form of a drug, which is typically more soluble than its crystalline form, therefore improving oral bioavailability [66]. In ASD, the amorphous drug is homogeneously distributed within an inert carrier, and materials such as Soluplus® [67] and hydroxypropyl methylcellulose acetate succinate (HPMCAS) [68] have been specifically developed to aid solubilization in amorphous solid dispersions, by working as the drug carrier and have been widely used [69–71]. One of the processes that produce ASDs is HME and, as we have seen before, this is usually a prior step to FDM printing. It is therefore not surprising that several poorly soluble drugs have been used in tablet printing. Works by Gioumouxouzis et al. [72], Skowyra et al. [44], Solanki et al. [73] and Wei et al. [74] are all examples where the drug in the final product was in its amorphous form, that way aiding solubility. Conceição et al. have shown a different approach in achieving a tablet of a poorly soluble drug, through using cyclodextrins in a semisolid extrusion process. With cyclodextrins, these form a complex with the hydrophobic drug molecules. In this case, drug-release was modified by altering cellulose ethers in the formulation that acted as a binder [75].

4.4.1.6 Choice of Excipients

As with APIs, several excipients have been used to produce 3D printed tablets. Either on their own or by combining with other excipients, many different polymers have been explored. One of the main disadvantages is the limited choices of materials. One of the most used polymers for printing pharmaceutical tablets and capsules is PVA [9, 10, 25, 44, 64, 76, 77]. PVA can be both drugs loaded using HME [9, 10, 28, 29, 64, 76] or by being soaked in a drug-containing solution [25, 31, 44, 50] in order to include the target drug. One issue that can be found, especially with filaments being soaked in a drug-containing solution is the limit of drug loading in PVA. This will consequently require more material and a bigger tablet in order to achieve the desired dose of a drug. Saviano et al. demonstrated how the particle size of the PVA polymer can affect processability and the mixing of polymer-drug [76]. Facing the hurdle of high temperatures in HME and FDM, Pereira et al. used water as a co-plasticizer in order to decrease the temperatures needed for extrusion and extrusion printing, without the need to add another polymeric material [64]. However, polymeric blends have been used, whether to improve processability and extrudability or to modify the drug-release profile with FDM printed tablets. The polymers used in FDM printing usually lead to extended release profiles. Kempin et al. [27] and Kollamaran et al. [78] have demonstrated how using a blend containing Kollidon® can produce immediate release tablets. But even though a polymer might improve extrudability by decreasing the required temperature, it might not decrease the required temperature for the printing process, as the process parameters are different. Therefore, this might still not be the process for drugs that suffer degradation with increased temperature [74]. Moreover, polymer blends are almost always necessary for a semi-solid paste-based extrusion. In this case, a much wider variety of materials are available as commonly used excipients can be turned into a paste for printing and a much higher drug load can be obtained. Different types of oral solid dosage forms such as orodispersible films made with cyclodextrins and HPMC [75] or with maltodextrin, sorbitol and hydroxyethylcellulose [54] become possible with a semisolid extrusion process.

4.4.1.7 Summary

In summary, there are two different material extrusion techniques that facilitate the manufacturing of oral solid dosage forms, FDM and semisolid extrusion printing. While for FDM printing a filament is the feedstock material, with the exception of direct powder printing [33, 34], for semisolid extrusion, a paste is used instead. Oral solid dosage forms can be manufactured in a one-step process or coupled with traditional manufacturing processes such as HME and coating. Polymeric materials are usually used as the carriers in FDM which tend to show an extended drug-releasing period. Commonly used pharmaceutical excipients can be used in semisolid extrusion and its releasing rate can be easily modified based on the choice of excipients. HME coupled with FDM can be a way to manufacture tablets where the drug is in its amorphous form which can aid solubilization for poorly soluble drugs. However, for

FDM, the high processing temperatures can be a hurdle for both the excipient and APIs. Some drugs can suffer from degradation at the processing temperature. For semisolid extrusion, printing can be done at room temperature and Cyclodextrin-formulations can be a way around to overcome the drug solubility challenge in material extrusion [75].

4.4.2 Material Jetting (Inkjet Printing)

Material jetting introduced a voxelated manufacturing concept where droplets of materials were ejected and solidified as voxels to stack up into a final product. Voxels of different materials can be mixed and matched to form functionalized composite with controlled distribution at a scale of dozens of micrometers, which gives the user plenty of freedom to design the material and therefore performance of the device as needed. External energy sources such as infra-red (IR) or ultra-vilot (UV) light might be needed to help the liquid ink solidify.

4.4.2.1 Material Jetting Systems

Material jetting (or inkjet printing) is an AM technique where individual droplets are selectively deposited onto a substrate [79]. The jetting process consists of three phases [80]: droplet generation, positioning and interaction of droplets with the substrate, and finally the solidification mechanism to form the target structure. There are two mechanisms of dispensing inks: continuous inkjet printing and drop-on-demand inkjet printing.

In continuous inkjet printing, ink is forced through a nozzle and a piezoelectric transducer controls the frequency of drops by applying a disturbance at regular intervals. Inks used are usually conductive and when passing through a charged electrode, their placement is controlled with the help of field plates. A catcher can be used to collect areas where no ink is required and those drops can be reused [81].

If greater control over placement for ink deposition is required, a drop-on-demand mechanism is used. Drops are only ejected when required. Two mechanisms for droplet generation in a drop-on-demand printhead include Thermo and piezoceramic controlled. Thermo-based printheads will rapidly heat a thermal resistor. This results in the surrounding ink being vapourised causing an air bubble to grow, forcing material out of the nozzles. When the heater is turned off, a transient pressure wave is generated allowing stable ink ejection [82]. Alternatively, a piezoelectric material can be used to force the ejection of a droplet (Fig. 4.8). A voltage is applied to the material and the resultant mechanical deformation causes ink ejection. A specific waveform is applied to control the movement of the piezoelectric material that will affect jetting behaviour. This form of drop-on-demand inkjet printing may be preferred when investigating inks that are thermally sensitive or degrade due to heat [83].

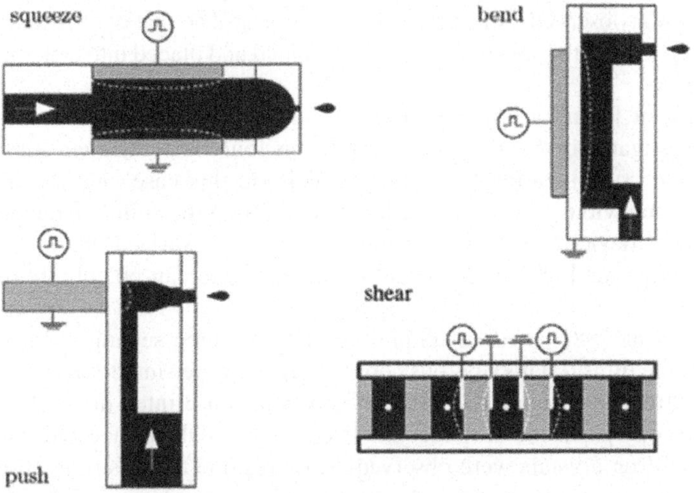

Fig. 4.8 Schematics of various DoD inkjet printers with piezoelectric material controlling drop ejection (reproduced from [83])

4.4.2.2 Film Dosage

Here, we looked into some early investigations of inkjet printing for potential oral delivery dosage forms. Examples include the jetting of small particles, individual drops, and thin films. Most involve the use of edible substrates to form a thin drug-loaded dosage.

Material jetting of liposomes to produce dosage forms was described in an early patent submitted by the Hewlett-Packard Company in 2005. The solution was loaded into a material reservoir and selectively jetted onto a substrate where it could potentially be checked for defects [84]. Meléndez et al. [85] also used this thermal inkjet printing technique. Prednisolone had been dissolved in a solution of ethanol, water, and glycerol to form a jettable ink. The ink was printed on Teflon-coated fiberglass substrates. Powder x-ray diffraction (XRD) and Raman mapping were found to have formed Form I and Form III of the API. This work did not rule out that thermal inkjet printing had a role to play in encouraging polymorphic formation [85]. Buanz et al. [86] used a modified Hewlett-Packard printer cartridge to dose aqueous drug (salbutamol sulphate) solutions onto oral films (clear acetate films). Both water and glycerine were added to the formulation to achieve better printability [86].

Material jetting can also be a viable route for poorly water-soluble APIs. Felodipine, with polymer excipient polyvinylpyrrolidone (PVP) and solvents ethanol and dimethyl sulfoxide (DMSO), was used by Scoutaris et al. [29]. High-pressure homogenization of folic acid was successfully printed in a nanosuspension form by Parkeike et al. 2011. The nanosuspension was found to be small enough to move through the nozzles without interference or blockage. Consistent formation of droplets was observed. An edible substrate was used so that once the nanosuspension was jetted onto the substrate, it could be folded and placed into a pharmaceutical capsule [39].

The idea of jetting a polymer-drug formulation on edible substrates has been further investigated by Sandler et al. [87]. It was found that different substrates can influence the crystallization behaviour of APIs (in this case, due to different ink penetration behaviour), further allowing inkjet printing the ability to create complex drug delivery devices. Additional parameters such as surface energy, roughness, capillary action, and chemical content of the substrate can potentially play a role [87].

Genina et al. [88] jetted rasagiline mesylate on three separate film substrates, orodispersible films (ODFs), porous copy paper, and water-impermeable transparent films, to further investigate the effect of substrates on printed doses. It was found that on the copy paper the drug recrystallized on the ODF surface and transparency film and no drug crystals were observed. This was due to the strong absorption of the ink's copy paper into the matrix of the substrate, caused by its high porosity. The copy paper also had the greatest linear association between the imprinted layers and the drug dose because of the high ink absorption [88]. More recently, inkjet printing of oral buccal films has also generated significant research interest [89].

4.4.2.3 Jetting into Pre-formed Excipients

In addition to the printing of edible film dosages, inkjet printing has been used as a loading tool to deposit APIs into excipient containers for oral delivery.

Fox et al. [90] inkjet printed drug topotecan and biologic insuling into planar oral micro-devices (Fig. 4.9). Because of the pH and enzymatic conditions that occur in the gastrointestinal tract, these devices were created to resolve issues such as low oral bioavailability. Due to its high throughput, while retaining high performance, inkjet printing was used. The method was shown to be reasonably accurate (e.g., in comparison to the 19.2 μg of insulin anticipated, an average of 21.5 μg of insulin was detected), with load variations attributable to droplet size variance. The ease of personalization was highlighted with the system's ability to be easily programmable. The method did not require any UV light or heat that would not negatively affect sensitive biologics and drugs. The authors illustrated the use of inkjet printing for various drug-containing micro-devices to create low-waste but high-capacity loading devices [90].

A photocurable ink to inkjet print the hydrophilic drug ropinirole HCl into preformed tablets and formulated with PEG dithiol and the drug, Norbornene moieties functionalized hyaluronic acid by Acosta-Vélez et al. [91]. A rapid step-growth polymerization reaction through thiolene chemistry led to a drug-loaded hydrogel when exposed to visible light with Eosin Y acting as the photo-initiator. Various parameters such as the loading of functionalized hyaluronic acid, the crosslinking ratio, and the light exposure time were tested for the tunability of mechanical properties. Dissolution of the preformed tablets revealed that 60% of

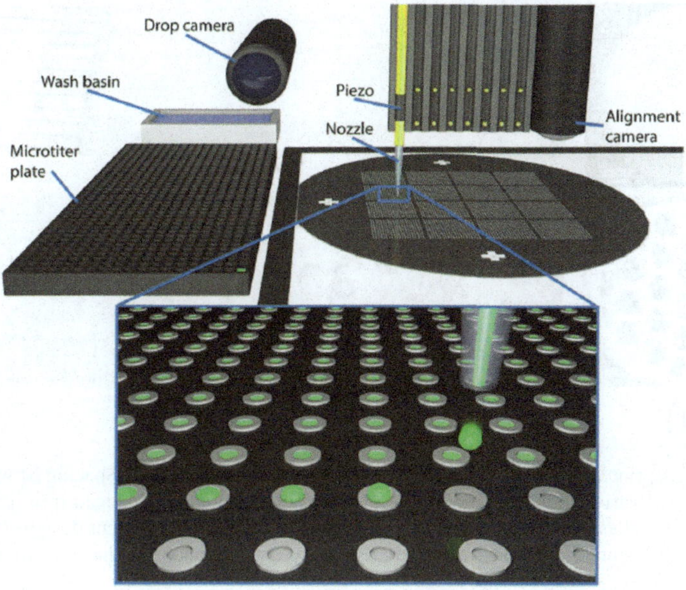

Fig. 4.9 Inkjet-based process for depositing drugs into micro-devices (reproduced from [90])

ropinirole HCl was released within 15 min [91]. Following a similar method, the authors achieved inkjet printing of a UV curable formulation for hydrophobic drugs. Naproxen and ibuprofen were formulated with individual inks. The bio-ink consisted of the monomer polyethylene glycol diacrylate (PEGDA), the photo-initiator Eosin Y, the co-initiator methoxide-polyethylene glycol-amine, and the plasticizer polyethylene glycol (PEG). The ink was deposited and capped off into preformed tablets. For characterization, the cured tablets were then extracted from the preformed models. SEM imaging was used to describe the microstructure of the printed formulation (with naproxen) cross-section. No crystals of naproxen were seen to indicate that the substance had dissolved in the polymer [92].

4.4.2.4 Freestanding Printed Oral Solid Dosage

The first fully inkjet-printed oral solid dosage tablet was achieved using a hot-melt inkjet method by Kyobula et al. [7] (Fig. 4.10a). By melting and mixing the components, a mixture consisting of 5 wt% fenofibrate in beeswax was formed. The mixture was loaded onto an LP50 PiXDRO (Meyer Burger) printer using a hot-melt chamber and heated to 90 °C for optimal printing. The tablets were printed on a substrate of polyethylene terephthalate (PET) at 40 °C with various internal structures to control drug-release (see Fig. 4.10a) and were thoroughly characterized [7].

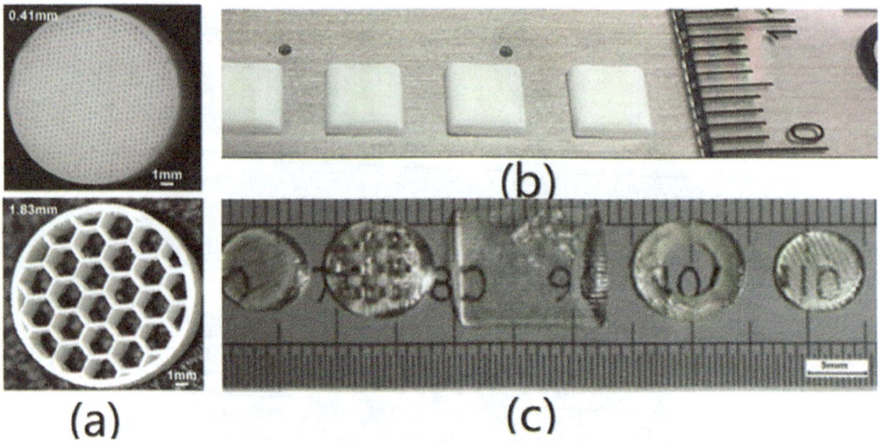

Fig. 4.10 **a** A printed tablet with internal solid and honeycomb structures. Spacing between honeycomb cells was varied to achieve different surface/volume ratios (reproduced from [7]), **b** PVP-thiamine HCl tablets (top) and drug-release profiles of tablets with different designs (reproduced from [93]), **c** UV-cured inkjet printed tablets with different designs (reproduced from [6])

Clark et al. [94] created free standing tablets using a UV-curable ink. The drug, ropinirole HCl, was formulated with PEGDA and used Irgacure 2959 as a photoinitiator (water was used as a solvent for the drug). As the peaks associated with the acrylate were not detected, Fourier-transform infrared spectroscopy (FTIR) showed that the printed tablet had been cured to a high degree. Confocal Raman spectroscopy revealed the existence of drug-rich and poor regions at the micron scale. As no proof of drug crystallinity was found, the drug was assumed to exist as a solid dispersion [94].

Cader et al. [94] used a water-based ink formulation to create freestanding tablets capable of a rapid drug-release. No additional solvents were involved in the formulation as all components were water-soluble. PVP and thiamine HCl were the polymer carrier and model API used in this investigation, respectively. Polysorbate 20 was used to reduce the high surface tension of water, allowing the ink to have an ideal Z printability value of 5.0. Glycerol was used as a plasticizer to prevent cracking and delamination during printing and allow for easier handling of the tablets. Raman spectroscopy and mapping showed that the drug existed in its ideal non-stoichiometric hydrate polymorphic form and was homogenously distributed throughout the tablet. Rapid drug-release of the tablets was seen and to illustrate dose variation, and different size tablets were printed [94].

UV-curable freestanding tablets were once again created by Clark et al. [6]. The poorly water-soluble drug, carvedilol, was formulated with Irgacure 2959 and a photocurable N-vinyl-2-pyrrolidone (NVP) and PEGDA matrix. It was shown that the printed drug (loaded at 10 wt%) was in its amorphous state. Highlighting inkjet printing's ability to create customized structures, the author created tablets of varying designs (ring, mesh, cylinder, and thin-film). By controlling the design of the tablet, different drug-release profiles were achieved [6].

4.4.2.5 Dosage with Multiple Inks

Material jetting can easily integrate different materials into a single device and therefore is considered as a convenient tool to integrate different functional inks for achieving dosage forms with complex drug loading, releasing profiles, or multiple APIs.

Oral dosage forms for the treatment of hypothyroidism were inkjet printed with oral drug combinations by Alomari et al. [95]. A two-cartridge thermal inkjet printer was used to print separate solutions of triiodothyronine and thyroxine on HPMC gel substrates cast on polyester sheets. Dose control was achieved by varying the length of prints (i.e. the pattern size). The authors have illustrated that inkjet printing is capable of dosing multiple drug-containing inks onto the same substrate [95].

A single device created using jet dispensing technology was illustrated by Scoutaris et al. [96]. The authors created a three-layer film for three antihypertensive agents: hydrochlorothiazide, amiloride HCl, and carvedilol (Fig. 4.11a). The first layer (layer 1) was printed with ink formulated with ethylcellulose, copovidone, and the APIs hydrochlorothiazide and carvedilol (dissolved in ethanol). The second layer (layer 2) was printed with only amiloride HCl, dissolved in an ethanol and water cosolvent system. The third layer (layer 3) consisted of 'dots' of carvedilol with copovidone dissolved in methanol. Organic solvents are not always advantageous due to their hazardous nature. However, these were necessary due to the nature of the formulation) [96].

Scanning electron microscopy (SEM) showed that layers, where the drug is mixed with a polymer, are smooth and transparent, suggesting drugs were incorporated in the polymeric matrix. Drug release was influenced by varying the ratios of copovidone and ethylcellulose in layer 1. By increasing the copovidone content (potentially acting as a pore forming agent), hydrochlorothiazide and carvedilol release was faster. The

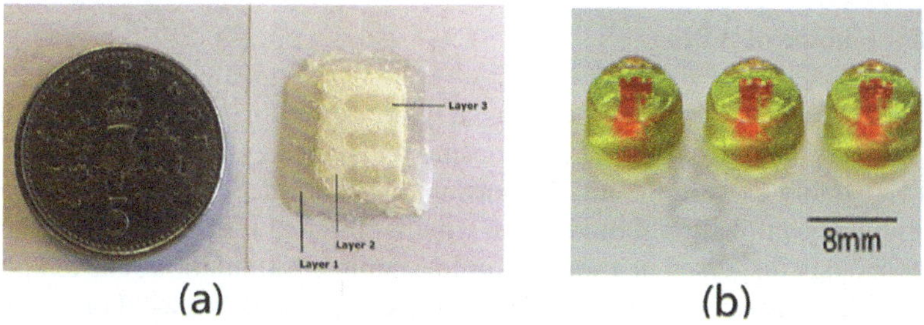

Fig. 4.11 **a** The three-layer film containing three antihypertensive drugs (reproduced from [96]). **b** Multi-material inkjet printed UV-cured tablets (reproduced from [97])

release of amiloride HCl was the fastest as this drug was not embedded in any polymer matrix, whereas hydrochlorothiazide and carvedilol release was sustained [96].

He et al. [97] (Fig. 4.11b) created a reactive pro-drug formulation to create customized designed tablets capable of delivering extended and controlled drug releases. The reactive prodrug used was able to polymerize into drug-attached macromolecules during the inkjet process. Hydrophilicity was tuned allowing hydrolysis to be hindered or facilitated allowing for a controlled drug release. This was achieved by controlling loadings of photocurable materials of 2-hydroxyethyl acrylate, 2-hydroxyethyl methacrylate, and 2-hydroxyethyl acrylamide as reactive linkers. Significantly high ibuprofen loading (58 wt%) was achieved by attaching the drug to 2-hydroxyethyl acrylate through a cleavable ester bond. In addition to controlling materials, the design freedom offered through inkjet printing allowed for an additional method of controlling drug release. This was done by varying the geometry of the tablets and by spatially controlling the position of the tablet composition. Critically, the loaded drug composition was seen to be homogenously distributed through the ink information and printed tablet without any indication of recrystallization and precipitation. This once again highlights inkjet printing's ability to create tablets with high physical stability with a high drug loading [97].

A thesis by Cader in 2019 was able to show the feasibility of creating a wholly inkjet printed tablet capable of delivering a drug release with a controlled mechanism. Using three inks, a multi-material reservoir system was created for one-directional drug release. It consisted of a rapidly dissolving core (a PVP-thiamine matrix, similar to Cader et al. [98]), an impermeable base and side-walls made of polycaprolactone (PCL) and a release controlling membrane on the top surface (a blend of PCL and PVP, with PVP acting as a pore forming agent). With the application of this release controlling membrane (25 printed layers) controlled release was achieved and zero-order kinetics were exhibited. By varying the PVP loading in the release-controlling membrane ink and different printing parameters, the thickness of the membrane and its pore area sizes were controlled, influencing releasing. Different drug-release profiles were achieved in this thesis for rapid, controlled extended, delayed, and bimodal (fast to slow) release [98].

4.4.2.6 Choice of APIs

As can be seen in the previous sections, a wide variety of APIs have been used to create inkjet printed dosages. When choosing APIs, the key criteria are the ability of the API to stably exist in the ink. No unwanted occurrences of phase separation or undesirable crystallization should occur, whether the API is dissolved in a solvent or homogeneously dispersed in a solution or in a curable formulation. The temperature at which inkjet printing occurs can also limit the type of APIs used. For example, in hot-melt inkjet printing by Kyobula et al., fenofibrate was able to withstand high temperatures without degrading [7]. High temperatures may also be required to reduce the viscosity of inks for more stable jetting behaviour [79]. Once a stable ink (in the correct environment) is achieved, the ink must be able to pass through the nozzle system with loading homogeneity. This is especially important when using APIs which are encapsulated in liposomes for example [84], where particle size may be larger than nozzle diameter. The final consideration for API selection is its ability to remain stable during the ink's interaction with the substrate and solidification stage of printing. If solvent evaporation is the mechanism of solidification, different drug polymorphisms may form [85]. Substrate heating will also have to be taken into consideration to ensure the heating does not degrade the drug. The drug's stability to UV light will need to also be considered in photocurable ink formulations that are used to create tablets [94]. In addition to drug stability, the printability of inks may be influenced by the API and its loading. How the drug affects surface tension and viscosity should always be investigated. It is however more likely that excipients will play a larger role in affecting printability than the API.

4.4.2.7 Choice of Excipients

Excipients will have a major role in the printability of the ink. In most inkjet formulations, the polymer excipient, which would make up the matrix of the tablet would have a large role in determining the viscosity of the ink. Depending on the type of API and the desired release rate, the choice of polymer excipient can be determined. For example, if fast drug release is desired, water soluble polymers such as PVP can be used [93]. Or if a slower drug release is required, ethylcellulose can be added to the formulation [96]. Varying ratios of polymer excipients can also control drug release, as illustrated by Acosta-Vélez et al., where authors controlled PEGDA loading to change porosity, influencing drug release [92].

The solidification mechanism of inkjet printing will also have an impact on determining what excipient or ink ingredients are required. In the case of solvent-based inks, the choice of solvents greatly affects surface tension and viscosity. In the case of Cader et al., water was used. With an initial surface tension value of 65 mNm^{-1}, polysorbate 20, a surfactant, was required to reduce it to 37 mNm^{-1} [93]. In addition, the time for evaporation can also affect printing behaviour. If evaporation occurs too quickly, nozzles may get blocked (as was seen when printing inks at high temperatures by Cader). In contrast, if evaporation is too slow, longer drying time will be required before additional layers can be printed [98]. The choice of solvent will also need to take into account safety factors. Toxic solvents like methanol that have been used in studies will need to be tracked to ensure that there is no toxicity associated with the finished tablet [99]. Hence, safer liquids like water may be used, but this does reduce the types of drugs that can be used, due to solubility limits. When using photocurable materials, similar care will have to be taken to ensure no remaining uncured and potentially hazardous material remains in the finished product.

4.4.2.8 Summary

Material jetting has been used on numerous occasions for oral drug delivery applications. Initially, this had been done onto edible substrates that would make up part of the dosage form, but more recent papers have moved on to freestanding and wholly inkjet-printed tablets. These have been successfully produced using solvent, UV-curable, and hot-melt formulations. A wide range of drugs and biologics have been jetted in numerous forms including microspheres, liposomes, nanosuspensions, and solutions. Drugs with both poor and high water solubility have been formulated into printable inks where high drug loading can be achieved. The existing studies have shown material jetting is a reproducible, uniform, and accurate method of pharmaceutical product production. The ability to control drug release has been achieved by varying the substrate, the ink formulation component's loading, device geometry, printing parameters, and spatial location of the APIs. Multi-material inkjet printing has been used to create dosages containing different APIs and well as multi-layered tablets, capable of delivering controlled drug release.

4.4.3 Binder Jetting

As the first AM technique successfully produced FDA approved tablet (Spritam®), the binder jetting technique is the one that is closest to a real 3D printed commercial product. In this section, we will look into how binder jetting works in oral solid dosage form manufacturing, the advantageous applications into binder jetting printed tablets, and how to choose the excipient and APIs for this technique.

4.4.3.1 Binder Jetting System

Binder jetting technique was firstly invented by the Massachusetts Institute of Technology in 1993. It contains a layer-wise powder supplying system that spreads fresh powder onto the printing platform and inkjet printhead(s) that selectively deposit liquid binder onto the pre-designed location of each fresh powder layer to stick the particles in the deposited region together. By repeating the jetting and powder recoating process, the final object can be formed. Similar to material jetting techniques, two types of drop-on-demand printheads (thermal and piezo) are normally used. However, unlike material jetting, the liquid ink does not need to construct the final solid structure of the dosage form, which gives more freedom on the choice of ink formulations. Compared with the material extrusion technique, the machine setup and overall cost will be much higher, which may lead to fewer research studies in the binder jetting oral solid dosage form field. Nonetheless, binder jetting is an AM technique that manages to balance manufacturing resolution, the flexibility of materials, and production scale. Therefore the AM technique is most likely to be used for mass production in the future. Apercia's Zipdose technology [100], for example, introduced a conveyer belt based powder recoating system that allows continuous tablet production and makes it capable of manufacturing a large batch of tablets.

4.4.3.2 Rapid Delivery Dosage

Oral solid dosage forms made by the binder jetting process will normally claim rapid delivery as one of the key advantages. This is owing to the outstanding porosity created by the powder bed. The tablet Spritam® [101] from Apricia has an average disintegration time of 11 seconds. In Yu et al.'s [102] work, their tablet released about 97.7% of the drug in the initial 120 seconds. By choosing different excipient powder and binder inks, the disintegration and releasing time could range between 11 seconds and 20 hours, but such a rapid designation and releasing speed within dozens of seconds can hardly be achieved by the other AM techniques. To ensure good powder fluidity and smooth recoating during the manufacturing process, loose powder is normally used which contains a lot of free void space. Although the recoating roller could provide some compression force to pack the powder to a certain level, the printing process still mainly relies on the deposited liquid binder to fill these voids and connect particles. The liquid will create a binding bridge to connect particles and the user can define the volume of liquid binder deposited per unit area. By reducing the binder volume, the void between particles will be less saturated and therefore, during the application, fluid in the environment (e.g. water) can quickly penetrate the tablet and achieve thorough contact with the binder bridge. This will accelerate the binder dissolving speed and as soon as the binder bridge was fully dissolved, the tablet will fully disintegrate.

4.4.3.3 Multi Binder/Active Dosage

Binder jetting technology can produce complex oral solid dosage forms by including multiple binder or APIs, based on which, the user can design complex releasing profiles of a single drug and achieve sequentially releasing of multiple APIs. The binder jetting technique has the potential to produce such a complex dosage by integrating multiple inkjet print-heads in one machine, each containing a binder or API. In 1996, Wu et al. [103] published a proof-of-concept for using the binder jetting process to produce multi-active devices containing both methylene blue and alizarin yellow as mock actives and PCL and PEO as excipients and shell, respectively [104]. However, the follow-up studies mostly focused on a single active ingredient but used different binder inks to design the release profile of a tablet. Katstra et al. showed that by using different binders and powdered excipients, the release profile can be delayed as needed [105]. Rowe et al. [106] published their work in the same year showing that within one tablet, by combining two different binders that have different disintegration times, a complex divided tablet can be achieved. This work demonstrated that the user can program the dosage's releasing profile as needed. As a proof of concept, they designed and printed a tablet that has a dual-pulse release profile where the drug is released sequentially in each section. Although only one drug was printed in these two studies, multi-active tablets, in principle, can easily be achieved by replacing one of the binder inks with a drug-loaded binder ink, or by incorporating an additional printhead to deposit drug loaded liquid.

4.4.3.4 Choice of APIs

Regarding the APIs, there were three different strategies that can be used for the binder jetting process. The first strategy is to mix the target drug into the excipient powder [102, 107, 108]. If the target drug is in powder form, this is the most straight forward strategy that does not need to be concerned about disturbing the target drug's chemical or physical properties. This strategy is also comparable with the traditional compress molding way of manufacturing tablets. However, the limitation of this strategy is that once mixed, the drug concentration in the powder is defined. Therefore, different drug loading can only be achieved by changing the overall volume of the dosage form. The releasing speed could still be engineered by choosing different binder ink or design structures with different infill as mentioned in the material extrusion section, but the design freedom of this strategy is restricted. This strategy is also not quite feasible for producing devices containing multiple APIs as this will require constantly changing the powder supply during the manufacturing process.

The other two strategies are mixing the APIs into the binder ink [107, 109–111] or having an independent drug-loaded ink [103, 112] in addition to binder ink. Both strategies provide the user more freedom to design the performance of the final device by enabling the user to engineer the spatial distribution of the target drug(s) in a tablet. With the help of the computational tool, a spatial drug distribution map can be generated based on user-defined drug-release profile(s). The target tablet can then be manufactured following the guidance of these designs. However, limitations still exist for these two strategies. Firstly, the ink shall follow the designing principles for inkjet printing, where the viscosity and surface tension of the it shall meet certain criteria. Most of the existing studies prepared the binder ink by dissolving selected APIs into certain organic solvents to form a liquid ink that meets the viscosity and surface tension requirements for inkjet. Once printed onto the powder bed, the organic solvent will evaporate and the dissolved APIs will precipitate out. The choice of the organic solvent is complicated as the toxicity of the solvent shall be considered as solvent residuals may exist in the end product. In the meanwhile, the volatility of the solvent is also important. If the solvent is too volatile, the ink will evaporate and solidify at the printhead nozzles and block the printhead; if the evaporation speed is low, it may require excessive evaporation time between layers. In addition, the interaction between the solvent and the target drug needs to be taken into consideration.

4.4.3.5 Choice of Excipients

For the binder jetting process, the powder bed will form the excipient. The powder shall meet certain physical property requirements to ensure a reliable recoating for each layer. The morphology and size distribution can decide the flowability of the powder and recoated layer thickness. These will then have an impact on product quality and structure resolution. Higher manufacturing resolution will provide the user with higher design flexibility and better control of API distribution in a tablet, both of which enable the production of oral solid dosage forms with complex releasing profiles. In current studies, the excipient powder that has been used includes PCL, PEG [103, 112], lactose [102], mannitol [107, 111], PVAc [109], PVP [109, 111], cellulose [110], hydroxypropyl methylcellulose [102, 113], and $CaSO_4$ [108, 112].

As many traditional dosages are made by compressing the molding of powder form excipients, there are many existing FDA approved candidates that could be considered. If the user would like to consider new excipients, it is important to consider whether they could be made into powder form with reliable flowability.

4.4.3.6 Summary

Binder jetting is a pseudo-area-based printing technique that provides a much higher production speed than point-based AM techniques such as material extrusion. It is more suitable for large-scale production where based on the patent [114], a convey belt based system can be applied to make it a competitive process versus mass production. The overall machine cost will be higher than that of material extrusion considering the printhead cost and controlling system. The key limitation of the binder jetting technique is normally only one type of powder can be used in each batch, and therefore limit the level of releasing control. Although research has been carried out to selectively deposit different powder either within one layer or in different layers [115–117], such strategy can significantly slow down the processing speed as well as the difficulties in recycling the unused powders, both of which will increase the overall cost.

4.4.4 Powder Bed Fusion

Powder bed fusion is a widely used AM technique. However, there were only a few studies investigating the possibility of using this technique for manufacturing oral solid dosage forms. In this section, we will look into these studies and discuss the limitation of this technique in dosage manufacturing.

4.4.4.1 Technique

The concept of powder bed fusion is very similar to that of the binder jetting technique. The fresh loose powder is sequentially spread onto a building platform layer by layer. When each layer of powder is spread, instead of using a binder to form solid bridges between particles, a focused energy beam (normally laser) is used to scan the target area to melt the particles partially. The melted particle shield will merge and form solid bridges once it cools down.

Compared with the binder jetting technique, because the target material will be heated up to the melting temperature which may cause excipient or APIs degradation, the chosen excipient and APIs need to be chemically stable at the melting temperature of the excipient powder. In addition, the powder bed fusion technique requires the excipient/API melt to have relatively low zero viscosity and low surface tension [118] at the processing temperature to ensure the melt will not flow away. The user also needs to consider the optical property of the powder; the deposited energy needs to be properly absorbed by the powder and converted into heat to melt the excipient particle. At last, for the powder bed fusion process, the API has to be mixed into the excipient powder and therefore the user will have less flexibility to design the final dosage's drug loading and releasing profiles. By considering all these limitations, the powder bed fusion technique was not widely considered as a potential method for 3D printing oral solid dosages.

4.4.4.2 Phased-Release Tablet

Similar to that of binder jetting, the releasing speed of the printed dosage form can also be controlled by varying the level of binding between the particles. In principle, if lower energy was deposited, the level of particle melting and bridging will be limited on the particle surface, which will end up with a more porous structure with accelerated soaking and drug-release speed [119, 120]. On the contrary, if higher energy was used, the dense structure will inhibit environment fluid penetration and reduce the drug release. Cheah et al.'s study confirmed this [121], in which they sintered the powder with four different laser power and five different scanning speeds. It was found that the region with a higher energy dose will form a denser structure and within one tablet, and the porosity of different regions can be engineered by controlling the energy dose. Therefore, a core–shell structure was designed by varying the energy dose, which showed a lower releasing speed than a uniform structure with a lower energy dose. A similar phenomenon was also observed by Leong et al. in [122]. Changing the infill level was another strategy for engineering the releasing profile of the final dosage. As mentioned in the previous section, dosage forms with less infill and a higher surface/volume ratio exhibit a faster releasing speed. Fina et al. [123] investigated the use of gyroid lattice structure printed by powder bed fusion technique and confirmed this strategy is also feasible for controlling the drug-release speed. By sequentially switching the feeding powder, Awad et al. [124] successfully achieve a multi-drug dosage form by using the powder bed fusion

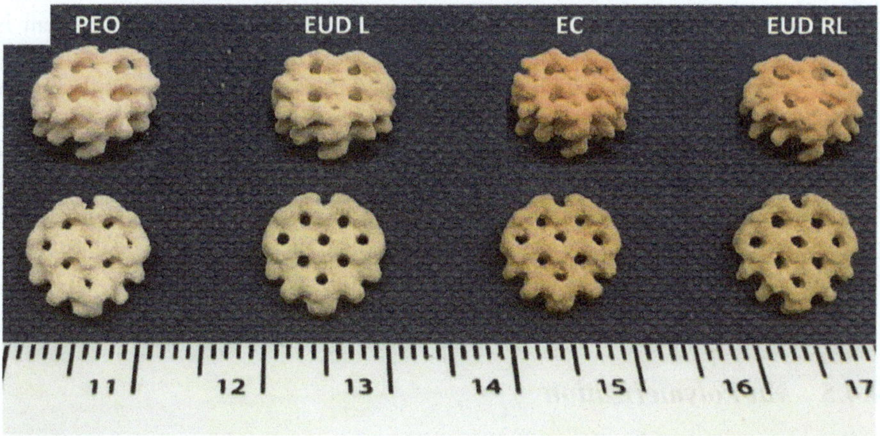

Fig. 4.12 Engineering the releasing speed by introducing different infill ratios (reproduced from [123])

technique. However, challenges still remain, which include how to efficiently coat different powders and how to recycle the unused powder, which will need to be solved before this could be used in production (Fig. 4.12).

4.4.4.3 Choice of Excipients and APIs

In the early attempts, commercial nylon powder [119, 120] was used as an excipient to investigate controlling the releasing profile of the printed dosage form through the strategy mentioned in 4.4.2. At this stage, the dye was used instead of APIs. Then PCL [122, 125–128] and PLLA [122] were introduced as excipient materials. Rather than dye, real APIs started to be added into the powder to produce real dosage forms that are targeting for the actual application. At this period, the APIs that were attempted include: Ibuprofen [127], Fluorouracil [128], progesterone [129]. In recent studies, excipient powder that has been commonly used in the pharmaceutical industry was also investigated and showed a positive result. Eudragit L100-55 [123, 130], Eudragit RL [123], Kollidon VA 64 [131], and Kollicoat IR [130] were all confirmed to be suitable for powder bed fusion directly or with little modification. In these studies, paracetamol was used as a model drug and it was confirmed that this drug did not degrade at the processing temperature of all the commercial excipient powders mentioned.

4.4.4.4 Summary

In summary, powder bed fusion is a possible routine for producing oral solid dosage forms. However, compared to the binder jetting process, it possesses some limitations in excipient and API choices especially when temperature sensitive ingredient is included. In addition, most of the current powder bed fusion techniques are using a laser to sinter the powder, which is a point-based processing method, while binder jetting is pseudo-area based printing which will provide more opportunity in scaling up. At last, the binder jetting technique also provides the choice of in-line design of the spatial distribution of the drug and is also capable of integrating multiple drugs into one dosage form, which will be challenging for powder bed fusion considering the powder cross-contamination and production cost. Therefore, based on the current processing concept, the powder bed fusion technique is relatively limited.

4.4.5 Vat Polymerization

Vat polymerization relies on photoreactive ink formulations which convert monomers into large molecules to construct solid structures. This technique can provide the highest manufacturing resolution (e.g. the two-photon printing technique and project micron stereolithography). Although it is not widely used in oral solid dosage forms, for some special therapeutical applications that require high processing resolution (e.g. microneedles), it still has a lot of potentials.

4.4.5.1 Technique

Vat polymerization equipment normally includes a resin vat containing free running photoreactive monomers or oligomers, a printing stage, and an energy source (e.g. UV, visible light, infra light) that will selectively trigger the photopolymerization reaction to form the solid object. The energy dose strategy (polymerization) could be point-based (stereolithography, two-photon polymerization), area-based (digital light processing) or even volume-based (computed axial lithography). The two-photon polymerization technique has the highest printing resolution (~150 nm) but is limited to manufacturing relatively small structures (millimeter range), while computed axial lithography has the highest manufacturing speed, but with restricted printing resolution (~50–100 μm). Most of the commercial equipment is based on stereolithography and digital light processing technique. For manufacturing oral solid dosage forms, the main challenge is the availability of compatible APIs and excipients. As photoreactive materials were not commonly used in traditional mass production of oral solid dosages, new materials and formulations will need to be developed to meet the processing requirement of vat polymerization. These new materials and formulation will, however, bring safety and regulation concerns.

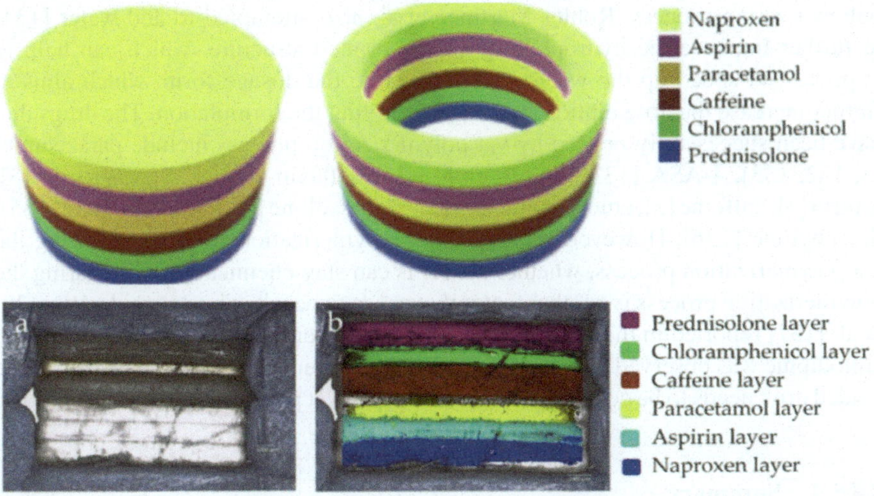

Fig. 4.13 Oral solid dosage printed by vat polymerization containing six different drugs (reproduced from [3])

4.4.5.2 Phased-Release Tablet with Single and Multiple APIs

Like the powder bed fusion technique, the target API will need to be added to the formulation. Therefore, the strategy of designing drug loading and releasing speed with a selected formulation can be changing the tablet volume and the infill of the tablet structure respectively. Martinez et al. [132] investigated the drug-release profile of vat polymerized dosage forms with different geometries and surface/volume ratios. In their study, they confirmed that a high surface volume ratio can significantly increase the loaded drug's releasing speed. It is also possible to manufacture a multi-material dosage form (or polypill) through vat polymerization. Robles-Martinez et al. [3] attempt to sequentially switch the liquid resin in the vat to create a dosage form with multiple APIs. A cylinder and ring shape tablet stacked up with six different blocks of drugs were successfully achieved. As a proof-of-concept work, both works have demonstrated that vat polymerization can provide the user design freedom for controlling the dosage form's releasing profile or manufacturing complex dosage forms that contains more than one API. Nevertheless, they are still at the lab scale and will face production speed and complexity challenges when scaling up (Fig. 4.13).

4.4.5.3 Choice of Excipients and APIs

PEGDA is the most preferable excipient monomer for existing studies [3, 132–134]. This monomer is hydrophilic which can accelerate the water penetration to help the loading drug diffuse out. It contains two acrylate functional groups and ensures relatively high reactivity, which is an important requirement for the vat polymerization process. Robles-Martinez et al. also attempted to add water [134] to further improve the hydrophilicity of the printed structure, which can help to improve and speed up the water penetration into the dosage form, which aims to further increase the drug elution speed by modifying the formulation. The drugs that have been successfully printed by vat polymerization process include paracetamol [3, 132, 133], 4-ASA [133], ibuprofen [134], riboflavin [134, 135], naproxen [3], aspirin [3], caffeine [3], chloramphenicol [3], prednisolone [3], L-ascorbic acid [135], theophylline [136]. However, because photopolymerization is required during the vat polymerization process, whether the APIs can stay chemically stable during the polymerization process is another factor that needs to be taken into consideration. Xu et al. [137] reported in their work that reaction between the PEGDA monomer with amlodipine was observed which confirmed that the chemical property stability of the loaded drug needs to be considered when designing vat polymerization formulations.

4.4.5.4 Summary

Vat polymerization is a technique that has a lot of potential in manufacturing drug delivery devices. Although in oral solid dosage form manufacturing, its manufacturing speed and flexibility of integrating multiple APIs into a single dosage is still restricted. Meanwhile, because of its outstanding manufacturing resolution means for some special applications, it would be the only feasible AM technique that can manufacture micron or even nano geometries. One of the biggest challenges that vat polymerization is faced with is the availability of feasible excipients and APIs and when will these new formulations pass a safety test and receive regulatory approval.

4.5 Summary and Outlook

AM in the pharmaceutical industry is not yet widely used, even though it is extensively used to research possible pharmaceutical applications. While continuing to be used and explored in research, it may not replace traditional manufacturing processes in the near future. Instead, it will likely be used for specific medications or applications, such as the previously mentioned orphan drugs.

Even though 3D printing provides the ability to personalize drugs and tailor them to patients and it could be a good manufacturing technique for applications such as niche and orphan drugs that are needed in much smaller quantities, the industry would still need to play a part in it. Just like with Spritam®, GMP conditions and regulations from the FDA, EMA, or agency that regulates a specific country/region were still followed, quality control measurements were still taken into account and that will always have to be a certainty. When dealing with people's lives and health, such measures are necessary, which is why regulations and guidelines exist. Contrary to all the traditional manufacturing techniques, the AM techniques are not regulated for pharmaceutical applications and setting the regulatory requirements can take a long time. The companies would need to acquire new equipment which adds into their costs, and the equipment manufacturers would have to change and modify them to ensure they could meet pharmaceutical standards. Moreover, process changes and evolution in the pharmaceutical industry tend to be difficult, as is the case with changing from the production of big batches to a continuous manufacturing process.

Nonetheless, the possibilities that AM offers to the printing of medicines are being continuously researched and that will undoubtedly lead to its application in the industry.

References

1. Arafat B et al (2018) Tablet fragmentation without a disintegrant: a novel design approach for accelerating disintegration and drug release from 3D printed cellulosic tablets. Eur J Pharm Sci 118:191–199
2. Lamichhane S et al (2019) Customized novel design of 3D printed pregabalin tablets for intra-gastric floating and controlled release using fused deposition modeling. Pharmaceutics 11(11):564
3. Robles-Martinez P et al (2019) 3D printing of a multi-layered polypill containing six drugs using a novel stereolithographic method. Pharmaceutics 11(6):274
4. Khaled SA et al (2015) 3D printing of tablets containing multiple drugs with defined release profiles. Int J Pharm 494(2):643–650
5. Khaled SA et al (2015) 3D printing of five-in-one dose combination polypill with defined immediate and sustained release profiles. J Control Release 217:308–314
6. Clark EA et al (2020) Making tablets for delivery of poorly soluble drugs using photoinitiated 3D inkjet printing. Int J Pharm 578:118805
7. Kyobula M et al (2017) 3D inkjet printing of tablets exploiting bespoke complex geometries for controlled and tuneable drug release. J Control Release 261:207–215
8. Khaled SA et al (2018) Extrusion 3D printing of paracetamol tablets from a single formulation with tunable release profiles through control of tablet geometry. AAPS PharmSciTech 19(8):3403–3413
9. Goyanes A et al (2015) Effect of geometry on drug release from 3D printed tablets. Int J Pharm 494(2):657–663
10. Goyanes A et al (2015) 3D printing of medicines: engineering novel oral devices with unique design and drug release characteristics. Mol Pharm 12(11):4077–4084
11. Gibson M (2016) Pharmaceutical preformulation and formulation: a practical guide from candidate drug selection to commercial dosage form. CRC Press
12. Banker GS, Siepmann J, Rhodes C (2002) Modern pharmaceutics. CRC Press
13. Zhang GG et al (2004) Phase transformation considerations during process development and manufacture of solid oral dosage forms. Adv Drug Deliv Rev 56(3):371–390
14. Saifee M et al (2009) Drug polymorphism: a review. Int J Health Res 2(4)
15. Cui Y (2007) A material science perspective of pharmaceutical solids. Int J Pharm 339(1–2):3–18
16. Láng P et al (2013) Polymorph screening of an active material. J Pharm Biomed Anal 84:177–183
17. Ventola CL (2014) Medical applications for 3D printing: current and projected uses. Pharm Ther 39(10):704

18. Walsh J et al (2018) Patient acceptability, safety and access: a balancing act for selecting age-appropriate oral dosage forms for paediatric and geriatric populations. Int J Pharm 536(2):547–562
19. Goyanes A et al (2017) Patient acceptability of 3D printed medicines. Int J Pharm 530(1–2):71–78
20. Buyukgoz GG et al (2020) Exploring tablet design options for tailoring drug release and dose via fused deposition modeling (FDM) 3D printing. Int J Pharm 591:119987
21. Perrie Y, Rades T (2012) FASTtrack pharmaceutics: drug delivery and targeting. Pharmaceutical Press
22. Wen H, Park K (2011) Oral controlled release formulation design and drug delivery: theory to practice. Wiley
23. Siepmann J, Siegel RA, Rathbone MJ (2012) Fundamentals and applications of controlled release drug delivery. vol 3. Springer
24. Siepmann J, Siepmann F (2012) Modeling of diffusion controlled drug delivery. J Control Release 161(2):351–362
25. Goyanes A et al (2014) Fused-filament 3D printing (3DP) for fabrication of tablets. Int J Pharm 476(1–2):88–92
26. Tabriz AG et al (2020) 3D printed bilayer tablet with dual controlled drug release for tuberculosis treatment. Int J Pharm 120147
27. Kempin W et al (2018) Immediate release 3D-printed tablets produced via fused deposition modeling of a thermo-sensitive drug. Pharm Res 35(6):124
28. Li Q et al (2017) Preparation and investigation of controlled-release glipizide novel oral device with three-dimensional printing. Int J Pharm 525(1):5–11
29. Chen D et al (2020) Preparation and in vitro evaluation of FDM 3D-printed ellipsoid-shaped gastric floating tablets with low infill percentages. AAPS PharmSciTech 21(1):6
30. Reddy Dumpa N, Bandari S, Repka MA (2020) Novel gastroretentive floating pulsatile drug delivery system produced via hot-melt extrusion and fused deposition modeling 3D printing. Pharmaceutics 12(1):52
31. Goyanes A et al (2015) 3D printing of modified-release aminosalicylate (4-ASA and 5-ASA) tablets. Eur J Pharm Biopharm 89:157–162
32. Cui M et al (2020) Fabrication of high drug loading levetiracetam tablets using semi-solid extrusion 3D printing. J Drug Deliv Sci Technol 101683
33. Fanous M et al (2020) Simplification of fused deposition modeling 3D-printing paradigm: feasibility of 1-step direct powder printing for immediate release dosage form production. Int J Pharm 578:119124
34. Fanous M et al (2020) Development of immediate release (IR) 3D-printed oral dosage forms with focus on industrial relevance. Eur J Pharm Sci 155:105558
35. Samaro A et al (2020) Screening of pharmaceutical polymers for extrusion-based additive manufacturing of patient-tailored tablets. Int J Pharm 586:119591
36. Khaled SA et al (2014) Desktop 3D printing of controlled release pharmaceutical bilayer tablets. Int J Pharm 461(1–2):105–111
37. Pietrzak K, Isreb A, Alhnan MA (2015) A flexible-dose dispenser for immediate and extended release 3D printed tablets. Eur J Pharm Biopharm 96:380–387
38. Dragoman MV (2014) The combined oral contraceptive pill-recent developments, risks and benefits. Best Pract Res Clin Obstet Gynaecol 28(6):825–834
39. Huffman MD, Xavier D, Perel P (2017) Uses of polypills for cardiovascular disease and evidence to date. Lancet 389(10073):1055–1065
40. Lu D-Y et al (2014) New perspectives of HIV/AIDS therapy study. Recent Pat Anti-Infect Drug Discovery 9(2):112–120
41. Tagami T et al (2018) Defined drug release from 3D-printed composite tablets consisting of drug-loaded polyvinylalcohol and a water-soluble or water-insoluble polymer filler. Int J Pharm 543(1–2):361–367
42. Daly R et al (2015) Inkjet printing for pharmaceutics–a review of research and manufacturing. Int J Pharm 494(2):554–567

43. Voura C et al,Printable medicines: a microdosing device for producing personalized medicines. The authors describe a new method for the production of personalised medicines involving the direct printing of drug solutions or suspensions containing APIs and excipients onto edible paper carrier materials utilising ultra-precise microdrop printing technology. The approach allows ultra-precise dosing of the API according to the specific requirements of each individual patient and is suited for highly active low-dose drugs. Pharm Technol Eur 23(1)
44. Skowyra J, Pietrzak K, Alhnan MA (2015) Fabrication of extended-release patient-tailored prednisolone tablets via fused deposition modelling (FDM) 3D printing. Eur J Pharm Sci 68:11–17
45. Goyanes A et al (2017) Development of modified release 3D printed tablets (printlets) with pharmaceutical excipients using additive manufacturing. Int J Pharm 527(1–2):21–30
46. Alhijjaj M, Belton P, Qi S (2016) An investigation into the use of polymer blends to improve the printability of and regulate drug release from pharmaceutical solid dispersions prepared via fused deposition modeling (FDM) 3D printing. Eur J Pharm Biopharm 108:111–125
47. Ilyés K et al (2019) The applicability of pharmaceutical polymeric blends for the fused deposition modelling (FDM) 3D technique: material considerations–printability–process modulation, with consecutive effects on in vitro release, stability and degradation. Eur J Pharm Sci 129:110–123
48. Genina N et al (2017) Anti-tuberculosis drug combination for controlled oral delivery using 3D printed compartmental dosage forms: from drug product design to in vivo testing. J Control Release 268:40–48
49. Beck R et al (2017) 3D printed tablets loaded with polymeric nanocapsules: an innovative approach to produce customized drug delivery systems. Int J Pharm 528(1–2):268–279
50. Ibrahim M et al (2019) 3D printing of metformin HCl PVA tablets by fused deposition modeling: drug loading, tablet design, and dissolution studies. AAPS PharmSciTech 20(5):195
51. Cheng Y et al (2020) 3D printing of extended-release tablets of theophylline using hydroxypropyl methylcellulose (HPMC) hydrogels. Int J Pharm 591:119983
52. Khaled SA et al (2018) 3D extrusion printing of high drug loading immediate release paracetamol tablets. Int J Pharm 538(1–2):223–230
53. Cui M et al (2019) Effect of novel internal structures on printability and drug release behavior of 3D printed tablets. J Drug Deliv Sci Technol 49:14–23
54. Elbl J, Gajdziok J, Kolarczyk J (2020) 3D printing of multilayered orodispersible films with in-process drying. Int J Pharm 575:118883
55. Holländer J et al (2018) 3D printed UV light cured polydimethylsiloxane devices for drug delivery. Int J Pharm 544(2):433–442
56. Seoane-Viaño I et al (2021) Semi-solid extrusion 3D printing in drug delivery and biomedicine: Personalised solutions for healthcare challenges. J Controlled Release
57. Goyanes A et al (2019) Direct powder extrusion 3D printing: fabrication of drug products using a novel single-step process. Int J Pharm 567:118471
58. Charoenying T et al (2020) Three-dimensional (3D)-printed devices composed of hydrophilic cap and hydrophobic body for improving buoyancy and gastric retention of domperidone tablets. Eur J Pharm Sci 155:105555
59. Fu J et al (2018) Combination of 3D printing technologies and compressed tablets for preparation of riboflavin floating tablet-in-device (TiD) systems. Int J Pharm 549(1–2):370–379
60. Huanbutta K, Sangnim T (2019) Design and development of zero-order drug release gastroretentive floating tablets fabricated by 3D printing technology. J Drug Deliv Sci Technol 52:831–837
61. Smith DM et al (2018) Pharmaceutical 3D printing: design and qualification of a single step print and fill capsule. Int J Pharm 544(1):21–30
62. Smith D et al (2018) 3D printed capsules for quantitative regional absorption studies in the GI tract. Int J Pharm 550(1–2):418–428
63. Siyawamwaya M et al (2019) 3D printed, controlled release, tritherapeutic tablet matrix for advanced anti-HIV-1 drug delivery. Eur J Pharm Biopharm 138:99–110

64. Pereira BC et al (2019) 'Temporary Plasticiser': a novel solution to fabricate 3D printed patient-centred cardiovascular 'Polypill' architectures. Eur J Pharm Biopharm 135:94–103
65. Gioumouxouzis CI et al (2018) A 3D printed bilayer oral solid dosage form combining metformin for prolonged and glimepiride for immediate drug delivery. Eur J Pharm Sci 120:40–52
66. Göke K et al (2018) Novel strategies for the formulation and processing of poorly water-soluble drugs. Eur J Pharm Biopharm 126:40–56
67. Soluplus—For better solubility & bioavailability
68. Ashland, AquaSolveTM hypromellose acetate succinate
69. Alshahrani SM et al (2015) Stability-enhanced hot-melt extruded amorphous solid dispersions via combinations of Soluplus® and HPMCAS-HF. AAPS PharmSciTech 16(4):824–834
70. Tajarobi F et al (2011) The influence of crystallization inhibition of HPMC and HPMCAS on model substance dissolution and release in swellable matrix tablets. Eur J Pharm Biopharm 78(1):125–133
71. Linn M et al (2012) Soluplus® as an effective absorption enhancer of poorly soluble drugs in vitro and in vivo. Eur J Pharm Sci 45(3):336–343
72. Gioumouxouzis CI et al (2017) 3D printed oral solid dosage forms containing hydrochlorothiazide for controlled drug delivery. J Drug Deliv Sci Technol 40:164–171
73. Solanki NG et al (2018) Formulation of 3D printed tablet for rapid drug release by fused deposition modeling: screening polymers for drug release, drug-polymer miscibility and printability. J Pharm Sci 107(1):390–401
74. Wei C et al (2020) Development of 3D printed tablets by fused deposition modeling using polyvinyl alcohol as polymeric matrix for rapid drug release. J Pharm Sci
75. Conceição J et al (2019) Hydroxypropyl-β-cyclodextrin-based fast dissolving carbamazepine printlets prepared by semisolid extrusion 3D printing. Carbohydr Polym 221:55–62
76. Saviano M et al (2019) Poly (vinyl alcohol) 3D printed tablets: the effect of polymer particle size on drug loading and process efficiency. Int J Pharm 561:1–8
77. Goyanes A et al (2015) Fabrication of controlled-release budesonide tablets via desktop (FDM) 3D printing. Int J Pharm 496(2):414–420
78. Kollamaram G et al (2018) Low temperature fused deposition modeling (FDM) 3D printing of thermolabile drugs. Int J Pharm 545(1–2):144–152
79. Magdassi S (2010) The chemistry of inkjet inks. World Scientific
80. Derby B (2010) Inkjet printing of functional and structural materials: fluid property requirements, feature stability, and resolution. Annu Rev Mater Res 40:395–414
81. Martin GD, Hoath SD, Hutchings IM (2008) Inkjet printing-the physics of manipulating liquid jets and drops. J Phys Conf Ser
82. Cummins G, Desmulliez MP (2012) Inkjet printing of conductive materials: a review. Circuit world
83. Wijshoff H (2010) The dynamics of the piezo inkjet printhead operation. Phys Rep 491(4–5):77–177
84. Gore M (2005) Systems and methods for preparation of pharmaceutical dosage using compositions containing aqueous vesicles. Google Patents
85. Meléndez PA et al (2008) Thermal inkjet application in the preparation of oral dosage forms: dispensing of prednisolone solutions and polymorphic characterization by solid-state spectroscopic techniques. J Pharm Sci 97(7):2619–2636
86. Buanz AB et al (2011) Preparation of personalized-dose salbutamol sulphate oral films with thermal ink-jet printing. Pharm Res 28(10):2386
87. Sandler N et al (2011) Inkjet printing of drug substances and use of porous substrates-towards individualized dosing. J Pharm Sci 100(8):3386–3395
88. Genina N et al (2013) Evaluation of different substrates for inkjet printing of rasagiline mesylate. Eur J Pharm Biopharm 85(3):1075–1083
89. Eleftheriadis G et al (2020) Development and characterization of inkjet printed edible films for buccal delivery of B-complex vitamins. Pharmaceuticals 13(9):203

90. Fox CB et al (2017) Picoliter-volume inkjet printing into planar microdevice reservoirs for low-waste, high-capacity drug loading. Bioeng Transl Med 2(1):9–16
91. Acosta-Vélez GF et al (2017) Photocurable bioink for the inkjet 3D pharming of hydrophilic drugs. Bioengineering 4(1):11
92. Acosta-Vélez GF et al (2018) Photocurable poly (ethylene glycol) as a bioink for the inkjet 3D pharming of hydrophobic drugs. Int J Pharm 546(1–2):145–153
93. Cader HK et al (2019) Water-based 3D inkjet printing of an oral pharmaceutical dosage form. Int J Pharm 564:359–368
94. Clark EA et al (2017) 3D printing of tablets using inkjet with UV photoinitiation. Int J Pharm 529(1–2):523–530
95. Alomari M et al (2018) Printing T3 and T4 oral drug combinations as a novel strategy for hypothyroidism. Int J Pharm 549(1–2):363–369
96. Scoutaris N et al (2018) Jet dispensing of multi-layered films for the co-delivery of three antihypertensive agents. Drug Deliv Transl Res 8(1):32–42
97. He Y et al (2020) A reactive prodrug ink formulation strategy for inkjet 3D printing of controlled release dosage forms and implants. Adv Ther 1900187
98. Cader HK (2019) Inkjet printing of solid oral dosage forms. PhD thesis
99. Buanz AB et al (2015) Ink-jet printing versus solvent casting to prepare oral films: effect on mechanical properties and physical stability. Int J Pharm 494(2):611–618
100. West TG, Bradbury TJ (2019) 3D printing: a case of ZipDose® technology—world's first 3D printing platform to obtain FDA approval for a pharmaceutical product. 3D and 4D printing in biomedical applications: process engineering and additive manufacturing, pp 53–79
101. Hsiao WK et al (2018) 3D printing of oral drugs: a new reality or hype? Taylor & Francis
102. Yu D-G et al (2009) A novel fast disintegrating tablet fabricated by three-dimensional printing. Drug Dev Ind Pharm 35(12):1530–1536
103. Wu BM et al (1996) Solid free-form fabrication of drug delivery devices. J Control Release 40(1–2):77–87
104. Sun Y, Soh S (2015) Printing tablets with fully customizable release profiles for personalized medicine. Adv Mater 27(47):7847–7853
105. Maroni A et al (2017) 3D printed multi-compartment capsular devices for two-pulse oral drug delivery. J Control Release 268:10–18
106. Rowe C et al (2000) Multimechanism oral dosage forms fabricated by three dimensional printing™. J Control Release 66(1):11–17
107. Yu D-G et al (2009) Novel drug delivery devices for providing linear release profiles fabricated by 3DP. Int J Pharm 370(1–2):160–166
108. Chang S-Y et al (2020) Binder-jet 3D printing of indomethacin-laden pharmaceutical dosage forms. J Pharm Sci 109(10):3054–3063
109. Wang C-C et al (2006) Development of near zero-order release dosage forms using three-dimensional printing (3-DP™) technology. Drug Dev Ind Pharm 32(3):367–376
110. Katstra W et al (2000) Oral dosage forms fabricated by three dimensional printing™. J Control Release 66(1):1–9
111. Lee K-J et al (2003) Evaluation of critical formulation factors in the development of a rapidly dispersing captopril oral dosage form. Drug Dev Ind Pharm 29(9):967–979
112. Shi K et al (2019) Drop-on-powder 3D printing of tablets with an anti-cancer drug, 5-fluorouracil. Pharmaceutics 11(4):150
113. Yu DG et al (2007) Tablets with material gradients fabricated by three-dimensional printing. J Pharm Sci 96(9):2446–2456
114. Yoo J et al (2014) Three-dimensional printing system and equipment assembly. Google Patents
115. Chianrabutra S, Mellor B, Yang S (2014) A dry powder material delivery device for multiple material additive manufacturing. University of Southampton, UK
116. Chivel Y (2016) New approach to multi-material processing in selective laser melting. Phys Procedia 83:891–898
117. Wei C et al (2020) Hybrid ultrasonic and mini-motor vibration-induced irregularly shaped powder delivery for multiple materials additive manufacturing. Addit Manuf 101138

118. Schmid M, Amado A, Wegener K (2015) Polymer powders for selective laser sintering (SLS). In: AIP conference proceedings, AIP Publishing LLC
119. Leong K et al (2001) Fabrication of porous polymeric matrix drug delivery devices using the selective laser sintering technique. Proc Inst Mech Eng H: J Eng Med 215(2):191–192
120. Low K et al (2001) Characterization of SLS parts for drug delivery devices. Rapid Prototyping J
121. Cheah C et al (2002) Characterization of microfeatures in selective laser sintered drug delivery devices. Proc Inst Mech Eng [H] 216(6):369–383
122. Leong K, Chua C, Gui W (2006) Building porous biopolymeric microstructures for controlled drug delivery devices using selective laser sintering. Int J Adv Manuf Technol 31(5–6):483–489
123. Fina F et al (2018) 3D printing of drug-loaded gyroid lattices using selective laser sintering. Int J Pharm 547(1–2):44–52
124. Awad A et al (2019) 3D printed pellets (miniprintlets): a novel, multi-drug, controlled release platform technology. Pharmaceutics 11(4):148
125. Leong K et al (2007) Characterization of a poly-ε-caprolactone polymeric drug delivery device built by selective laser sintering. Bio-Med Mater Eng 17(3):147–157
126. Salmoria G et al (2012) Development of functionally-graded reservoir of PCL/PG by selective laser sintering for drug delivery devices: this paper presents a selective laser sintering-fabricated drug delivery system that contains graded progesterone content. Virtual Phys Prototyping 7(2):107–115
127. Salmoria G et al (2016) PCL/ibuprofen implants fabricated by selective laser sintering for orbital repair. Procedia CIRP 49:188–192
128. Salmoria G et al (2017) 3D printing of PCL/Fluorouracil tablets by selective laser sintering: properties of implantable drug delivery for cartilage cancer treatment. Drugs 4:6
129. Salmoria G, Klauss P, Kanis L (2017) Laser printing of PCL/Progesterone tablets for drug delivery applications in hormone cancer therapy. Lasers Manuf Mater Process 4(3):108–120
130. Fina F et al (2017) Selective laser sintering (SLS) 3D printing of medicines. Int J Pharm 529(1–2):285–293
131. Fina F et al (2018) Fabricating 3D printed orally disintegrating printlets using selective laser sintering. Int J Pharm 541(1–2):101–107
132. Martinez PR et al (2018) Influence of geometry on the drug release profiles of stereolithographic (SLA) 3D-printed tablets. AAPS PharmSciTech 19(8):3355–3361
133. Wang J et al (2016) Stereolithographic (SLA) 3D printing of oral modified-release dosage forms. Int J Pharm 503(1–2):207–212
134. Martinez PR et al (2017) Fabrication of drug-loaded hydrogels with stereolithographic 3D printing. Int J Pharm 532(1):313–317
135. Karakurt I et al (2020) Stereolithography (SLA) 3D printing of ascorbic acid loaded hydrogels: a controlled release study. Int J Pharm 119428
136. Kadry H et al (2019) Digital light processing (DLP) 3D-printing technology and photoreactive polymers in fabrication of modified-release tablets. Eur J Pharm Sci 135:60–67
137. Xu X et al (2020) Stereolithography (SLA) 3D printing of an antihypertensive polyprintlet: case study of an unexpected photopolymer-drug reaction. Addit Manuf 33:101071

Chapter 5
Additive Manufacturing of Metal Implants and Surgical Plates

Di Wang, Yongqiang Yang, and Changjun Han

5.1 Background

The demands for metal implants and surgical plates are presently motivated by health problems and medical needs. Metal implants have been widely used in the human body throughout tissue repair and replacement. Surgical plates can help surgeons to better orient during the accurate operation. Metal implants and surgical plates have been conventionally produced by investment casting or forging. However, the increasing requirements of patients and surgeons for implant customization are hindered by the limitation of design complexity and batch production of traditional manufacturing methods of metal implants.

Additive manufacturing (AM) is an emerging technology to bring about revolutionary innovation to manufacturing industries by providing great flexibility for the design and fabrication of complex and/or customized parts through a layer-by-layer strategy [1–3]. Compared with traditional manufacturing methods, AM can provide a promising alternative for the fabrication of personalized implants with designed geometries. Aided by computed tomography (CT) or magnetic resonance imaging (MRI) scans, reverse engineer technology can quickly reconstruct three-dimensional (3D) models of the human body. These models can then be processed by AM into parts for disease analysis and surgical plan customization.

Titanium and its alloys, 316L stainless steel and Co–Cr alloy, known to have excellent biocompatibility, have been widely used in various AM metal implants

D. Wang (✉) · Y. Yang · C. Han
School of Mechanical and Automotive Engineering, South China University of Technology, Guangzhou 510641, China
e-mail: mewdlaser@scut.edu.cn

Y. Yang
e-mail: meyqyang@scut.edu.cn

C. Han
e-mail: cjhan@scut.edu.cn

K. Zhou (ed.), *Additive Manufacturing*,
https://doi.org/10.1007/978-3-031-04721-3_5

[4]. Moreover, metal matrix composites produced by AM have great potential for orthopedic applications due to their special chemical and biomedical properties. In addition, it is important for the metal implant to mimic mechanical characteristics of tissue to avoid many undesirable effects, caused by dissimilar mechanical properties between the implant and the tissue. AM can produce porous implant with compatible mechanical properties to actual tissue and improved capability in bone ingrowth to extend the life of the prosthesis [5]. In terms of intraoperative navigation, AM has been demonstrated to be conducive to accurate surgery and reducing tissue damage [6].

Representative AM processes for metal implants and guide plates include selective laser melting (SLM) and electron beam melting (EBM), etc. SLM and EBM processes use a high-energy beam (laser or electron beam) to selectively melt powder layers. SLM and EBM have become the mainstream manufacturing technologies for implants and guide plates due to high printing precision and excellent performance of manufactured parts. In general, AM of metal implants and guide plates can realize the desired personalized shape and structure to match human bones, the mechanical properties to avoid stress shielding, and superior biological properties.

This chapter introduces metal materials for additively manufactured implants, AM of metal implants, and AM of surgical guide plates. After a brief introduction of background, the performances of common metals materials in AM (Fe-based alloys, titanium and its alloys, Cobalt-based alloys, tantalum, magnesium and its alloys, and zinc and its alloys), will be introduced and compared. These performances include relative density, mechanical properties, chemical and biomedical properties, etc. The potential applications of materials with regards to AM implants will also be mentioned. Then, application cases related to AM of metal implants and surgical guide plates were presented, including the design, fabrication and post-processing of implants and plates. Challenges and future perspectives on AM for metal implants and surgical guide plates will be given in the end.

5.2 Metal Materials for Additively Manufactured Implants and Surgical Plates

Metal materials used for implants should be biocompatible and have high mechanical strength, low elastic modulus similar to bone, excellent corrosion and wear resistance, and low costs [4]. Another important factor is the possibility of obtaining porous structures because the structures affect the mechanical strength and the biological properties of the new tissues. Meanwhile, the rough inner and outer surface of the porous structure is suitable for the growth of bone tissues, which can accelerate the process of osseointegration.

There are many types of metallic materials that can be used for fabricating metal implants. In the International Standards for Organization standards (ISO5832) under

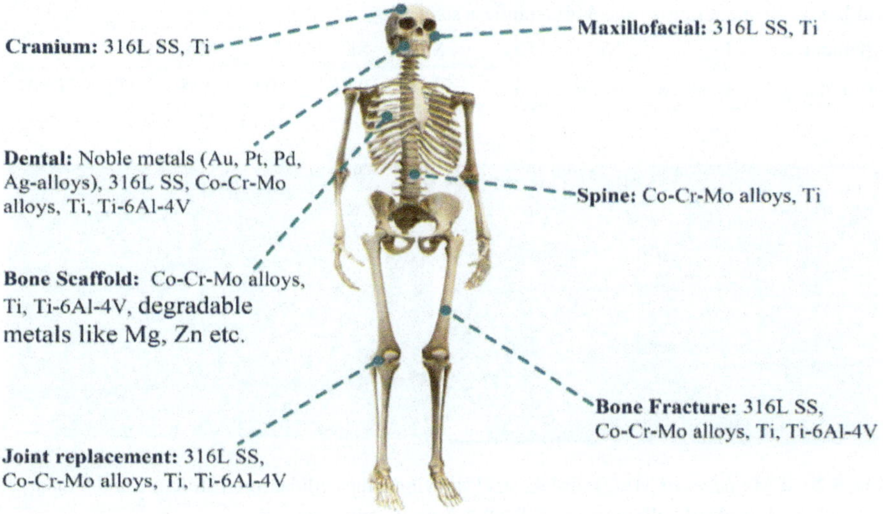

Fig. 5.1 The clinical application of common metal implants. Reprinted with permission from a work by Bai et al. [7]

the general title "Implant for surgery–metallic materials", 12 types of metallic materials can be divided into three categories: titanium-based, cobalt-based and steel-based alloys. The common metal implant materials include 316L stainless steel, Ti6Al4V and CoCrMo alloys. The clinical applications of common metal implants are shown in Fig. 5.1 [7]. In this section, three types of common metal materials including their microstructure and properties under different AM techniques are introduced. At the same time, the other three types of metal implant materials are briefly introduced.

5.2.1 Fe-Based Alloys

5.2.1.1 316L Stainless Steel

Stainless steel is an Fe-based alloy with a percentage of nickel and chromium ranging between 11 and 30 wt%. The most commonly used stainless steel alloy in the human body as an implant is the 316L, due to its good biocompatibility and corrosion resistance. Powder bed fusion (PBF) of 316L stainless steel powder is ideal for personalized implants. The composition of the powder meets the requirements of ASTM F138-13a [8]. The specific components of the powder are shown in Table 5.1 [8]. SLM-printed 316L stainless steel in SLM is mainly introduced.

The manufacturing process of metal parts via SLM undergoes rapid melting and cooling with cooling speeds of up to 1×10^3 °C/s, the crystal nucleation rate is greatly

Table 5.1 Element ratio of the 316L stainless steel alloy

Element	Fe	Ni	Cr	Mo	Si	C	Mn	P	S
Ratio (wt%)	Balance	12.06	17.53	2.16	0.86	0.03	0.38	0.02	0.01

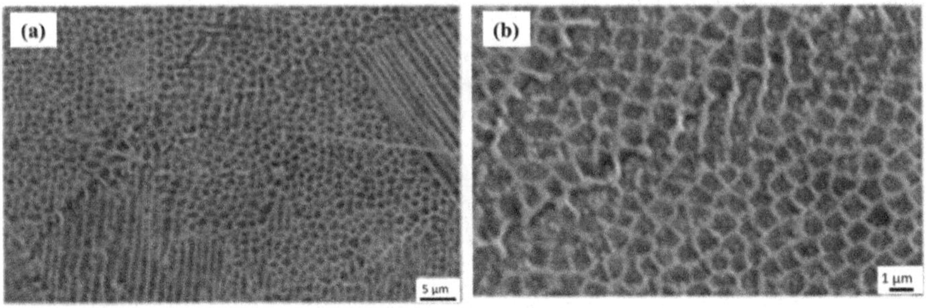

Fig. 5.2 SEM images of 316L stainless steel showing **a** the cellular microstructure and **b** enlarged view of (**a**). Reprinted with permission from a work by Sun et al. [10]

increased and grain size is suppressed at the same time. Therefore, microstructure mainly consists of microcellular crystals and columnar crystals of 1–10 μm [9]. Sun et al. [10] investigated the microstructure of 316L stainless steel and observed the typical cellular sub-grain microstructures from the scanning electron microscopy (SEM) images shown in Fig. 5.2. Epitaxial growth was evident as grains can grow across melt pool boundaries. It was found that there was no obvious difference in the cellular size among the SLM samples printed with power energy density from 98.70 to 108.57 J/mm^3. The average cell size was measured to be ~1 ± 0.5 μm.

Relative density is the ratio of the actual density of the part to the theoretical density of the material. The relative density is an essential indicator for the quality of the part fabricated by AM. Tolosa et al. [11] studied the mechanical properties of austenitic stainless steel 316L manufactured by SLM and achieved a relative density of 99.90%. Laleh et al. [12] produced a near completely dense 316L stainless steel with a relative density higher than 99.5% via SLM and evaluated its erosion-corrosion behavior in a jet impingement system.

The SLM process of 316L stainless steel has been relatively mature. Compared with casting and forging, the yield strength and tensile strength of the printed 316L components are significantly improved, but the ductility is reduced. It is mainly attributed to the formation of a fine cellular dendritic structure under rapid solidification during the SLM process. This results in a higher strength of the printed 316L parts than those of the cast and forged 316L stainless steel parts that experience lower cooling rates. A comparison of the properties of 316L stainless steel parts manufactured by SLM, casting and forging is shown in Table 5.2.

Besides, Čapek et al. [15] fabricated a highly porous (87 vol.%) 316L scaffold via SLM, whose mechanical properties were similar to those of trabecular bone, and the cytocompatibility of the printed 316L scaffold was improved. Highly porous 316L

Table 5.2 A comparison of the properties of 316L stainless steel parts manufactured by SLM [13], casting [9] and forging [14]

Property	SLM	Casting	Forging (at 330 °C)
Yield strength (MPa)	480	≥175	150
Tensile strength (MPa)	600	≥480	405
Elongation (%)	30	40	62

Table 5.3 Mechanical properties of porous SLM 316L and nonporous SLM 316L [15]

Property	Porous SLM 316L	Nonporous SLM 316L
Tensile modulus of elasticity (GPa)	0.12 ± 0.02	193 ± 3
Tensile yield strength (MPa)	3.46 ± 0.25	567 ± 15
Ultimate tensile strength (MPa)	14.55 ± 1.56	635 ± 23
Compressive modulus of elasticity (GPa)	0.15 ± 0.03	187 ± 5
Compressive yield strength (MPa)	3.01 ± 0.13	497 ± 24
Flexural modulus of elasticity (GPa)	0.20 ± 0.05	–
Flexural yield strength (MPa)	3.82 ± 0.42	–

scaffolds prepared by SLM are suitable for repairing trabecular bone defects. The mechanical properties of porous SLM 316L and nonporous SLM 316L can be seen in Table 5.3.

Materials used in metal implants must meet specific requirements for chemical and biomedical properties. Biological properties can be evaluated in corrosiveness, biocompatibility, in vitro degradability, protein adsorption capacity, etc. Kong et al. [16] systematically compared anti-corrosion properties and biological compatibility of 316L SS produced by traditional technique and SLM under different laser beam power. The results showed that the SLM-printed 316L samples have higher corrosion resistance and biocompatibility under high laser power. The cells' concentration and proliferation also increased with increasing laser beam power.

316L stainless steel is rarely used for long-term load-bearing applications due to its poor fatigue performance and relatively low wear and corrosion resistance. Therefore, metallic biomaterials have gradually been replaced by cobalt alloys and titanium alloys with better biocompatibility, fatigue, wear, and corrosion properties. Stainless steel is still employed for internal fixation implants in fixation systems due to its low cost than those of other alloys [17].

5.2.1.2 Biodegradable Fe-Based Alloys

Biodegradable metallic materials are often used in some medical implants, such as bone scaffolds-aided bone repair, due to their high strength and stiffness. It will be absorbed and degraded gradually in the human body, avoiding the risk of removing the implant again.

Fe-based alloys, such as pure Fe and Fe–Mn, are some of the most studied types of biodegradable metallic materials. Fe-based alloys do not release hydrogen gas during biodegradation [18], and possess various advantages including excellent mechanical properties, low costs, and good biocompatibility. AM porous structures with the increased surface area can effectively improve the degradation rates of their materials that degrade slowly. Li et al. [19] demonstrated for the first time how SLM could be used to fabricate topologically ordered porous iron. However, to mimic the natural graded or gradient structures of human bones, AM porous biomaterials must call for graded designs. Li et al. [20] reported the direct printing of a functionally graded biodegradable porous metal (i.e., pure iron) using SLM. Moreover, Fe–Mn alloy was used to produce bone scaffolds by SLM [21], and Mn was added to control the high degradation of Fe.

5.2.2 Titanium and Its Alloys

Titanium and its alloys have good mechanical properties, excellent corrosion resistance, and good biocompatibility. Although they are more expensive than stainless steel and have poorer wear resistance than cobalt–chromium alloys, they adapt well to bones when implanted into the human body [22]. Therefore, they are the preferred metal implant materials.

5.2.2.1 Pure Titanium

Pure titanium (Ti > 99%) is one of the most commonly used titanium materials for implant applications. Pure titanium mainly possesses the α phase with a hexagonal close-packed (hcp) structure. Pure titanium has excellent biocompatibility because a stable oxide layer is formed spontaneously on its surface [23]. However, the wear resistance of pure titanium is poor. It is essential to improve the wear properties of pure titanium because osteolysis and aseptic loosening may occur due to poor wear resistance [24].

Pure titanium for biomedical applications can be manufactured by various AM techniques, such as SLM [25, 26], EBM [27], laser engineered net shaping (LENS, also known as laser-based direct energy deposition) [28], etc. Laser processing parameters can influence the resulting microstructure of SLM-produced pure titanium samples, changing from a lamellar α phase to an acicular martensitic α' phase (Fig. 5.3). Moreover, the yield strength, ultimate tensile strength, compressive prop-

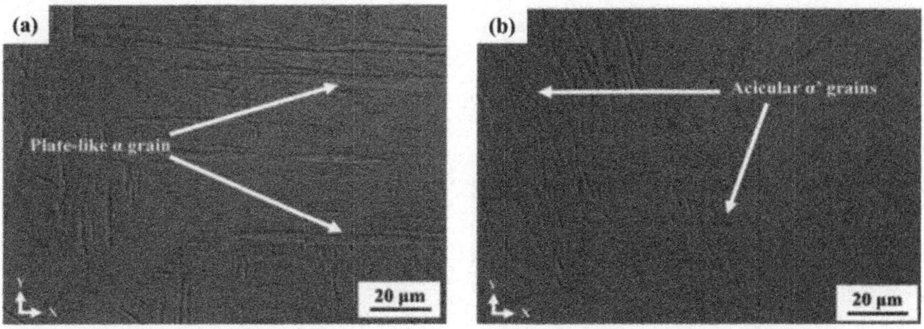

Fig. 5.3 Cross-sectional (Y–X) optical microstructure of the SLM-manufactured pure titanium samples: **a** the plate-like α phase and **b** the acicular α′ phase. Reprinted with permission from a work by Attar et al. [29]

erties and microhardness of SLM-produced pure titanium are 555 MPa, 757 MPa, 1136 MPa and 261 HV, respectively, which are superior to the corresponding properties for typically processed and deformed pure titanium samples. Such significant improvements in mechanical properties can be attributed to the grain refinement coupled with the acicular α′ phase resulting from the high cooling rate occurring in the SLM process [29].

The fabrication of porous titanium scaffolds is still a research hotspot. Han et al. [30] prepared pure titanium Schwartz diamond functionally graded porous scaffolds (FGPSs) with a wide range of graded volume fraction (from 5% to 20%) by SLM, revealed the effect of graded volume fraction on the mechanical property of FGPS, and developed a mathematical model between graded porosity and property of FGPS, which provides a way to design and manufacture superior graded scaffolds for bone implant applications.

5.2.2.2 Ti6Al4V

The most common titanium alloy for AM in orthopedic implants is the Ti6Al4V alloy, which can be used in various implants, including joints, spine, bone traumas [31, 32], etc. Ti6Al4V is an α–β titanium alloy, which exhibits both the high strength of the α phase and the low elastic modulus of the β phase, in which aluminum (Al) and vanadium (V) exist as an α phase stabilizer and a β phase stabilizer, respectively. The three most developed techniques to fabricate Ti6Al4V alloy are SLM, EBM and direct energy deposition (DED). The composition of the Ti6Al4V alloy meets the requirements of ASTM F136-13 [33]. The specific components of the powders are shown in Table 5.4.

The microstructure of Ti6Al4V parts varies with the applied AM techniques. Rafi et al. [34] investigated the microstructure of Ti6Al4V prepared by SLM and EBM. Figure 5.4a shows the optical microstructure of SLM-produced Ti6Al4V alloy. The SLM-printed Ti6Al4V alloy exhibited a complete martensitic (α′) microstructure.

Table 5.4 Element ratios of the Ti6Al4V alloy

Element	Ti	Al	V	Fe	Si	C	O	N	H
Ratio (%)	Balance	6.040	3.940	0.020	0.012	0.013	0.108	0.013	0.001

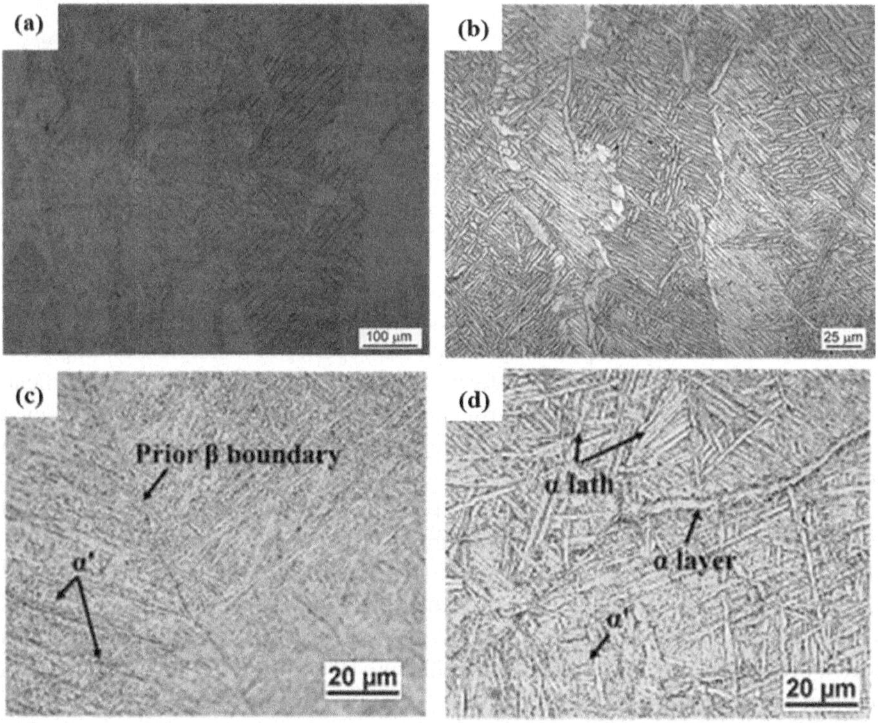

Fig. 5.4 Optical micrographs showing microstructure in Ti6Al4V produced: **a** SLM and EBM **b**. Reprinted with permission from a work by Rafi et al. [34] **c** LENS in low power condition and **d** LENS in high power condition. Reprinted with permission from a work by Zhai et al. [35]

The optical micrographs of EBM-produced Ti6Al4V are shown in Fig. 5.4b. The microstructure is mainly composed of α-phase and a small amount of β within the prior β columnar grains oriented along the build direction. The α-phase possesses a lamellar morphology with β phase surrounding the α lamellae boundary. Zhai et al. [35] studied the microstructure of Ti6Al4V alloy prepared by LENS. Ti6Al4V printed with low power shows martensitic microstructure with acicular α′ phase (about 0.6 μm thick) at high magnification in Fig. 5.4c. Comparatively, Ti6Al4V printed with high power shows regions of parallel α lath (about 1.7 μm thick), and regions of acicular α′ phase, leading to a mixed microstructure of martensitic α′ and α + β lamellae in Fig. 5.4d.

In these cases, prior β columnar grain boundaries are clearly visible. The difference in microstructure is induced by the differences in cooling rate when the β transforms to α as it cools through the transus temperature. Since the SLM has cooling rates on the order of 10^6 K/s, this results in the transformation of α to α′. Different from SLM, during EBM the building chamber is held at 650–700 °C throughout the building process, which does not allow the transformation of α to α′. When fabrication is completed, the building chamber is slowly cooled to room temperature, which results in the formation of α platelets. For LENS, high power fabrication results in a slower cooling speed. Therefore, α layers are observed in the high-power condition, but not in low power case.

High relative density has been achieved in both the SLM- and EBM-printed Ti6Al4V alloy. Van Hooreweder et al. [36] printed Ti6Al4V with a relative density of 99.7% via SLM with a bi-directional scanning strategy and optimal process parameters. Ataee et al. [37] prepared Ti6Al4V scaffolds by EBM technology, of which the density was more than 99%.

The comparison of mechanical properties of Ti6Al4V alloy parts manufactured by common AM technologies is shown in Table 5.5. The tensile strength of EBM-produced Ti6Al4V is lower than that of SLM-produced Ti6Al4V. This is due to the difference in the microstructure of the parts produced by the two processes. A harder martensite phase was observed in SLM due to a high cooling rate, while a softer α phase due to a lower cooling rate was obtained from EBM. Different AM processes lead to various microstructure in the printed Ti6Al4V, which affects the mechanical properties significantly.

Titanium alloys can provide a favorable surface for osteoblast adhesion and osteogenic differentiation [39]. The biocompatibility of Ti6Al4V is excellent, especially when it directly contacts tissue or bone. Wang et al. [40] evaluated and compared the biocompatibility of Ti6Al4V fabricated by EBM and SLM, including cytocompatibility, blood compatibility, skin irritation and skin sensitivity. The results showed that both the EBM- and SLM-printed Ti6Al4V exhibited good cell biocompatibility and blood compatibility, and showed no dermal irritation and skin allergic reaction.

The Young's modulus of titanium and its alloy does not match that of natural bone, far exceeding the 15–30 GPa of natural bone [41]. The difference between titanium and bones in mechanical properties of titanium and its alloy cannot be successfully transferred to the adjacent bone tissue through the implant, resulting in the stress shielding phenomenon. Therefore, low-modulus titanium alloys (β-titanium alloys,

Table 5.5 A comparison of the properties of Ti6Al4V parts manufactured by different AM processes

Process	Yield tensile strength (MPa)	Ultimate tensile strength (MPa)	Elongation (%)	Source
SLM	-	1204	7.8	[38]
EBM	950	1020	14	Arcam
LENS	973	1077	11	Optomec

etc.) are considered to have better biomechanical advantages [42]. Moreover, AM of gradient porous Ti6Al4V structure is also a promising solution [43].

5.2.2.3 Titanium–Niobium Alloys

Titanium–niobium alloys have high biomechanical strength and low elastic modulus closer to human cortical bone. They have been widely used as implant materials for dentistry, orthopedic and artificial hip joints.

Chlebus et al. [44] studied the possibility of using Ti–6Al–7Nb for medical implants. The alloy has higher corrosion resistance and biocompatibility than Ti6Al4V because it replaces vanadium with niobium in its chemical composition. In terms of microstructure and mechanical properties of Ti–6Al–7Nb, Xu et al. [45] studied the changes of microstructure and mechanical properties of SLM-produced Ti–6Al–7Nb alloy before and after heat treatment, as shown in Table 5.6. The microstructure of the alloy without heat treatment was composed of columnar prior β grains with fine acicular α′ martensite. After being solidified at 1055 °C and aging at 540 °C, the columnar structure of the original β grain is completely transformed into the equiaxed structure via phase transformation and grain growth, and the α′ martensite is dissolved into β, the finely layered $\alpha + \beta$ structure is formed by air cooling. The resultant microstructure had reduced strength and hardness but increased ductility. The reduction in yield and ultimate tensile strength would be positive to minimize stress shielding of orthopedic implants. The improved elongation meets the requirements for biomedical applications.

In terms of biocompatibility, Rotaru et al. [46] tested the biocompatibility and evaluated the osseointegration of the Ti–6Al–7Nb alloy used in the manufacturing of personalized implants with SLM technology and compared the growth viability of osteoblastic-like cells on different Ti–6Al–7Nb alloy samples (plain, coated with hydroxyapatite or SiO_2–TiO_2) implanted into the cranial bone of Wistar rats. All results suggested that the Ti–6Al–7Nb alloy with or without infiltration was well biologically tolerated, and hydroxyapatite infiltration can enhance the osteoconductive property.

However, Al is a toxic ion and should not be released inside the human body. Thus, alloys such as Ti–27.5Nb [47] and Ti–42Nb [48], biomedical materials with excellent properties, are used as alternatives to Ti-6Al-7Nb. Zhao et al. [49] in-situ fabricated a Ti-25Nb alloy by SLM using different laser energy densities. The acicular-shaped

Table 5.6 A comparison of the properties of Ti–6Al–7Nb parts manufactured by SLM and SLM with heat treatment [45]

Method	Yield strength (MPa)	Tensile strength (MPa)	Elongation (%)
SLM	1082	1160	9.7
SLM with heat treatment	817	940	11.5

grains can be observed at low energy densities (70–80 J/mm^3), corresponding to α' phase (Fig. 5.5a, b). As the energy density increases to 90 and 100 J/mm^3, the martensite grains grow up to the coarsened lath shape (Fig. 5.5c, d). Furthermore, the cellular-shaped subgrains appear at the energy density of 110 J/mm^3 (Fig. 5.5e). Therefore, with an increase in the energy density, the microstructure of the Ti–25Nb alloy varies from acicular-shaped grains to coarsened lath-shaped grains and lath-shaped grains + cellular-shaped subgrains.

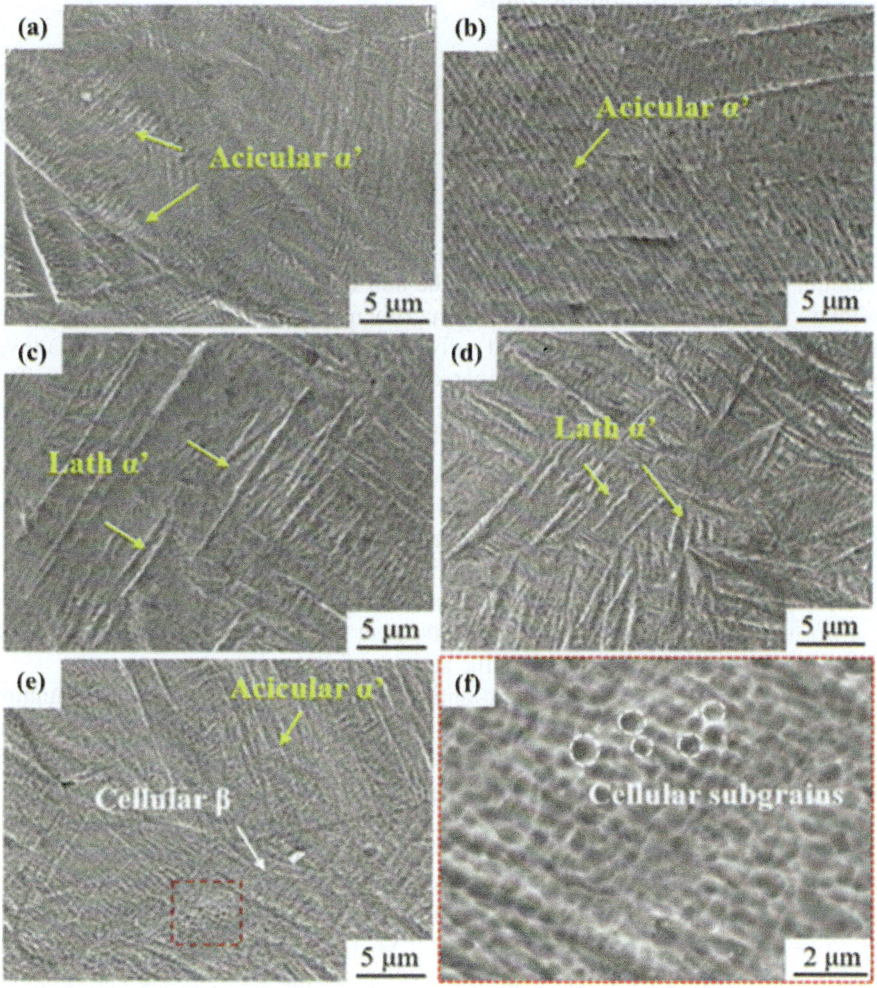

Fig. 5.5 Microstructures of the SLM-produced Ti–25Nb samples at different energy densities in the x–y plane observed by SEM: **a** 70 J/mm^3, **b** 80 J/mm^3, **c** 90 J/mm^3, **d** 100 J/mm^3, **e** and **f** 110 J/mm^3. Reprinted with permission from a work by Zhao et al. [49]

For porous scaffolds, Liang et al. [50] studied in-situ synthesized titanium–niobium scaffolds using SLM. The addition of Nb enhanced the formation of Nb_2O_5 and associated surface hydrophilicity. Moreover, Nb_2O_5 activated the anti-inflammatory macrophage and promoted osteogenic differentiation. Therefore, titanium–niobium porous scaffolds presented reinforced osseointegration and accelerated osteogenesis.

5.2.2.4 Titanium–Tantalum Alloys

For titanium alloys, tantalum is a promising additive for biomedical applications. The biocompatibility of tantalum, which is non-toxic and harmless to the human body, is better than titanium. In addition, tantalum, namely, a β phase stabilizer of titanium alloys, can reduce the elastic modulus of titanium alloys. Therefore, titanium-tantalum alloys are expected to be used in dental implants and orthopedic surgery [51].

From the perspective of medical application, the most significant advantage of titanium-tantalum alloys lies not only in the non-toxicity of alloy elements but also in their corrosion resistance. Corrosion experiments in environments simulating the human body (e.g., artificial acidified saliva) showed that titanium-tantalum alloys had higher corrosion resistance compared with currently used alloys such as Ti–6Al–7Nb, which can stabilize the titanium passivation layer, inhibit the dissolution of Ti and Ta ions, and provide a biological neutral layer in corrosive body fluids [52].

There is little literature on titanium-tantalum alloys produced by SLM. In 2019, Zhao et al. [53] fabricated titanium-tantalum alloys by SLM with different Ta contents (0–25 wt%). With the increase of Ta content, the SLM-produced titanium-tantalum alloys exhibited the microstructure evolution from the lath α grain to acicular $\alpha' +$ primary cellular β grains, accompanying with the gradual suppression of martensite transformation (Fig. 5.6). The β-stabilized effect of Ta promoted the formation of the β (Ti, Ta) solid solution phase in the alloy.

Sing et al. [54] fabricated titanium-tantalum alloy samples with 50 wt% tantalum by SLM, and the results showed that the alloy consists of the β phase. Moreover, the yield strength, the ultimate tensile strength, the elongation and Young's modulus of SLM-produced titanium-tantalum alloy were 882.77 MPa, 924.64 MPa, 11.72% and 75.77 GPa respectively, which can be a potential material for biomedical applications due to its high strength to modulus ratio compared to Ti6Al4V and pure titanium. In 2018, the group [55] reported the effect of laser process parameters on part quality. The results showed that fully dense titanium-tantalum parts with a relative density of $99.85 \pm 0.18\%$ can be obtained from SLM. In 2019, Soro et al. [56] successfully produced porous Ti–25Ta with SLM. The mechanical properties were suitable for bone replacement applications, showing significantly reduced elastic modulus, ranging from 14 to 36 GPa depending on the level of porosity.

In addition to the titanium alloys introduced above, the preparation of titanium matrix composites by SLM is also the focus of research. So far, the studies on the SLM of titanium matrix composites are mainly focused on Ti–TiC and Ti–TiB

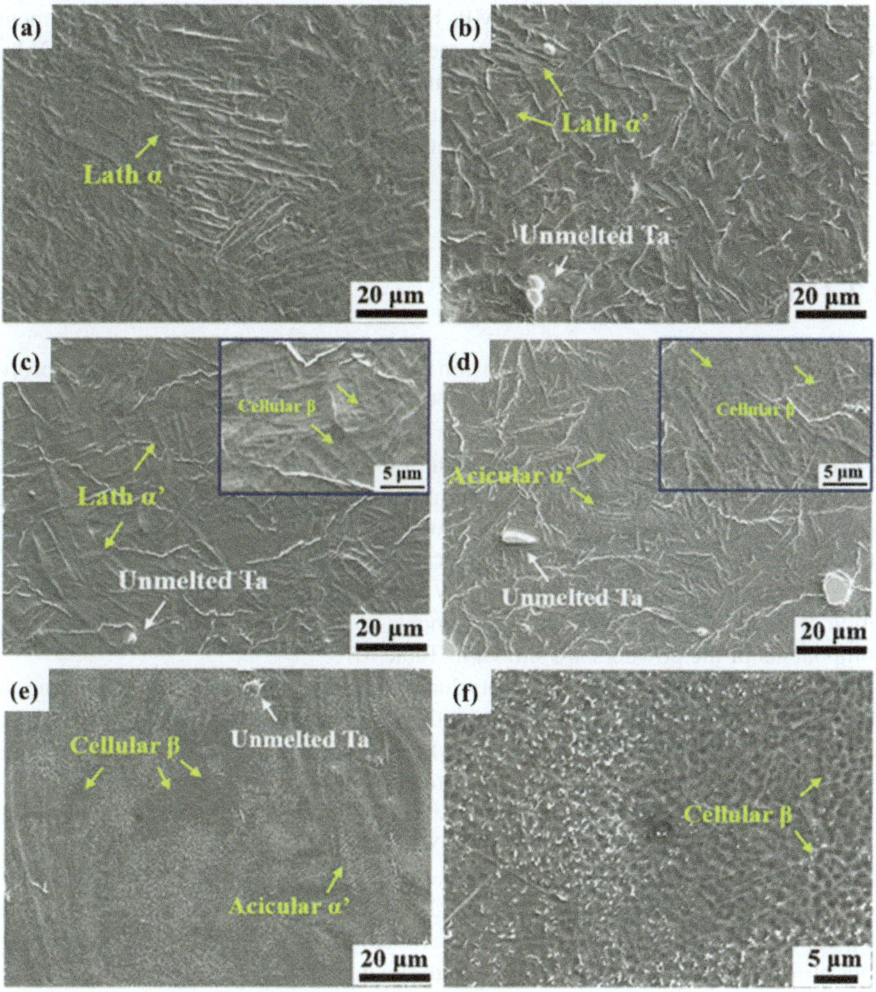

Fig. 5.6 Microstructure of SLM-produced titanium-tantalum alloys with different Ta contents. **a** Ti, **b** Ti–6Ta, **c** Ti–12Ta, **d** Ti–18Ta, **e** and **f** Ti–25Ta. Reprinted with permission from a work by Zhao et al. [53]

composites [57, 58]. Titanium/hydroxyapatite (Ti/HA) composite has the potential for orthopedic applications due to its favorable mechanical properties, excellent biocompatibility and bioactivity. However, there are few studies on the preparation of this material by SLM. Han et al. [59] in-situ prepared the pure titanium and nanoscale HA (Ti-nHA) composites by SLM for the first time. The results showed that relatively long lath-shaped grains of pure titanium evolved into short acicular-shaped and quasi-continuous circle-shaped grains with varying contents of nHA. The elastic modulus of the Ti-nHA composites was higher than that of pure titanium due to the effect of grain refinement. With the addition of 2% nHA, the ultimate tensile strength significantly reduced to 289 MPa but still meets the application requirement of bone implants. Moreover, they fabricated functionally graded materials (FGMs) of titanium/hydroxyapatite (Ti/HA) with quasi-continuous ratios by SLM for the first time [60]. Based on the results of these studies, it can be suggested that the SLM technique could be utilized to fabricate the Ti/HA biomedical material composites for bone implant applications.

5.2.3 Cobalt-Based Alloys

Medical Co–Cr alloys have good corrosion resistance and excellent mechanical properties. One of the typical alloys is the CoCrMo alloy, which has good mechanical properties, corrosion resistance and wear resistance. In the early 1930s, the cast CoCrMo alloy has been used in dental and biomedical fields [61]. Nowadays, the CoCrMo alloy is widely used in surgical implants, cardiovascular stents, and dental repair. Especially its wear resistance is much better than that of titanium alloys and stainless steel, making it more suitable for active joints [62].

CoCrMo alloy can be manufactured by SLM and BEM technology. The composition of metal powder meets the requirements of ASTM F75-12 [63]. The specific components of the powders are shown in Table 5.7.

The microstructure of the SLM-produced CoCrMo alloy is nearly the same as that of 316L stainless steel [16]. Zhang et al. [64] obtained the SEM images of the SLM-produced CoCrMo alloy. The size of melting pools is 30–70 μm, as shown in Fig. 5.7a. In order to analyze the microstructure of melting tracks, the high magnification SEM images were achieved, as shown in Fig. 5.7b. Two shapes of the grain are observed: columnar cell and cellular cell. Xiang et al. [65] presented the microstructure of the CoCrMo alloys produced by the EBM. Figure 5.7c, d show SEM images of the columnar grains from both X–Z and X–Y planes. Each columnar grain consists of parallel dendritic structures with short secondary arms. Fine γ-Co columnar grains consisting of dendrites were formed due to the dendritic segregation of Cr and Mo elements. Moreover, the formation of carbides was attributed to the segregation of carbon into interdendritic regions and grain boundaries. Carbides were observed to be formed at both columnar grain boundaries and interdendritic regions.

CoCrMo alloy parts manufactured by AM have a high density. Tan et al. [66] prepared the CoCrMo parts with a relative density higher than 99.2% by EBM. Ghani et al. [67] produced the full dense CoCrMo parts by SLM and DMLS with default process parameters, and the yield relative density for both Co–Cr–Mo parts produced by SLM and DMLS is 99.3%.

Table 5.7 Element ratio of the CoCrMo alloy

Element	Co	Cr	Mo	Si	Mn	Fe	N	C	Ni
Ratio (%)	Balance	29.40	6.00	0.80	0.75	0.26	0.19	0.15	0.09

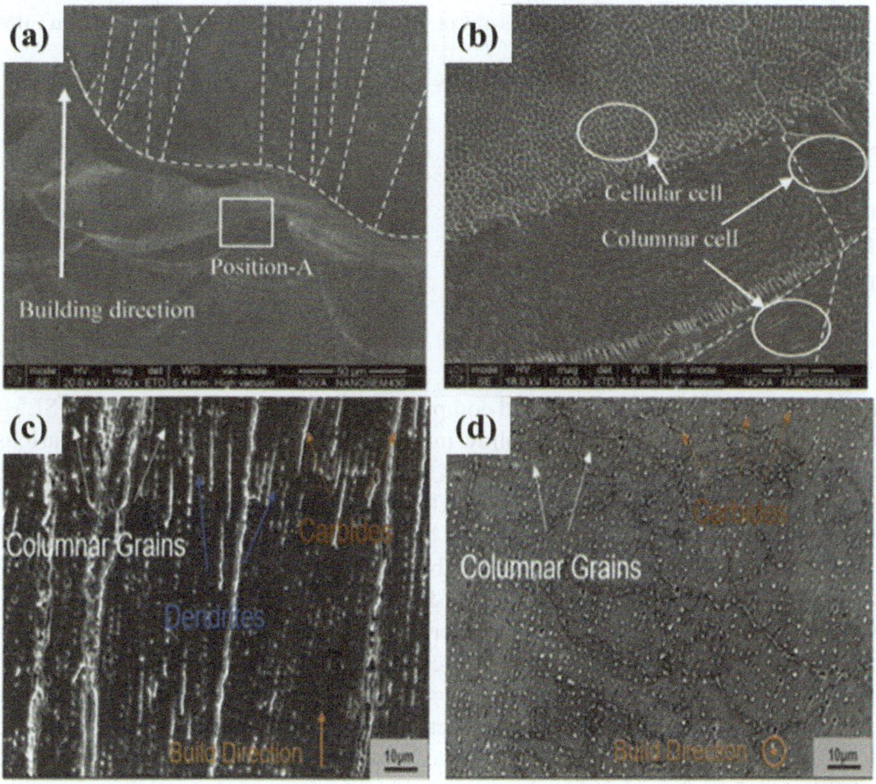

Fig. 5.7 SEM image of the SLM-printed CoCrMo alloy: **a** microstructure of the melting track, **b** the enlarged image of position A. Reprinted with permission from a work by Zhang et al. [64]. SEM images of the EBM-printed CoCrMo alloy showing columnar grains and carbides: **c** in the X–Z plane, **d** in the X–Y plane. Reprinted with permission from a work by Xiang et al. [65]

As shown in Table 5.8, compared with the ASTM F75 standard, the tensile strength, yield strength, and Rockwell hardness of CoCr alloys manufactured by SLM and EBM are higher. A fine and homogeneous structure is formed due to the rapid occurring melting and solidification processes. Besides, the crystalline grains are restricted from growing by the grains nearby. The fine sizes of the grains together with a large number of crystal boundaries hinder dislocation which lead to greater tensile and yield strength [68]. The high cooling rate in SLM generates finer grains. The structure of the SLM-produced parts consists mainly of the oversaturated Co-base phase and the carbide-dispersed Co-base phase [69]. This is especially true for cases, in which the carbides are substantially refined and dispersed in the matrix structure, thereby resulting in greater tensile strength, yield strength, and hardness.

At present, there are a few studies on the biocompatibility of CoCrMo alloys made by SLM, especially the hemolysis rate under different corrosion conditions. Zhang

Table 5.8 A comparison of the properties of CoCrMo parts manufactured by SLM [69] and EBM [68] with ASTM

Material category	Yield strength (MPa)	Tensile strength (MPa)	Elongation (%)	Rockwell hardness (HRC)
SLM	825	1070	7.2	38
SLM with heat treatment	660	1210	14.3	36
EBM	730	–	–	47
EBM with HIP	560	960	–	34
ASTM F75-12	450	655	>8	25

et al. [70] determined both the blood compatibility and the ion release rates of SLM-manufactured and cast CoCrMo alloy implants following submersion in various corrosive solutions, and investigated the surface structure and morphology of the samples before and after the experiment. The results showed that the surface of SLM-manufactured parts was more uniform and less porous than that of cast parts, resulting in a lower ion release rate and more favorable blood compatibility. Hedberg et al. [71] studied the corrosion sensitivity of CoCrMo dental alloys manufactured by SLM. It shows that rapid cooling and strong temperature gradients during laser melting result in the formation of a fine cellular structure with cell boundaries enriched in Mo (Co depleted), and suppression of carbide precipitation and formation of a martensitic ε (hcp) phase at the surface. These properties were shown to decrease the SLM alloys' corrosion and metal release susceptibility compared with the cast alloy. Porous bone substitutes take more risks of postoperative infection due to their much larger surface area. Han et al. [72] prepared for gentamicin-loaded silk fibroin coatings on AM porous CoCrMo bone substitutes via electrophoretic deposition technique, which improved the biological functionality of porous structures made of medical metals.

5.2.4 Others

5.2.4.1 Tantalum

Tantalum is a biocompatible material with unique mechanical properties. Porous tantalum is the most common type of tantalum in implants application. By designing the porosity, it can achieve the elastic modulus similar to bones, reducing stress shielding [73]. Compared with other medical metal materials, porous tantalum has the typical characteristics of high volumetric porosity, low elastic modulus and high surface friction coefficient, which is an ideal bone substitute material [74].

Porous tantalum implants have been very positive in terms of fabrication using SLM. Wauthle et al. [75] manufactured a highly open porous (80%) pure Ta implant by SLM. Guo et al. [76] used SLM to manufacture porous Ta scaffolds and found that the porous Ta scaffold not only had a regular pore shape and connectivity but also had controllable elastic modulus and compressive strength. Moreover, the osteogenesis and osseointegration results in vitro and in vivo were improved compared with the porous Ti6Al4V scaffold manufactured using the same technology. These results demonstrate that the porous Ta scaffold manufactured by SLM is potentially useful for orthopedic clinical application.

The pore size of porous tantalum is 400–600 μm, the porosity is 70–80%, and the elastic modulus is about 3 GPa, which is between human cancellous bone (0.1–1.5 GPa) and cortical bone (12–18 GPa) [77]. Thijs et al. [78] experimented using pure tantalum for fabrication by SLM and achieved a relative density of 99.6%. The yield strength of SLM Ta is also reported to be higher than those obtained via conventional methods.

5.2.4.2 Magnesium and Its Alloys

Magnesium and its alloys are one of the common degradable metals. Magnesium has mechanical properties matching natural bone. Besides, magnesium and its alloys also have bone conductivity, promote bone growth and cell adhesion, and will not induce human inflammatory responses [73]. Therefore, they are commonly used to fabricate cardiovascular stents and bone screws [79, 80].

A disadvantage of magnesium is its rapid degradation. Nowadays, researchers worldwide have made positive progress in the biodegradability of magnesium alloys produced by SLM. Li et al. [81] successfully fabricated WE43 (wt% yttrium and 3 wt% rare earth elements) scaffolds by SLM. The constructed scaffolds showed enhanced biodegradability, and the biodegradation rate of 0.17 mL/cm^2·day over four weeks was lower than the values reported for most cast or extruded WE43 magnesium alloy. Xu et al. [82] improved the biodegradation resistance of the ZK30-Cu alloy by grain refinement resulting from rapid solidification in the SLM process. Moreover, the alloy also showed a good antibacterial ability due to the addition of copper.

The mechanical properties of magnesium samples produced by SLM, such as compressive strength, tensile strength, yield strength and elastic modulus, are better than those produced by traditional technologies. Therefore, it is conducive to applying magnesium in the field of plastic surgery [83]. Ng et al. [84] studied the effects of different laser processing parameters on magnesium's microstructure and mechanical properties during SLM. The results show that the average hardness of magnesium decreases from 0.95 to 0.59 GPa with the increase in laser energy density, while the corresponding elastic modulus is between 27 and 33 GPa.

5.2.4.3 Zinc and Its Alloys

Zinc is an indispensable element in human nutrition. Zinc can be used in orthopedic implants with good biocompatibility [85]. Compared with magnesium, the degradation rate of zinc is relatively moderate [86], thus zinc and its alloys constitute a new generation of biodegradable metal materials for biomedical implants.

A huge problem in the preparation of Zn by SLM is that zinc's melting point and boiling point are relatively low, and the interval between them is so small that the unstable and low-density parts may be obtained [87]. To achieve the high density of Zn prepared by SLM, Wen et al. [88] developed a reasonable process window to obtain the zinc metal with a high density of 99.90%. Besides, Li et al. [89] applied SLM to fabricate porous zinc with a topologically ordered diamond structure. An integrative study was conducted on the static and dynamic biodegradation behavior (in vitro, up to four weeks), the evolution of mechanical properties with increasing immersion time, electrochemical performance, and biocompatibility of the AM porous zinc. The results show that their mechanical properties surprisingly increased and indirect cytotoxicity revealed good cellular activity up to 72 h.

It is difficult for pure zinc to meet the clinical mechanical requirements of the weight-bearing bone. Still, it can be combined with other elements to form a zinc-based alloy to ensure mechanical properties. There is little research on the mechanical properties of SLM-produced porous scaffolds of zinc alloys. Qin et al. [90] fabricated the porous scaffolds of Zn-xWE43 (x = 0, 2, 5 and 8 wt%) by SLM with the mixed powder of pure zinc and the WE43 Mg alloy. The molding quality, microstructure and mechanical properties were studied in this study. It shows that the maximum tensile strength of Zn-5WE43 is 335.4 MPa, but the elongation is only 1%. The compressive strength and Young's modulus of Zn-5WE43 porous scaffolds are 73.2 MPa and 2480 MPa respectively, while pure Zn porous scaffolds are 22.9 MPa and 950 MPa respectively, which shows that AM Zn–Mg alloy has a broad prospect in biodegradable applications.

The comparison of the properties and applications of different biomedical metals is illustrated in Table 5.9. With the development of materials science, there are more and more composite metal materials. For the orthopedic implants fabricated by AM, we can completely control the materials according to the specific requirements of the surgical conditions of the patient, no longer affected by the limitation of traditional manufacturing processes. Selecting appropriate metal materials according to bone repair area can integrate the mechanical properties with the biological properties of implants, which is the development trend of implant materials [7].

5.3 Additive Manufacturing of Metal Implants

Recently, the rapid development of AM technology has triggered the transformation of manufacturing technology and promoted advanced design concepts. AM technology has been widely used in the medical field because of its flexible design and

Table 5.9 The properties of different biomedical metals

Material	Elastic modulus (GPa)	Yield strength (MPa)	Tensile strength (MPa)	Main application	Source
316L stainless steel	153.8	445.7	629.8	Joint replacement Bone defect repair Dental implant	[91]
Ti6Al4V alloy	109.2	1110	1267	Joint replacement Bone defect repair Bone scaffold	[92]
CoCrMo alloy	210	960	1400	Joint replacement Dental implant Bone scaffold	EOS
Tantalum metal	–	450	739	Joint replacement	[93]
Magnesium metal	–	254	296	Bone scaffold	[94]
Zinc metal	23	114	134	Bone scaffold	[85]

processing methods and extremely short design and processing cycle. Moreover, AM technology can also realize the personalized customization of artificial implants and surgical auxiliary tools, which significantly improves the surgical technology and therapeutic effect.

AM metal implants can be divided into standardized implants and personalized implants. The application of AM technology in orthopedics is constantly developing, including joint implants, trauma implants, spinal implants, dental implants, etc.

5.3.1 *Articular Implants*

AM technology plays an important role in joint implant systems. Bone tissue (such as cartilage) is not renewable in adulthood, which usually is healed and repaired by implants after injury. However, the health condition of each patient depends on various people. Standardized production of implants is prone to a poor matching degree between implants and the human body, and higher requirements for surgeons. The personalized implant was defined in the late 1990s. With the development of advanced manufacturing technology, advanced design concepts, and medical technology, AM technology can manufacture surgical models of patients to assist doctors in customizing surgical plans and meeting the individual needs of different patients to realize the manufacture of personalized implants.

5.3.1.1 Knee Joint

The human knee joint has the most complex structure and function of human joints. Total knee arthroplasty is the most effective treatment for osteoarthritis and knee deformity. However, the movement characteristics of the knee joint are always different among patients, and every patient has different requirements for the knee joint. With the expansion of the application field of AM technology, the design of personalized implants based on AM and computer-aided design (CAD) technology provides opportunities for the realization of personalized knee implants.

The fabrication of knee implants via reverse engineering and AM technology has been explored [95]. The design of the knee implant includes two steps: reverse design and forward design. The reverse design process mainly consists of reconstructing the lower limb vertical line, selecting personalized osteotomy position, and extracting anatomical position features. The forward design process mainly includes the personalized design of prosthesis contour and joint surface.

Since the contoured osteotomy surface of the fixed prosthesis cannot meet the patients' individual needs, the match of the osteotomy surface of the commercial knee implant is generally poor. While forwarding design, computer simulation of bone-cutting is used to extract the contour shape of each cutting plane on the distal femur and tibial platform, and then to perform parameter fitting on the contour of the patient's osteotomy surface, which can improve the matching of the contour of the implant, increase the coverage rate while reducing the degree of over-coverage of the implant body, and ensure the fluency of implants outline in appearance. Figure 5.8 shows the optimization of the osteotomy profile of the knee's personalized femoral and tibial implant sections.

As shown in Fig. 5.9a, the total knee prosthesis consists of the personalized knee prosthesis' two components, femoral prosthesis and tibial prosthesis fabricated by AM technology, and the tibial gasket fabricated by light-curing non-metal AM technology. As shown in Fig. 5.9b, the knee implant had a good reconstruction effect on the lower limb line and flexion motion by combining the knee implant experiment conducted on the cadaver with X-ray photos.

AK MEDICAL is the first company in China to apply AM technology for bone joint and spine replacement implants and obtain a CFDA registration license for metal implants. At present, it has studied various knee implant products according to user requirements, including A3 knee prosthesis, ACCK knee prosthesis, and A3GT knee prosthesis.

For A3 knee prosthesis, there are 12 models of bone snatching and 13 models of the shin bone platform, better matching the patient's anatomical structure, close to the patient's tailor-made. It is suitable for knee resurfacing surgery. ACCK knee prosthesis can partially limit the internal and external flipping and rotational activities of the collarbone by increasing the height and widening of the tibia gasket, thus partially replacing the lateral ligament function and improving joint matching. The ACCK knee prosthesis system provides a variety of different types of pads for the far end of the gland, the posterior relationship of the collarbone and the shin bone

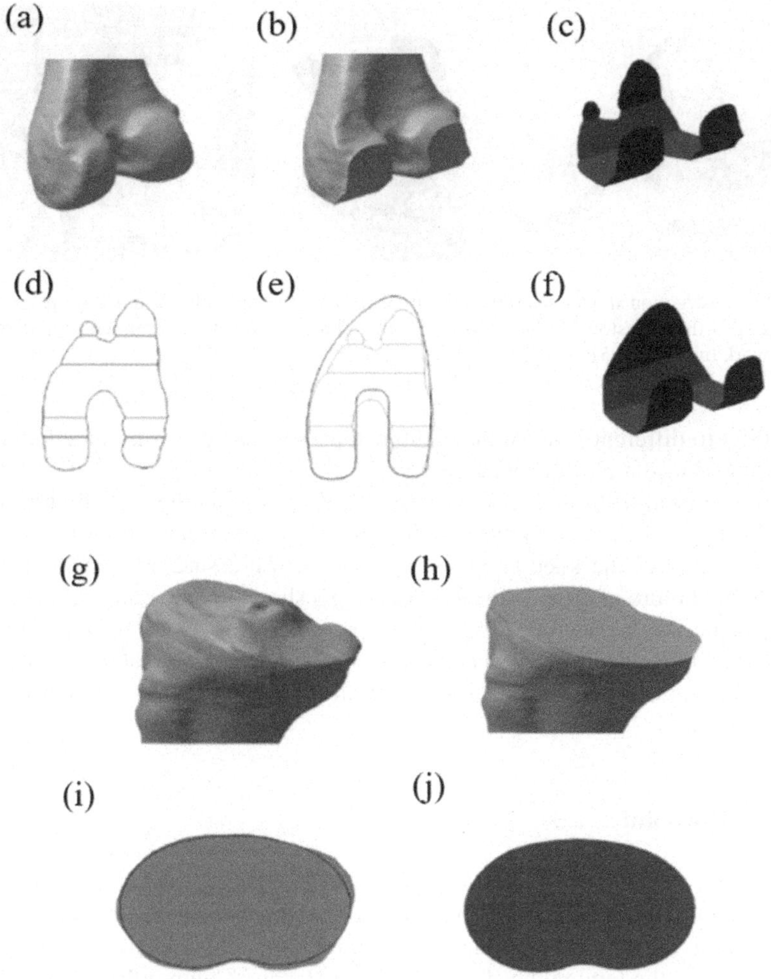

Fig. 5.8 The optimization of the osteotomy profile of the knee's personalized femoral and tibial implant sections: **a** determine the femoral osteotomy surface, **b**) simulate bone cutting, **c** extract the osteotomy profile, **d** determine the contour curve, **e** smooth the curve, **f** optimize the contour curve, **g** before tibial osteotomy, **h** after tibial osteotomy, **i** extract the osteotomy profile, **j** optimize the tibial profile. Reprinted with permission from a work by Wang [95]

pad to deal with bone defects, as well as the stress caused by concentric and eccentric extension rod dispersion gaskets and the limitations of the man of the psalm. For revision knee arthroplasty and complex primary knee arthroplasty. The A3GT knee prosthesis retains the full-size, high-bending, and all-in-one advantages of the A3 knee prosthesis, and the new CR-type prosthesis design allows the CR/AS prosthesis to be selected during surgery or switch to a PS/PS-plus prosthesis at any time.

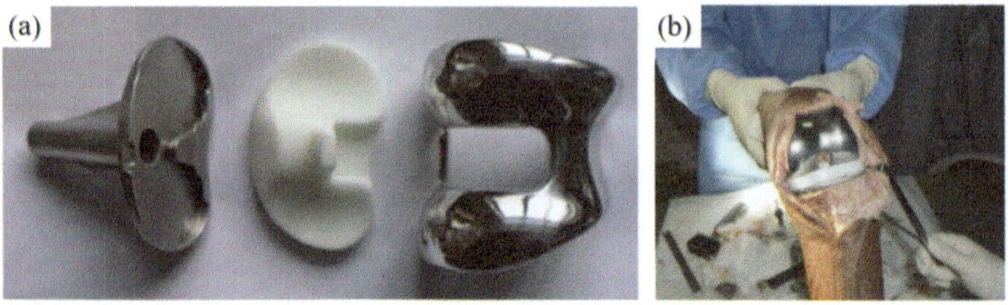

Fig. 5.9 Construction and application of personalized total knee prosthesis: **a** AM for personalized total knee prosthesis customization and **b** prosthetic cadaver verification. Reprinted with permission from a work by Wang [95]

According to different lesions, the combined prosthesis system is suitable for total knee joint surface replacement surgery.

Korkmaz et al. [96] designed an anatomical prosthesis that was highly in line with the shape of the knee joint and generated less stress and strain to maximize the adaptability of the knee bone and generate less pressure and tension to avoid the possible failure of the prosthesis. The results show that the contact stress of the tibial prosthesis is reduced by 50%, and the principal stress of the tibial prosthesis is lower. The finite element analysis results show that the anatomical design approach has advantages in terms of joint surface wear, aseptic loosening, and normal joint biomechanics.

5.3.1.2 Hip Joint

The hip joint is a ball-socket joint composed of the femoral head of the human body and the acetabulum of the pelvis. The femoral head is spherical, and the acetabulum is spherical and concave. The total hip joint comprises four components: the acetabulum cup, the lining, the femoral head, and the femoral stem. Commonly, the inner lining is used with materials such as ultra-high molecular polyethylene, and the femoral head is used with materials such as metal or alumina ceramics. The acetabulum cup and the femoral handle are biologically fixed with the bone. The inner lining is fixed in the acetabular cup, and the femoral head is sheathed on the femoral handle, and the two move to form a joint.

Xiao [97] designed the acetabular cup and femoral stem of the artificial hip joint implant regarding the anatomical shape and structure of the hip joint. The acetabular cup fixed with the pelvis is a substitute for the acetabulum on the pelvis. The femoral stem implant is inserted into the bone marrow cavity, and the head rotates with the acetabulum cup or the original acetabulum to achieve the extension and rotation of the medullary joint.

During the manufacturing process of the actual metal implant, a 3D model of the hip joint is first obtained from medical CT image data through Mimics software, as

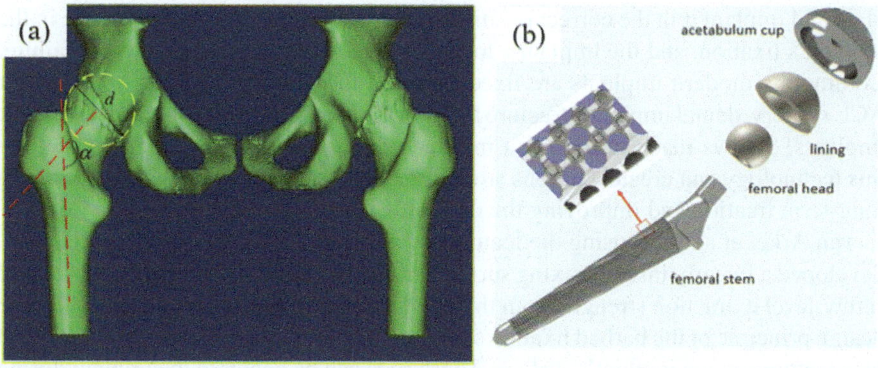

Fig. 5.10 Design process of the artificial hip joint: **a** a 3D model of hip joint extracted by Mimics and **b** porous artificial hip joint model. Reprinted with permission from a work by Xiao [97]

shown in Fig. 5.10a. The model is then processed to create a personalized artificial hip model. A perforated design is performed for the acetabular cup and the lateral femoral stem to create the model, as shown in Fig. 5.10b. The porous acetabulum and femoral stem make up the total hip to ensure its mechanical properties and shape stability, which meet the functional requirements of the joint. Finally, the model data are sliced and the artificial hip joint is printed by SLM technology.

In 2017, the artificial hip joint produced by AK MEDICAL became the first metal implant approved by the CFDA in China. Based on the previous generation products, the AK ML second-generation hip product has been launched in 2020. The ML TH biological femoral stem follows the classic flat wedge femoral stem design, with medial-lateral compression and anterior-posterior three-point fixation to provide immediate stability for the femoral stem. The proximal coating adopts Supravit patented coating technology. The handle body is sprayed under a vacuum environment, without bubbles, ensuring the coating is compact and uniform and preventing the generation of crack source and coating peeling. Based on 300 μm pure titanium coating, 50 μm hydroxyapatite coating is added to the proximal femoral stem to promote early bone reconstruction, which ensures the effect of osteointegration in the middle and late period, and reduces the risk of postoperative stem subsidence and loosening.

5.3.2 *Traumatic Implants*

Commonly, mild/severe fractures without missing bone can be healed by external fixation, or plate and nail fixation. Traumatic implants printed by AM can help patients heal better.

In general, if the spinal cord is severely injured, a spinal fusion is required. To fuse the vertebrae, the surgeon must carefully insert the bone screws to fix the guide plate and implant it in the correct position. Bone screws play an important role in the implant's fixation, and the implant's loosening is the main cause of surgical failure. Commonly, modern implants are fixed by press fit or screws. Screws are used for ACL surgery, dental implants, fracture fixation, and early intervention cartilage repair implants. Screws made by AM can improve the fixability of the implant, because this technology can create a porous structure in which bone can grow, thus ensuring long-term fixation and improving the implant's initial stability.

van Arkel et al. [98], using the features of design freedom of parts made by AM, developed a linearly inserted fixing surface that can be fixed in the bone and obtained screw-level extraction strength from the push-fit device. As shown in Fig. 5.11a, the design principle of the barbed fixation surface is to create small, hook-like directional bias features on the implant's surface. These rods can be bent into the implant during insertion to reduce interference with the bone and provide low push-in resistance

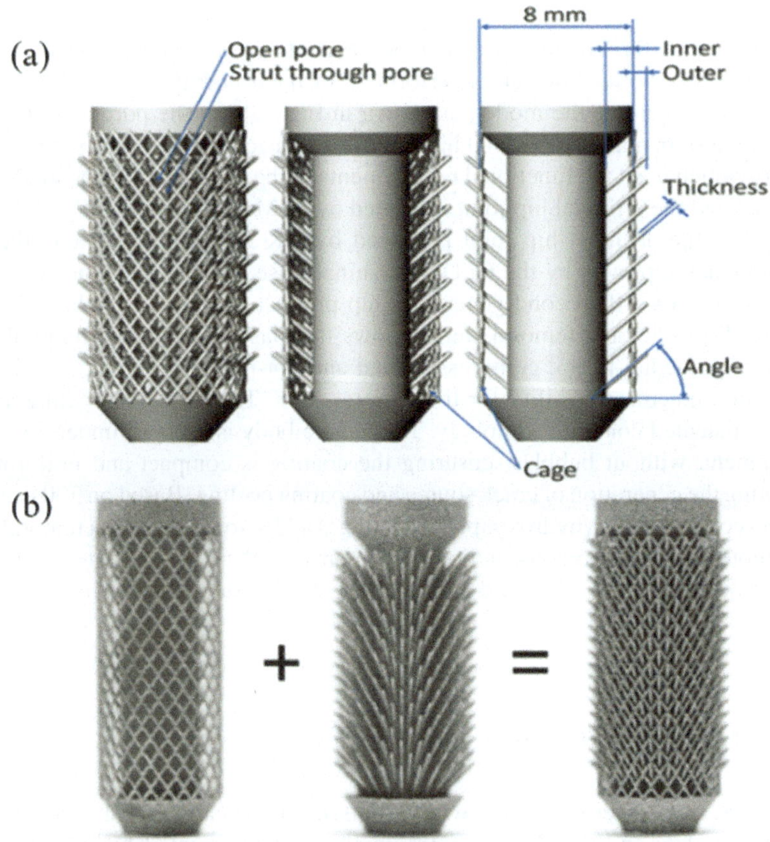

Fig. 5.11 Design optimization of bone screws: **a** the screw structure diagram and **b** the schematic diagram of optimization structure of barb fixation. Reprinted with permission from a work by van Arkel et al. [98]

when they are biased in direction. Then they can bend outward and grip the bone as they are pulled out to increase the interference and improve the stability.

As shown in Fig. 5.11b, the screws were printed by an SLM device when optimized. Experimental tests on the distal femur of the cadaver showed that the surface of the screws could be inserted more easily than the traditional press-fit technique because of halving the insertion force and doubling the anchoring force. It has two advantages. First, the lower insertion force requires less implant impingement while maintaining immobilization, facilitating minimally invasive surgery, and reducing the risk of impingement fractures. Second, the increased anchorage force provides a new option for less invasive implant design by using low-profile fixation techniques. With push-fit technology's inherent advantages, barb fixation can simplify existing procedures while enabling new surgical treatments.

5.3.3 Spinal Implants

During spinal surgery, it is necessary to completely remove the lesion site, and use the corresponding structural support for structural reconstruction to maintain the stability and integrity of the spinal structure. The spine is not only surrounded by important nerves and blood vessels, but the spinal canal also contains the spinal cord. Therefore, it is necessary to ensure the operation's accuracy to ensure the patient's safety. With the continuous development of SLM technology, its forming accuracy and effect are improving. It can directly shape a complex curved surface and a porous structure to meet spinal surgery demands.

5.3.3.1 Vertebral Implant

Chen et al. [99] used CAD and SLM technology to shape bionic porous Ti6Al4V spinal implant. A prosthesis was surgically implanted to connect the atlas and the third cervical vertebra, and the spinal structure was reconstructed in a patient with complete axial excision.

The CT scan data of the cases were imported into the Mimics software and stored in STL format, then imported into the GEOMAGIC software to eliminate the model noise, and further processed to obtain the 3D solid model, as shown in Fig. 5.12a. A bionic porous structure is constructed between the target binding surfaces, which is conducive to the growth of human tissue in the implants and the patient's recovery. The bionic porous structure was designed to be a secondary porous structure with pore sizes of 0.75 mm and 1.5 mm respectively, as shown in Fig. 5.12b, c. The biomimetic porous structure alone cannot well achieve the goal of stability of the spine reconstructed by the spinal implant. Therefore, it is necessary to construct a shell supporting the porous structure and a fixing screw hole between the upper and lower vertebrae, as shown in Fig. 5.12d. The overall design of the implant is

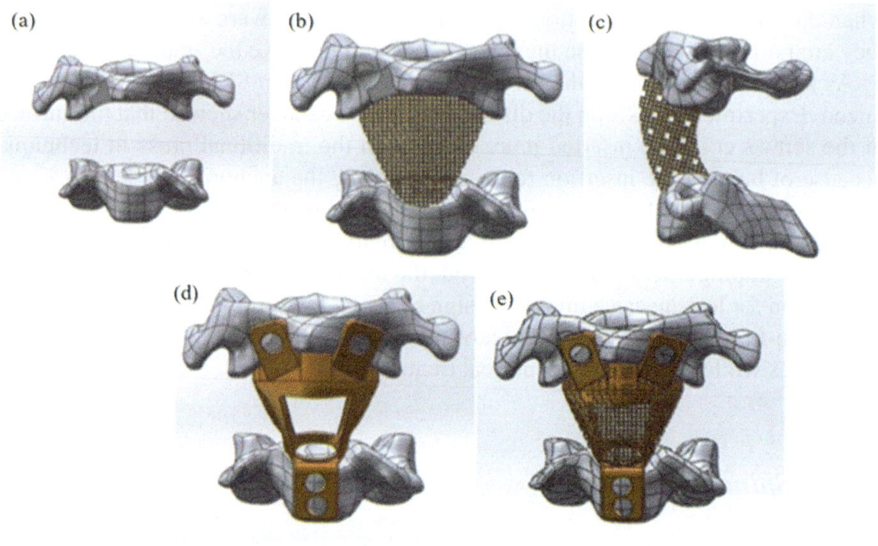

Fig. 5.12 Design of spinal implant: **a** the target spine model, **b** the bionic porous main structure, **c** the ionic porous main structure (side), **d** the outer frame of the prosthesis, and **e** the design effect of the prosthesis

combined the porous structure with the supporting part of the shell to form the final implant, as shown in Fig. 5.12e.

Finally, the implant model was made by SLM equipment, as shown in Fig. 5.13a, b. After molding, to verify the matching degree of the spinal implant, the surface of the prosthesis was sandblasted and polished, and the printed spinal implant was simulated by the matching resin model, as shown in Fig. 5.13c.

The CAPRI Cervical 3D Expandable Corpectomy cage system was developed by K2M, a medical technology company, and was CE-certified for use in surgical

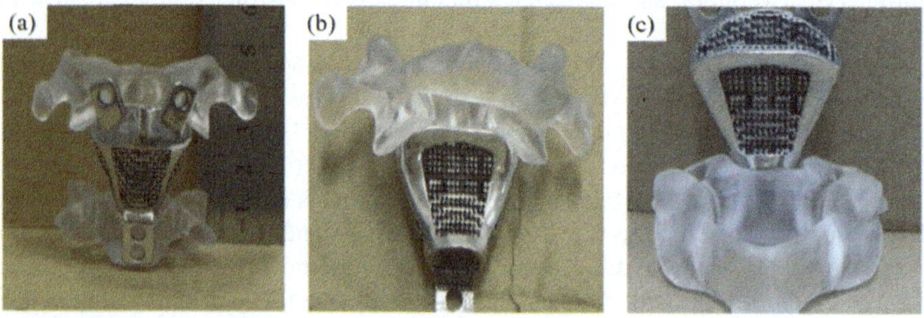

Fig. 5.13 Fitting situation of the prosthesis and resin models: **a** coordination between the prosthesis and the resin model, **b** fitting surface of the prosthesis and atlas, and **c** fitting of the prosthesis to the third vertebra

procedures. It is an implant that stabilizes the spine after vertebral column resection (VCR). The surgeon uses pedicle screws and rods to realign the spine, then the implant must replace the removed vertebrae and stabilize it in the new position. CAPRI is a porous implant adapted to bone graft growth, allowing the combination of complex internal geometries and customed rough surface structures. CAPRI implants allow for control of expansion in situ by precise height and lordosis angle and can be placed surgically along multiple angles to achieve skeletal integration, and they are successfully used in the operation of Rigshospitalet in Denmark.

4Web Medical, an Orthopedic implant company, has added an independent anterior spine truss system (ASTS-SA) to its AM spine implant portfolio. The company's latest adjuvant surgical device features an enhanced two-step locking system that prevents spinal screw regression and provides the surgeon with greater technical support during surgery. Because the company's TRUSS implant technology was used to create the ASTS-SA, it has the same enhanced healing properties as other transplant series. The addition of an integrated screw to the implant allows ALIF surgery without additional fixation or electroplating.

5.3.3.2 Infundibular Thoracic Implants

The funnel chest is famous for the characteristic depression of the chest. The patient's sternum and its adjacent costal cartilage sink, forming a funnel deformity. The most common cause of the funnel chest may be abnormal development of the sternum, ribs, or diaphragm. The commonly used surgical treatment methods are minimally invasive surgery and open surgery. Minimally invasive surgery involves inserting a plate into the body, and open surgery involves attaching the sternum to the correction target using a wire or stent.

Bellia-Munzon et al. [100] used computed tomography and 3D reconstruction technology to determine the overall features of the sternum and thoracic cavity and created STL files through software processing. According to the thorax's position, the metal prosthesis' curvature was determined, and the anterior and posterior pressure on the side of the implant was calculated.

The designed implant was fabricated by the SLM technique. Before introducing the implant, the implant was placed on the patient's chest to determine the final orientation and position. Use the tunnel dissector to create a retrosternal tunnel and pull out the tape through the tunnel. Tape is attached to the selected implant and pulled back through the retrosternal tunnel to the position previously identified by the CT scan. The implant is turned over with the convex surface toward the sternum. The process is repeated several times until the implant is successfully achieved.

In 2018, a British patient with a chest tumor had to have part of her sternum and three ribs removed during surgery. Without the implant, the patient would not survive. The hospital designed the implants from CT scans of the patient's chest, and Renishaw made them. Titanium alloy with a grid structure reduces its own weight and is beneficial to rapid wound healing. Good biocompatibility also avoids serious rejection reactions. This is the first case of a chest prosthesis manufactured and implanted in a patient in the UK.

Titanium alloy as a safe and effective material for chest wall reconstruction and repair, has a good clinical effect on personalized and anatomical chest wall reconstruction and repair. Zhang et al. [101] imported the 3D reconstructed STL format file into the EOS M280 printer (EOS GmbH, Krailling, Germany) for implant production. The printer uses medical-grade titanium alloy powder and laser sintering technology to produce implants. After routine cleaning and autoclave sterilization, the implants are used for the surgery.

5.3.3.3 Interbody Fusion Cages

Interbody fusion cages are successful methods for achieving fusion by enabling bone grafts to accommodate arthrodesis and restore biomechanical stability of the anterior column. They have been used to assist bone fusion for various spinal diseases, including degenerative disc disease, spondylolisthesis, trauma, tumor reconstruction, scoliosis, and so on. AM interbody fusion cages can be subdivided into cervical fusion cages and lumbar fusion cages.

Mobbs et al. [102] treated a patient with a vertebra abnormality via an interim plant fusion implant. Preoperative CT and MRI studies were used to construct a 3D model of the patient's anatomical structure and to design a customized AM titanium implant, as shown in Fig. 5.14. A standard L-4/L-5 anterior approach was performed, and the pathological level was confirmed by X-ray before disc and endplate removal. A custom-made AM titanium implant was packaged with a bone graft and inserted in a precise fit. Anterior plate fixation was performed to secure the motion segment, and additional posterior percutaneous pedicle screws were used to stabilize the L-4/L-5 segment further.

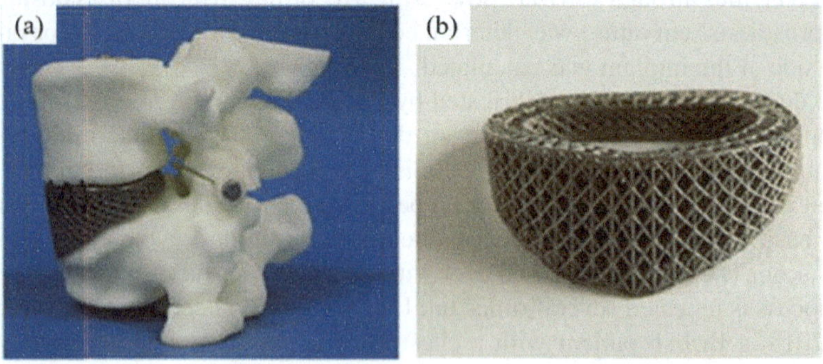

Fig. 5.14 Construction and application of personalized interbody fusion cages: **a** a plastic model of the half vertebra of an AM prosthesis and **b** prosthesis before sterilization and implantation. Reprinted with permission from a work by Mobbs et al. [102]

Osseus Fusion Systems, a Texas medical device company, has received approval from the US Food and Drug Administration (FDA) for its AM spine implant, ARIES. The implant is designed to help patients relieve back pain and shorten spinal recovery time. The ARIES interbody fusion device is printed in titanium and optimized for bone fusion and biological fixation. They are manufactured using PL3XUS, Osseus' proprietary AM technology, which uses powder bed fusion technology to manufacture titanium parts. ARIES has a porosity of 80%, allowing spinal tissue to grow through the construction. In addition, the device is equipped with a mesh that allows better fixation and diffusion of bone cells, resulting in faster healing, improving bone growth, and enhancing X-ray visibility. Each ARIES implant is available at various heights and lordosis (curvature of the spine inward) angles and is suitable for a variety of patient anatomies.

5.3.4 Dental Implants

With the improvement of people's economic living standards, the oral medical market develops rapidly. In prosthodontics, most of the implants are made by hand. The processing method is mainly casting which is time consuming; and it requires repeated imprinting and molding, which is cumbersome and brings great inconvenience to patients. Its accuracy and quality depend on various operators. As an advanced manufacturing technology, AM can well replace manual labor to produce personalized and accurate dental implants. Combined with 3D scanning and other technologies, such as computer software algorithms, it is widely used in dental bridge crowns, lingual brackets, dental implants, and other aspects.

5.3.4.1 Dental Bridge and Crown

Mai [103] adopted reverse engineering and CAD design methods to complete the design of the personalized dental crown fixed bridge. SLM molding technology was used to fabricate personalized CoCr alloy crown fixed bridge prosthesis with high molding accuracy. According to the characteristic structure of the tooth crown and fixed bridge, data processing (such as smooth surface, optimization of placement position, and support addition) was completed. Optimized process parameters were used to process the solid restorations with high surface quality. Reverse scanning was used to detect and evaluate whether the restorations have good molding accuracy. The wearing effect is obtained by grinding, sandblasting, porcelain, and other treatments.

Researchers produced metal crowns for the endangered walruses at the Quebec Aquarium. The gap between the crown and the ivory is between 0.25 and 0.45 mm. The veterinarian of the operation can choose the best crown according to the situation, which will neither weaken the adhesion of the adhesive nor prevent it from adhering closely to the ivory. The dental crown is made of cobalt-chromium alloys, which is more suitable for walruses because it can resist friction and is not easy to oxidize in saltwater.

5.3.4.2 Lingual Bracket

The lingual orthodontic technology is the most widely used "invisible" orthodontic technology, which is beautiful, effective, and suitable for many oral diseases. However, the traditional common lingual brackets are universal products produced by standardization, which are less suitable for different patients' teeth. The AM technology can create personalized brackets with higher matching with individual teeth according to the morphological characteristics of a patient's teeth.

Sun [104] used a 3D laser scanner to collect and reconstruct the dental model data. Based on the digital model, the CAD of personalized lingual brackets was carried out using 3D software, and then the personalized metal lingual bracket was directly manufactured through the SLM rapid forming machine.

During designing and manufacturing personalized lingual brackets, the patient's digital dental model was obtained through reverse engineering technology, and the defects of the model were also processed. Then the archwire plane was created on the model, as shown in Fig. 5.15a. The brackets and the base of the band ring are designed according to the lateral profile of the tongue of each tooth, as shown in Fig. 5.15b. The bottom of the bracket completely fits the side of the tongue, which makes the bracket have the function of "self-locking" and can be positioned directly on the tooth, thus improving the bracket's bonding effect and positioning accuracy. Finally, the STL file of the bracket was imported into Magics for SLM processing and manufacturing. The personalized tongue bracket after manufacturing is shown in Fig. 5.15c.

The personalized lingual invisible appliance, developed by Ruitong Biological Technology Co., Ltd., has obtained the medical product registration certificate and then has been put into the market. The brackets, archwires, and installation trays of Ruitong Biotechnology Co., Ltd. are digitally customized, eliminating a large number of technicians from doctors, shortening the treatment time, and improving the accuracy and efficiency of the treatment.

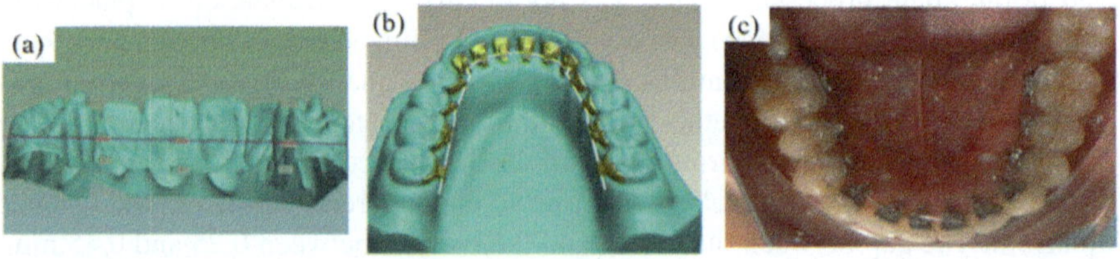

Fig. 5.15 Construction and application of personalized lingual brackets: **a** creation of the archwire plane, **b** schematic diagram of half brackets, and **c** personalized tongue brackets made by SLM. Reprinted with permission from a work by Sun [104]

5.3.4.3 Implantation of Dental Studs

The SLM process can form dental implants with the same complex surface morphology as the natural teeth, which can meet patients' personalized dental repair demands and have incomparable advantages over the traditional forming process. The most common way to repair and replace teeth is to implant artificial nails in the alveolar to replace the tooth's root as a fixed part and then cover the abutment and crown to form the whole tooth.

Wang et al. [105] obtained the 3D model of the human dental crown through reverse engineering and designed the 3D model of certain dental crowns according to the dental nail data provided by the Stomatological Hospital of Jilin University, as shown in Fig. 5.16. To ensure the normal physiological activities of the nail crown artificial implant after being implanted, the printed implant should meet certain mechanical and physiological conditions.

The Magics software uses the STL triangular surface file converted from the 3D model for parts placement, support addition, slicing, and scanning path planning. For crown parts, the concave lower surface should be matched and fixed with the base. To achieve the high precision requirements of the inner surface, the face that needs to be matched must be placed up. For the tooth nail parts, the most important thing is forming the threaded part, which is placed vertically. The pre-processed file was imported into the SLM equipment for printing. After 4 h, the solid parts of the crown and nail were formed. To further reduce the occlusal surface roughness of the crown and maintain a certain smoothness, a necessary shot peening treatment should be carried out on the occlusal surface after removing the support. After the shot peening treatment, the crown surface will undergo a certain degree of deformation strengthening, increasing the hardness of the crown surface and improving the wear resistance of the surface.

The Italian company Leader has produced conventional commercial implants (Tixos Titanium Implants) using SLM. Implants made by AM technology save raw

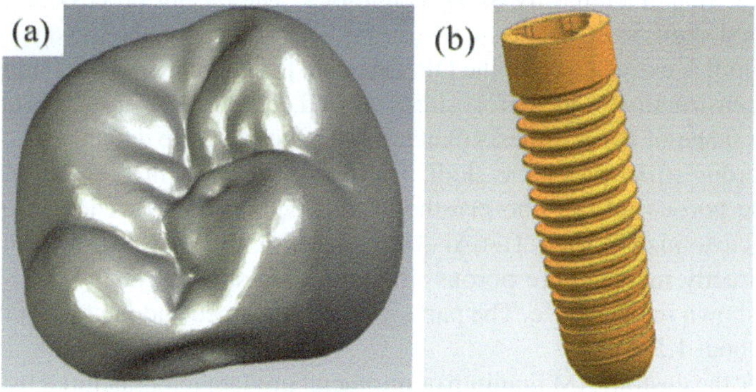

Fig. 5.16 Dental implant models: **a** a 3D model of the dental crown and **b** a 3D model of the dental nail. Reprinted with permission from a work by Wang [105]

materials, reduce costs, do not need complex surface treatments, and have surface pores of 100–200 μm. Animal experiments show that the bond-bonding rate of the direct laser printing surface is similar to that of the acid-etched sandblasting surface, and both are significantly better than the smooth pure titanium surface.

CBCT was used to obtain the dental root image data, and EBM technology was used to make personalized root implant teeth. The fabrication accuracy can be up to 0.2 mm. The scanning electron microscope showed a large number of trabecular-like spherical processes and about 100–200 μm of groove structure. After acid etching, the hydrophilicity was greatly increased, and the hydrophilic contact Angle was close to zero. The cylindrical implants made by the EBM technique were implanted into the tibia of miniature pigs. One month later, the tissue sections showed that a large number of new bones formed a bone combination with the implant's surface and grew into the gap on the implant's surface, and the osseointegration rate was up to 40–50%.

5.3.5 Other Applications

5.3.5.1 Cranial Repair

The skull is composed of the cerebral cranium and the cranial surface. The skull loss is a common operation in plastic surgery. The traditional repair method is that a doctor manually cut plastic custom rail network restoration according to the defect position of CT or MRI image information. When involving complicated anatomic areas or a large area of bone defect repair, the appearance of restoration is difficult to control. The prosthesis manufactured by SLM technology can be more accurately fitted to the injured part of the skull and improve the success rate of the surgery.

Xiao [97] developed a modeling system of porous structure, based on which the porous structure was designed for the skull prosthesis model, and the porous implant was manufactured by SLM technology. The design process of a typically personalized skull prosthesis is shown in Fig. 5.17a. Firstly, the 3D data file of the defected skull is extracted from the DICOM data format of CT or MRI images. Then the 3D reconstruction of the defect site is carried out via a left–right mirror image to obtain the shape of the prosthesis matching the skull and the defect site.

The porous structure of the skull prosthesis is shown in Fig. 5.17b. By optimizing the porous design, the prosthesis with appropriate mechanical properties could be customized. Then Ti–6Al–4V material with good biocompatibility was used to directly manufacture porous titanium alloy skull repair by the SLM technique, as shown in Fig. 5.17c. The parts were well-formed and the size of the porous unit was about 1.5 mm.

In 2016, the custom AM titanium craniofacial implant manufactured by Bioarchitects of the US obtained the 510(K) license from the US FDA [106]. According to BioArchitects, the AM titanium craniofacial implants are mainly designed to repair non-weight-bearing bones in patients' heads and faces.

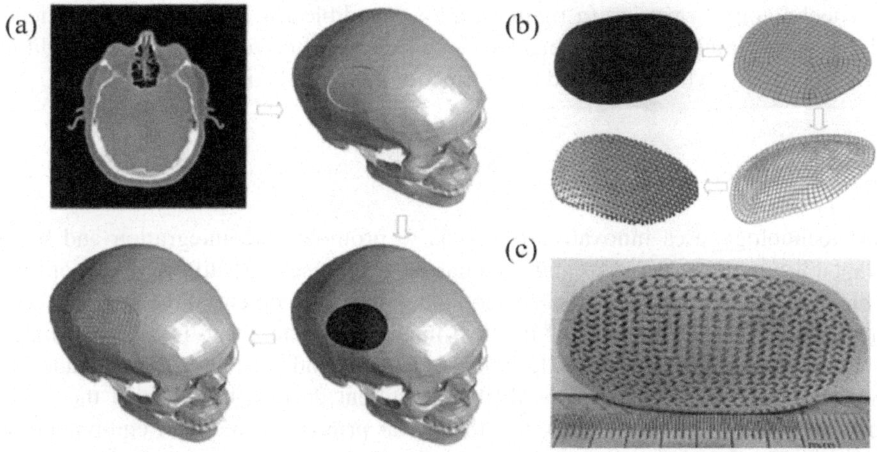

Fig. 5.17 Design and construction of personalized porous cranial: **a** a personalized porous cranial complex design process, **b** a porous structure design of skull prosthesis and **c** processed porous skull prosthesis. Reprinted with permission from a work by Xiao [97]

5.3.5.2 Maxillofacial Prosthetic Implants

The mandible is arched, consisting of a horizontal mandibular body and a vertical mandibular ramus. When viewed from the side, the mandible is L-shaped. The maxillofacial bones are the only movable bones in the middle of the human maxillofacial. Trauma, tumor, and other diseases often bring different degrees of trauma to the human mandible, resulting in the appearance of mandible defects. Autologous bone transplantation is the most widely used method for the repair of maxillofacial defects. The specific operation is usually to take the patient's autogenous bone of a certain length and shape and fix it to the defect site with the prefabricated titanium plate of maxillofacial surgery to finish the repair. Although this method will not produce rejection reactions, the autologous bone resources are limited, which causes damage to the bone extraction site and is not conducive to the repair of the facial appearance. The SLM technique can design and shape jaw implants and prostheses with a highly personalized appearance. At the same time, porous metals' appearance also guarantees the bone ingrowth performance of metal implants and prostheses.

Xiao [107] carried out 3D reconstruction and reverse modeling of the human mandible model, and a finite element mechanical simulation of the human mandible under midline occlusal movement. According to the normal stress distribution in the middle occlusal state of the mandible, the internal porous structure of the mandible prosthesis was designed. Then, the mandibular prosthesis was manufactured by the SLM technique, and the customized porous mandibular prosthesis was installed on the resin model to test the matching of the implant.

AM technology can directly print implants of bone defects (such as the upper, mandible, orbital, nasal, and ear bones) for defect reconstruction. Nickels [108] used

AM technology to process a titanium alloy mandible for an 83-year-old patient. The new "mandible" was implanted immediately after removing the mandible intraoperatively, which achieved good functions and aesthetic effects.

5.3.5.3 Ankle

AM technology uses innovative materials to promote osteointegration and bone adaptability to create complex and personalized prostheses with different and controllable porosity. Therefore, it has been applied in the replacement of the internal prosthesis. Based on the analysis of medical images, Liverani et al. [109] used image processing software to obtain 3D models of bones and ligaments. They made 3D models of replacement ankles with three different designs to better fit the ankle surface. Finally, the CoCR ankle prosthesis was printed by the SLM equipment, as shown in Fig. 5.18.

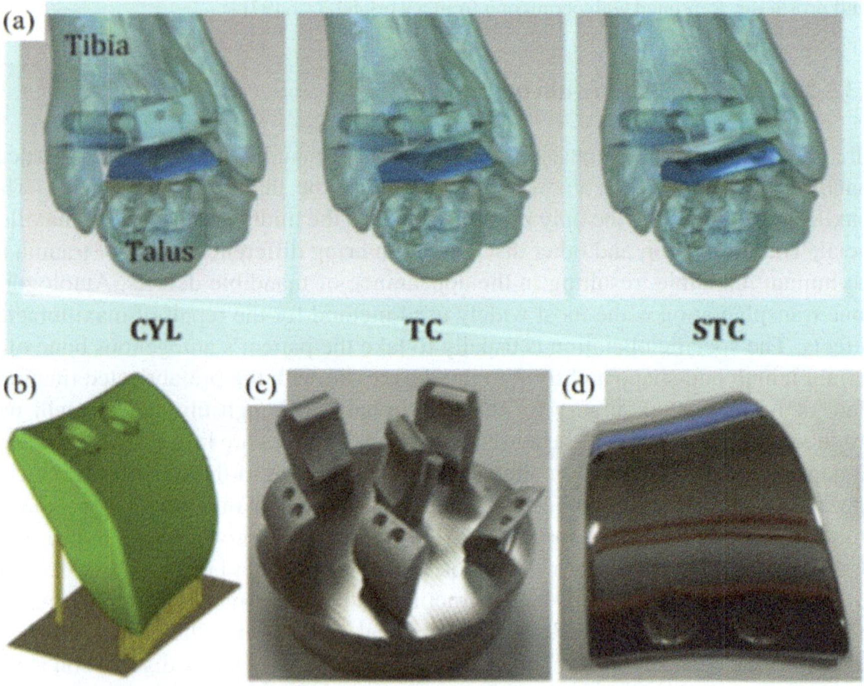

Fig. 5.18 3D models of ankle prosthesis and its manufacturing drawing: **a** 3D models of the replaced ankle with the three different designs, respectively CYL (left), TC (center), and STC (right), **b** an example of orientation in AutoFab of the talar component of the STC prosthesis, **c** final arrangement of the STC, TC and CYL prosthesis components on the platform, and **d** an example of the same talar component after the polishing. Reprinted with permission from a work by Liverani et al. [109]

5.3.5.4 Oar Bones

Mashudu Tshifularo's team at the School of Health Sciences at the University of Pretoria, South Africa, performed the world's first operation to transplant a bone in the middle ear using an AM component at Steve Biko Hospital, leaving a 40-year-old man with conductive hearing loss restoring hearing. But the operation failed, and one of the main reasons is thought to be the incorrect size of the prosthesis. Part of the working principle of hearing is to transmit vibrations from the eardrum to the cochlea through three small bones in the middle ear (called ossicles), which is a very small structure. Osteoconductive hearing loss occurs when the ossicle is damaged. Using AM technology, anatomical relationships in space can be reproduced accurately to create customized prostheses for patients with conductive hearing loss. Titanium alloy is used to produce this prosthesis, which is biocompatible and to replace damaged bone by an endoscope during surgery.

5.3.5.5 Trabeculae

In addition to designing and fabricating implants that match the external shape of the defect, AM can also design and fabricate the internal structure of the implant, known as "metal trabecular bone", such as the "titanium bone" structure of the implant of the Italian Leader company. Trabecular metal produced by EBM has also been widely used to design orthopedic and joint implants. The design of trabecular bone structure can guide bone tissue to grow into the implant, reduce the implant's elastic modulus and weight, and increase its friction coefficient and stability after implantation. At the same time, the addition of growth factors or cells as scaffolds can further promote osteogenesis.

5.3.6 Certificated Implant Products

According to the demands of patients, AM technology can customize the shape of the implant, accurately control the complex microstructure of the implant, and realize the dual adaptation of the shape and mechanical properties of the implant to the human body, thus it is very popular in the human implantation. With the rapid development, many products have been applied in the clinic.

At present, the US FDA has approved more than 100 AM orthopedic implants, including domestic products. At the same time, domestic R&D and application have also achieved many results and accumulated a lot of experience in material selection, certification process formulation, cooperation with the competent department review. The domestic regulations and standards of customized medical devices have also made rapid progress. The technology of AM implantable devices has been brought into the scope of supervision, and the relevant charging standards are constantly being clarified.

AM acetabular cup is the earliest metal AM implant to obtain a medical device registration certificate in China. At present, there are two kinds of home-made AM acetabular cup products that have been approved by the National Medical Products Administration (NMPA) of the State Drug Administration and obtained the medical device registration certificate, including the hip prosthesis and acetabular parts of Beijing Aikang Yicheng Medical Equipment Co., Ltd., and the trabecular hip prosthesis including acetabular cup parts produced by Just Medical Equipment (Tianjin) Co., Ltd.

Materialize's Mimics Innovation Suite (MIS) can transform medical image data into high-quality 3D models, and carry out analysis, measurement, design, and other work on this basis. MIS is the first software tool certified by the US FDA and EU CE and can be used in the clinical environment. In July 2020, MIS was certified by China's NMPA, which added vitality for Materialise further to promote the development of China's medical industry.

The AM degradable weaving engineering scaffold technology platform and the first commercial product of Johnson's subsidiary Tissue Regeneration System (TRS) have been annotated by the US FDA [110, 111]. In promoting the commercialization of the technology, the TRS uses the company's partner business model to work with medical, dental, and AM companies to develop tissue engineering products they are interested in. Once the co-developed product is approved by the US FDA, the partner will commercialize the product, and the TRS will act as the manufacturer of the approved product.

Based on the work experience, researchers have developed a small drill using AM technology, which can successfully help professional athletes with anterior cruciate ligament injuries. Researchers completed the printing of the Pathfinder ACL guide plate with Laser Design and registered it as a Class 1 medical device on the US FDA. This invention has been patented in the name of Danamed, and now Pathfinder has begun to sell on the market.

In July 2017, Emerging Implant Technologies GmbH (EIT), a German medical device company, developed the spinal support implant and was approved by the US FDA. The implant, made of SLM augmentation technology, achieves 80% porosity of the component and contributes to the cell growth necessary for bone repair, and has been used in 10,000 cases in 15 countries around the world.

In 2018, Exactech introduced the sessile shoulder, a bone-preserving prosthesis designed for anatomical total shoulder arthroplasty, consisting of a sessile cage, humeral head, and cage pelvis, providing intraoperative flexibility and an ideal bone-preserving prosthesis. In addition, to improve the probability of biological fixation, it also adds a laser AM porous bone cage structure to achieve bone infiltration and growth, which makes the implantation more convenient and reduces the operation time and blood loss.

In 2018, Michigan-based medical technology company Stryker's AM Tritanium (titanium alloy) TL arc lumbar spine received a 510(k) license from the US FDA, and the AM interbody fusion device has gradually gained market acceptance. Stryker's AM Sahara Transversal Extensible Implant received FDA 510K approval. With its layered titanium 3D technology, the transverse extensible fusion implant allows surgeons to achieve up to 30 degrees of spine correction in patients with skeletal maturity due to its passive expansion capability. Layered titanium 3D technology can create structures that were previously impossible using traditional methods. The titanium powder was selectively sintered with a high-energy laser beam to print Sahara implants and retain the rough surface structure. Clinical trials have shown increases in protein expression compared to smooth titanium surfaces.

Recently, the state has issued a series of policies, such as the regulations on the Supervision and Administration of customized Medical Devices (trial), the guiding principles for Technical Review of Registration of Medical Devices for Personalized augmentation of Bone, Joint and Oral hard tissue, and the pilot Action Plan for Manufacturing Standards (2020–2022), to support the development of AM technology. China Institute of Food and Drug Control, Medical device Standard Management Institute and other units are also actively promoting the formulation of AM implants related standards, AM metal implants industry will be standardized development.

5.4 Additive Manufacturing of Surgical Guide Plates

5.4.1 Guide Plate for Spinal Surgery

The surgical guide plate is a personalized guide plate that is designed by computer based on the individual data of patients, and it has been widely used in hip joint, knee joint, and spinal surgery. For some complex orthopedic diseases, it is not only necessary to make an accurate diagnosis and detailed operation plan before the operation, but also to carry out an accurate operation during the operation.

In spinal surgery, pedicle internal fixation is the most basic operation, and the accurate placement of screws is always difficult, which depends on the variation of anatomical morphology, the size, and direction of pedicles. Therefore, the research on improving the accuracy of nail placement in internal fixation has been concerned. At present, the more mature methods of nail placement are bare-hand nail placement, X-ray fluoroscopy-assisted nail placement, computer fluoroscopy navigation nail placement, etc.

Bare-handed nail placement surgery requires operators who have a sufficient understanding of complex anatomical structures and have sufficient experience in nail placement. X-ray fluoroscopy-assisted nail placement has a high screw dislocation rate. The equipment of computer fluoroscopy navigation nail placement and the application are high-cost, similarly and the hardware requirements of the hospital are high.

With the continuous development of digital medical technology, AM technology has been gradually applied to spinal orthopedic surgery and shows its advantages. This technology has the characteristics of optimizing preoperative planning, providing individualized and precise treatment. To assist pedicle screw implantation, it has been applied in spinal surgery to make a navigation template.

Zhang et al. [112] applied the Co–Cr–Mo alloy surgical guide plate formed by SLM to cervical vertebra surgery. First, the original data are obtained by CT or MRI, as shown in Fig. 5.19. After software processing, a 3D digital model identical to the patient's bone tissue was established, and the surface anatomical shape of the target bone tissue was extracted. Then, a complementary reverse template with the best nail channel is designed in the software. Finally, the data is input into the AM equipment to produce the template with a positioning guide hole.

Besides, the guide handle was lengthened at the design to facilitate the fixation of the guide on the vertebral lamina during the operation and prevent the displacement of the guide when drilling. At the same time, it also reduces the occlusion of the surgical field of vision, and the effect is more ideal. Finally, the Co–Cr–Mo material guide directly was manufactured by the SLM, as shown in Fig. 5.20a–d. The metal guide plate was used in cervical nail placement surgery, as shown in Fig. 5.20e. It showed postoperative internal fixation was good, and the reduction of atlantoaxial dislocation was satisfactory. During six months of follow-up, there was no recurrence of dislocation and the bone graft was fused as well.

Zhang et al. [113] designed an AM navigation template based on the bone surface after removing soft tissues in the 3D model. The navigation template was used to achieve accurate placement of cervical pedicle screws on the body. The cervical spine was first scanned by CT, and a 3D model of the cervical spine was established. The screw trajectory was designed to pass through the central axis of the pedicle. Then, by removing the soft tissue from the bony surface in the 3D model, the navigation template was designed and finally printed using an American Replicator23D printer, as shown in Fig. 5.21.

After designing and printing the navigation template, the soft tissue, at the upper and lower edge of the cadaveric lamina and the junction of the spinous process and the lamina, was removed by the surgeon in order to dock the navigation template. Then, the interspinous ligament was cut at the midpoint of the spinous process space and

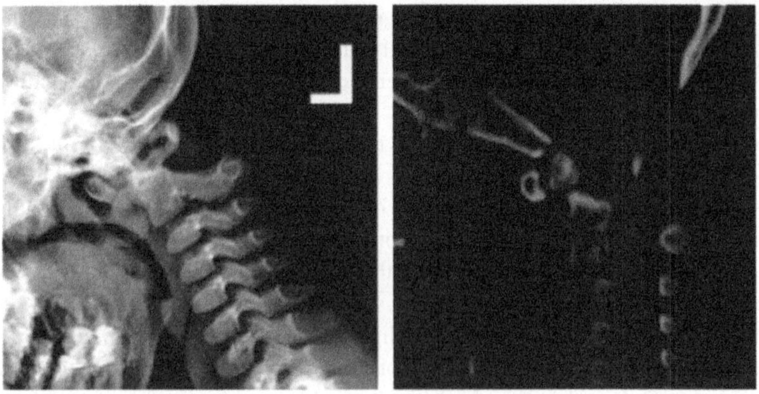

Fig. 5.19 Preoperative X-ray and CT examination show that the axial odontoid process was free with atlantoaxial dislocation

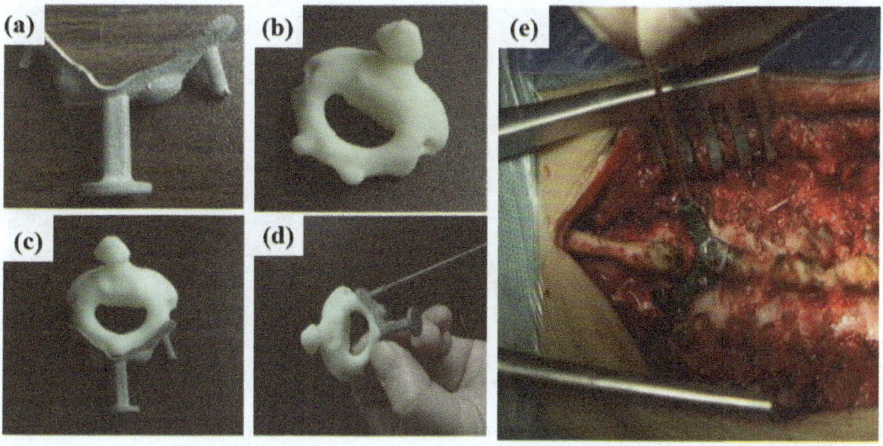

Fig. 5.20 **a** An AM metal nail guide plate, **b** and AM physical drawing of axis model, **c** a guide plate fitting with the axis model, **d** accuracy of the guide plate was tested by a simulated operation before the operation, and **e** clinical application of personalized surgical guide plate

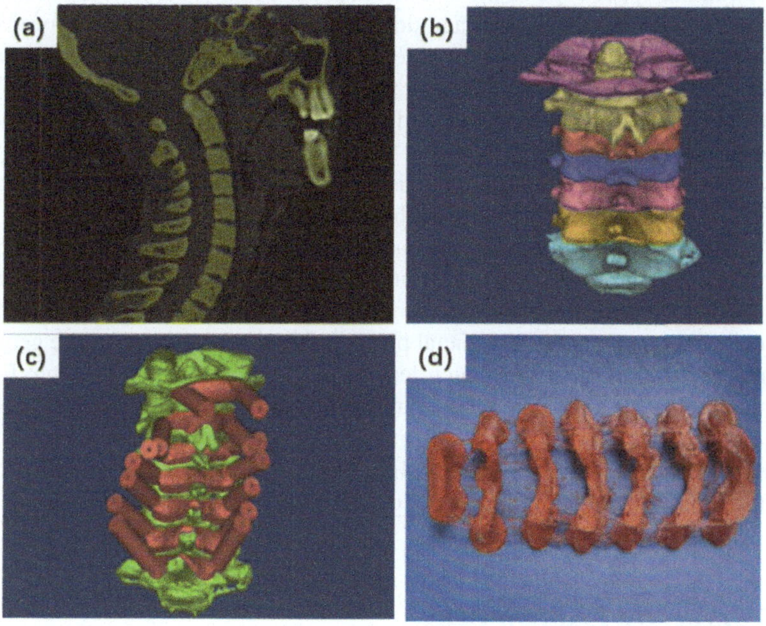

Fig. 5.21 **a** Growing region, **b** 3D reconstruction, **c** C1–7 navigational template models, and **d** 3D navigational templates. Reprinted with permission from a work by Zhang et al. [113]

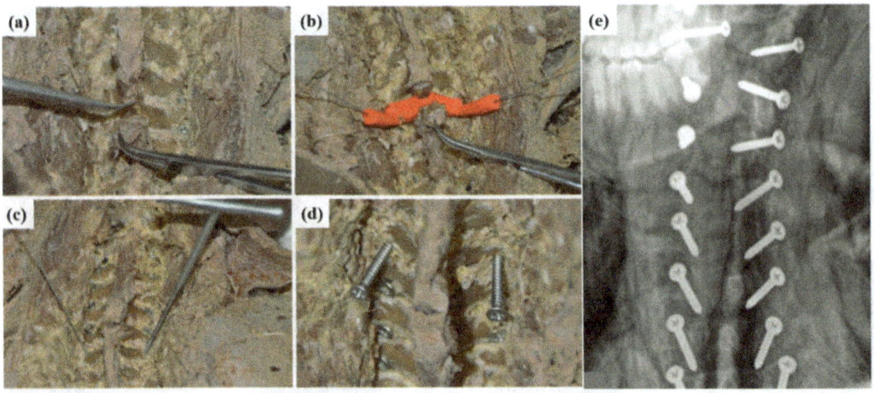

Fig. 5.22 **a** Removal of the interspinal ligaments, **b** docking of the navigational template, **c** opening the cortex with a hollow drill, **d** insertion of the screw, and **e** postoperative X-rays. Reprinted with permission from a work by Zhang et al. [113]

dissected to the upper edge of the spinous process. Finally, the navigation template was inserted through the incision and docked to make sure that the platform would not be displaced by push. The accurate positioning of cervical pedicle screws in cadavers using the navigation template was realized, proving that the guiding tool can effectively prevent intraoperative drift and accurately locate cervical pedicle screws, as shown in Fig. 5.22.

The use of AM navigation template is not affected by the adjacent vertebrae but also has some limitations. Firstly, from the aspect of data collection, the design of the guide plate is based on the patient's CT data with a certain precision, which has certain requirements for the equipment conditions of the hospital. Secondly, from the aspect of the guide plate's design, it is necessary to fully communicate between the person with medical knowledge and the metal printing designer to determine the geometric parameters of the guide plate, which leads to the increase in design time. Finally, in terms of the accuracy of the guide plate, due to the complex anatomical structure of the spine, the fitting surface of the guide plate produces certain deformation in design and processing, which leads to the deviation of the position and direction of the screw hole and affects the accuracy of nail placement.

5.4.2 Fracture Fixation Plate

In recent years, statistics show that the number of orthopedic hospitals and outpatients is increasing in China. The rate of traumatic fracture is gradually increasing, and therefore the number of such patients is gradually increasing accordingly, and the types of diseases are becoming more and more complicated and diversified. In the surgical treatment of traumatic fracture, difficulties, such as finding the direction of the fracture line, determining the location of the fracture block and the fixation scheme, adjusting the position of nails and steel plates, etc., are often encountered.

For example, for some complex acetabular fractures, the traditional reduction methods and nail placement can only be completed by doctors who have experience and surgical skills. Moreover, the standard plate used in the traditional treatment of complex pelvic acetabular fractures is poorly matched and can be not bent in the place with holes before the operation, and which limits fit between the plate and the bone surface. That seriously reduces the effect of operation and prolongs the time of operation. Therefore, the surgical guide plate with accurate screw angle and direction can be produced by AM, which can help the operator to reset and screw placement accurately and improve the accuracy of surgical operation.

Wang et al. [32] and He et al. [114] did a lot of research on the application of SLM in medicine, especially in orthopedics including process optimization, mechanical property testing, and biocompatibility of medical implants. For example, a male patient with a comminuted intra-articular fracture of the distal femur caused by a car accident was successfully applied. Mai et al. [115] studied the personalized design method of pelvic acetabular fracture, the design flow of personalized plate, the exploration of process window, and the biocompatibility of Ti6Al4V alloy formed by SLM. Based on the successful application of personalized plate in the operation of pelvic acetabular fracture, the eight cases of the design process were summarized.

Firstly, the patients were scanned by spiral CT, and the processed images were stored in DICOM format. The pelvic CT scanned image data were imported into Mimics15.0 digital medical processing software by lossless compression, and the 3D digital model of the affected pelvis was exported and repaired in STL format. Then, the previously intact pelvic model in STP format was used to extract these areas from the pelvic surface model as the initial design surface by the isometric surface function of SolidWorks. The edge of the personalized plate was modified and the boundary was smooth, the final model of the plate was got. Finally, compared with the damaged pelvic model, the screw position was determined in some areas of the initial surface of the plate to repair the damaged pelvis, as shown in Fig. 5.23a–f. The SLM technique was used to print the disc plate, which fitted perfectly to the patient's fracture site. The operation was very successful, and the patient's pelvic and acetabular fractures recovered well, as shown in Fig. 5.23g–j.

The systematic theory of parameterized design of pelvic acetabular fracture plate is obtained, which can quickly and efficiently obtain the design method of personalized pelvic acetabular fracture plate. Besides, it was proved that the Ti6Al4V alloy produced by SLM had excellent corrosion resistance, non-cytotoxicity, excellent mechanical properties, and biocompatibility, which provided a basis for applying SLM molded Ti6Al4V alloy in orthopedic implants.

Zheng et al. [116] designed the locking compression pediatric hip plate (LCP-PHP; Cindy, Switzerland), it can be applied to all kinds of orthopedic surgery in children. At present, it is mainly used to treat hip varus, hip valgus, or femoral neck fracture. During the operation, the navigation template of each case matched the marker of the proximal femur. Three screws can be accurately inserted into the femoral neck to stabilize the fracture with LCP-PHP. The process was as follows:

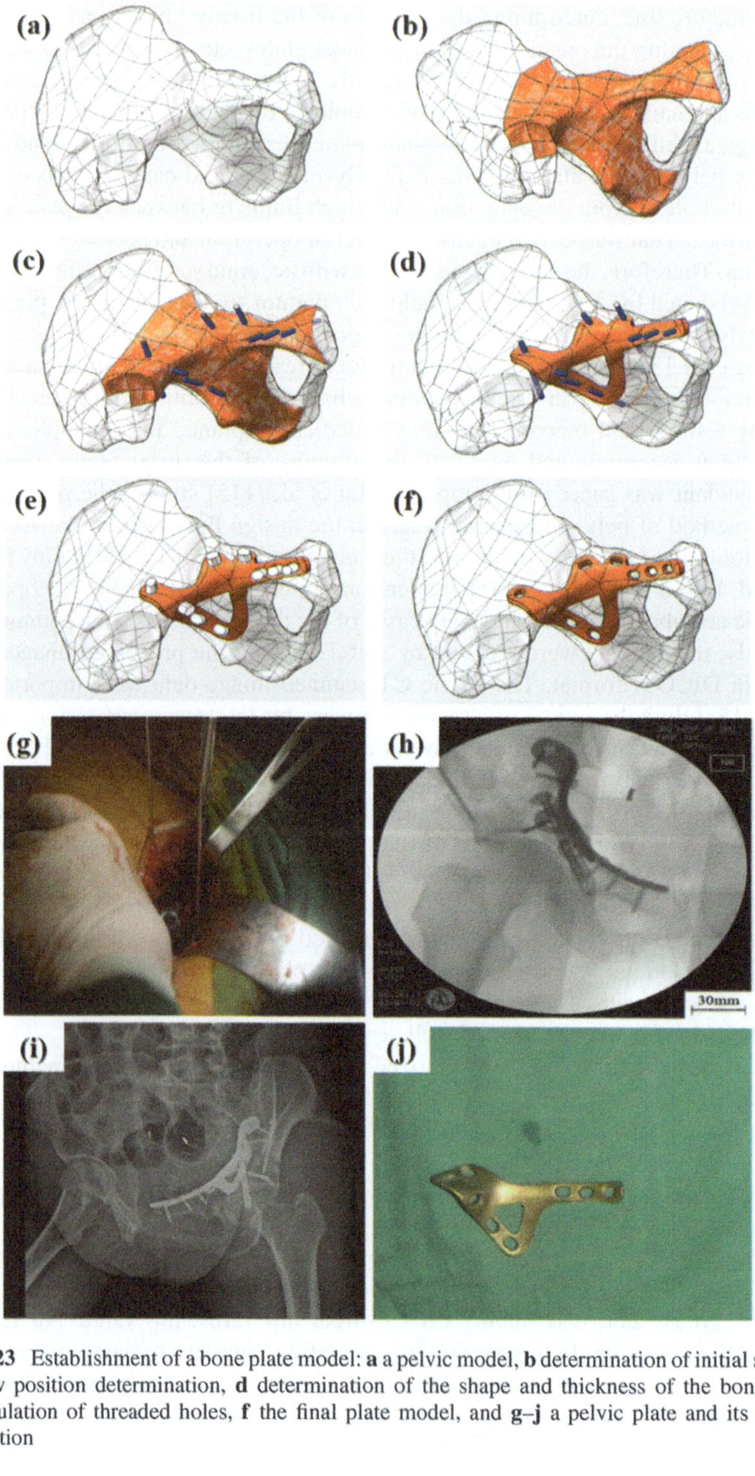

Fig. 5.23 Establishment of a bone plate model: **a** a pelvic model, **b** determination of initial surface, **c** screw position determination, **d** determination of the shape and thickness of the bone plate, **e** formulation of threaded holes, **f** the final plate model, and **g–j** a pelvic plate and its clinical application

The computed tomography data of children with fractures were imported into Mimics 17.0 software (Materialise NV, Leuven, Belgium) for 3D reconstruction of the proximal femur. Then, the local coordinate system of the hollow screw was slowly move and rotate to see if the screw was in the femoral neck and the axial and sagittal plane of the 3D interface. Optimization of screw path and screw length resulted in a high degree of matching between the printed polymer models of the reduced proximal femur and the navigation template. Before the operation, the model and the navigation template were sterilized by plasma, the navigation template was attached to the matching area of the proximal femur and the guide pin was inserted into the designed depth to maintain the fracture plate fixation. The radiogram obtained three weeks after the operation was effective.

Therefore, according to the special demands of complex fracture surgery and the patient's condition, computer-aided technology and SLM technology can be used to design and manufacture a fully fitted personalized bone plate, to improve the reduction quality of fracture, reduce the operation time, and bring revolutionary development for the realization of truly minimally invasive, personalized and accurate operation.

5.4.3 Bone Cutting Guide Plate

During the operation, the doctor's personal experience determines the operation of traditional orthopedic osteotomy surgery. In the resection process, surgical instruments (such as bone drills and bone knives) are difficult to locate and remove accurately, which increases the uncertainty of the operation time and results, and brings greater pain and health risks to patients. To overcome the problems of a complex operation, poor accuracy, and individual differences in traditional surgery, a personalized surgery board is very necessary.

The osteotomy's 3D model was printed for preoperative planning. The range of internal fixation fusion, decompression, and osteotomy was determined to remove the focus more accurately and improve the success rate of the operation. The template molded by SLM technology can locate accurately and was carry out high-temperature disinfection directly, which avoids debris production in the drilling process.

Yang et al. [117] conducted in-depth research on the direct manufacturing of 316L stainless steel parts by SLM technology. 316L stainless steel has the advantages of low cost, enough hardness, good biocompatibility, convenient disinfection, and more accurate positioning. Therefore, it is especially suitable for the manufacture of personalized medical auxiliary devices such as surgical templates.

The STL format obtained from Mimics is inputted into Pro/E, then the surface of the surgical template is constructed, offset, merged, and materialized. On this basis, the guide cylinder, the connecting device between the two pieces, and two retention holes for screw fixation are drawn according to the design intent. The designed effect diagram is shown in Fig. 5.24a. The CAD model is transformed into an STL format file, and the wireframe model is repaired. The processing position is placed, and

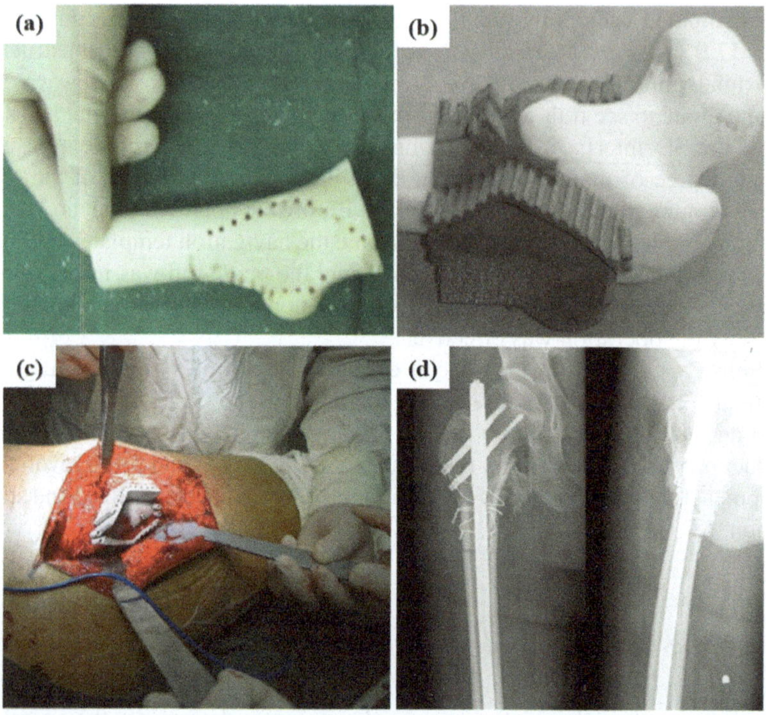

Fig. 5.24 **a** Assembling simulation before surgery, **b** trimming allograft bone in vitro, **c** placed model during operation, and **d** CT examination after the operation

the support is set through Magics12.0. The final layering process selection of the surgical template is completed, and the template model in CLI format is output to prepare for the next processing. The CLI format file generated by the slice is inputted into the SLM rapid prototyping system. After being adjusted to the best processing parameters, and the system is started for processing. The molded parts can be closely attached to the lesion site, as shown in Figs. 5.24b.

The template was successfully applied to the removed tumor of the upper femur of a male patient in the Army General Hospital of Guangzhou Military Region. The in vitro trimmed allogeneic bone was implanted into the resected area to help doctors to complete the operation more accurately, quickly, safely, and smoothly. Postoperative CT showed that the effect of the operation was good, as shown in Fig. 5.24c, d.

He [118] also studied the surgical reduction template for comminuted fracture. According to the patient's high-precision CT tomography data, a positioning template for surgical reduction of comminuted fracture was designed, as shown in Fig. 5.25a, b. In one case, for the need of the operation, four templates completely were attached to the bone surface and a circular guide tube was designed to be used together, and

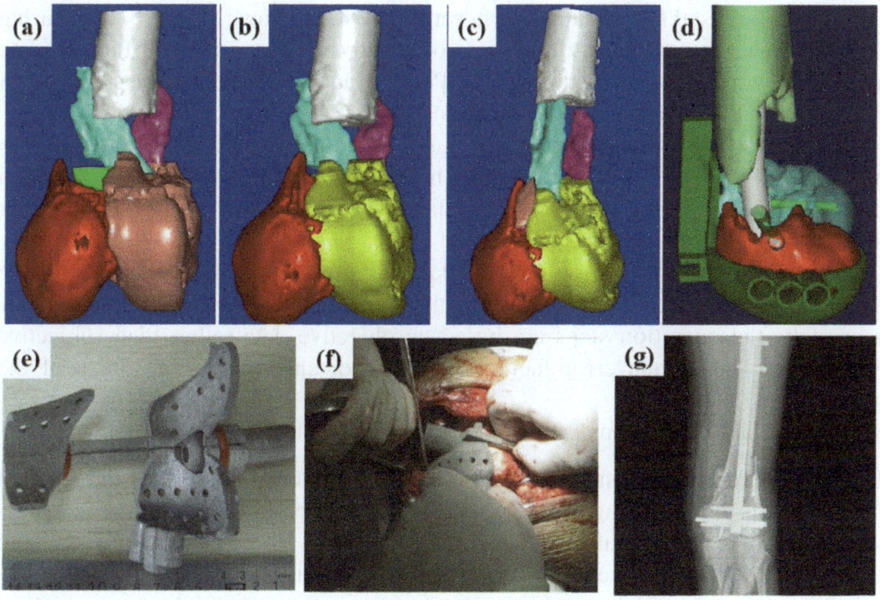

Fig. 5.25 **a** Preprocessing of CT image of the femur, **b** reduction of the articular surface, **c** bone fragment reduction, **d** surgical template, **e** designed surgical template, **f** placing template during operation, and **g** postoperative CT examination

a connecting device was also designed to fix the various parts of the template. The ligation ring was used to connect during the operation.

The soft tissue was removed by threshold segmentation and regional growth function, a clear figure of the femur was obtained, and a 3D fracture model was established. After completing the structural design of the guide plate, anatomical reduction of the labeled displaced bone piece was performed, the surgical reduction process was simulated, and the anatomical structure was restored, as shown in Fig. 5.25c, d. The processing results were transformed into STL format that can be recognized by the design software of Pro/E. After the completion of virtual reduction, the STL format femur model was imported into the 3D design software Pro/E to restore the appropriate position of the model at the fracture site of the bone joint, and the joint surface was removed through operations (such as selection surface, offset, merging, materialization), and then the necessary accessories were designed according to the preoperative virtual reduction results and the operation plan.

The five surgical templates cooperate well mutually by the optimized process of SLM rapid prototyping, as shown in Fig. 5.25e. The total assembly length is 143.2 mm, the hardness of the sample is HB188, the design size is taken as the standard, the error of the formed parts is less than ±0.5 mm and surface roughness is 30 μm. The dimensional accuracy, surface quality, and cooperation between various parts can fully meet the requirements of clinical reduction surgery. The molded surgical template was applied to a male patient with a comminuted intra-articular fracture of the distal femur caused by a car accident, as shown in Fig. 5.25f. During the operation, the operation effect was good, and the anastomosis of the template anatomically reduced the fractured piece. Postoperative CT examination showed that the reduction effect was good, as shown in Fig. 5.25g.

At the same time, the AM guide plate, used by Belgian AM service Materialise to help doctors plan osteotomy for children, has been approved by the US FDA, which means that children with congenital bone diseases or bone injuries will soon be better treated.

The individualized osteotomy guide plate designed by SLM is used to guide the positioning and osteotomy operation of the osteotomy guide plate. To osteotomy accurately, the operation steps are based on objective data rather than subjective experience. The human error factors are reduced, ensuring the accuracy of the operation.

However, AM osteotomy guide plate still has the following shortcomings:

1. In clinical application, due to the limitation of the visual field of the surgical incision, the matching surface between the osteotomy guide plate and bony structure is often small, which is poor and erratic. There are displacement and angulation during osteotomy, leading to the deviation of the osteotomy surface.
2. The materials widely used to make AM osteotomy guide plates are medical polymer materials, metal powder materials, photosensitive resin materials, ABS resin materials, etc. As the cytotoxicity of materials and the high production cost of the metal powder materials, it is not conducive to popularization and application.
3. This technology's clinical popularization and application are affected by the high cost and long cycle of AM osteotomy guide plate.

5.5 Conclusion and Outlook

This chapter summarizes the application of AM in metal implants and metal guide plates. The main conclusions are summarized as follows: Firstly, the materials and the properties of metal implants are introduced, mainly including three kinds of common metals materials (Fe-based alloys, Ti-based alloys, Co-based alloys) and three other types of metals materials, which can help to select different metal implant materials according to different medical demands. Secondly, comparing the AM cases of metal implants and guide plates studied by our group with other medical cases in the market, the application requirements, design process, printing process final clinical application of these cases are introduced, which provides different solution ideal for AM of different types of implants and guide plates.

With the development of AM technology, the application of AM in metal implants and metal guides has become more and more mature and extensive, especially having the advantage of complex structures and personalized customization to meet the demands of patients. So far, due to the limitation of biomaterials, manufacturing factors, and structural characteristics, there is still a gap between the ideal condition and the actual metal implant or guide plate, whose properties are still defective. Therefore, there are still research challenges to be overcome, in areas such as:

New metal biomaterials development: Selected biomaterials generally must meet the requirements of the biocompatibility and mechanical properties of implants. Some new metal materials, such as the porous tantalum, the degradable metal, the shape memory alloy, are better implant materials because their properties better match bone properties. To improve the toughness of implants, nickel can be added to titanium alloy. However, excessive nickel ions will cause toxicity to cells. Therefore, according to the idea of alloying, different metal mixture powders are combined by powder bed processing technology to obtain new implant materials, to eliminate the negative effects of toxic ions. In addition, PBF technology can combine expensive metals with other elements to improve expensive metal alloys' physical and chemical properties.

New structural design methods: Recently, the porous structure has been widely used in metal implants. Compared with other implants, porous implants have sufficient strength and stiffness, and can effectively avoid stress shielding. On the other hand, porous implants possessing good biocompatibility are suitable for cell penetration and growth. So far, the traditional design methods of porous structure mainly include constructing solid geometry based on UG, CATIA, and other CAD software, and designing images based on Micro CT, etc. However, the methods cannot effectively control and estimate the structure's mechanical properties and have the disadvantages of a complex implementation process, a large amount of data, and low computational efficiency. At South China University of Technology (SCUT), a porous implant was obtained by the mapping of a shaped control node. The method significantly reduces the implant's weight, facilitates the intermeshing fixation of the implant with the host bone and provides good biomechanical properties. Moreover, natural bone's porosity and topological changes can be restored via new modeling algorithms and connection strategies to design a bionic gradient structure.

Coordination of other processing methods: The parts produced by SLM technology have the characteristics of relatively poor surface quality and low surface precision, while traditional machining has the characteristics of high accuracy. Therefore, good manufacturing results can be obtained by the combination of two processing methods. Besides, researching and using new surface modification technologies can change the surface microstructure, promote the growth of bone cells and the formation of hydroxyapatite, improve the antibacterial properties of implants, and obtain metal implants with good biocompatibility.

The application of four-dimensional (4D) printing technology: 4D printing technology is a new technology developed by combining AM technology with intelligent material structures. Intelligent material structures realize their structural change with time under the excitation of the external environment based on AM, thus adding a time dimension to 3D objects. As one of the promising technologies in the medical field, 4D printing has great application value in intelligent medicine. Orthopedics, dentistry, dermatology, etc., will provide a broad space for applying 4D printing technology.

References

1. Han C, Fang Q, Shi Y et al (2020) Recent advances on high-entropy alloys for 3D printing. Adv Mater 32(26):1903855
2. Li B, Zheng H, Han C et al (2021) Nanotwins-containing microstructure and superior mechanical strength of a Cu-9Al-5Fe-5Ni alloy additively manufactured by laser metal deposition. Addit Manuf 39:101825
3. Li B, Han C, Wee JLC et al (2021) Interface formation and deformation behaviors of an additively manufactured nickel-aluminum-bronze/15-5 PH multimaterial via laser-powder directed energy deposition. Mater Sci Eng A 142101
4. Buj-Corral I, Tejo-Otero A, Fenollosa-Artés F (2020) Development of AM technologies for metals in the sector of medical implants. Metals 10(5):686
5. Bandyopadhyay A, Espana F, Balla VK et al (2010) Influence of porosity on mechanical properties and in vivo response of Ti6Al4V implants. Acta Biomater 6(4):1640–1648
6. Parthasarathy J (2014) 3D modeling, custom implants and its future perspectives in craniofacial surgery. Ann Maxillofac Surg 4(1):9
7. Bai L, Gong C, Chen X et al (2019) Additive manufacturing of customized metallic orthopedic implants: materials, structures, and surface modifications. Metals 9(9):1004
8. ASTM F138-13a (2013) Standard specification for wrought 18 chromium-14 nickel-2.5 molybdenum stainless steel bar and wire for surgical implants (UNS S31673). ASTM Int
9. Bai Y, Fu F, Xiao Z et al (2018) Progress in selective laser melting equipment, related biomedical metallic materials and applications. J Zhejiang Univ Sci A 19(2):122–136
10. Sun Z, Tan X, Tor SB et al (2016) Selective laser melting of stainless steel 316L with low porosity and high build rates. Mater Des 104:197–204
11. Tolosa I, Garciandía F, Zubiri F et al (2010) Study of mechanical properties of AISI 316 stainless steel processed by "selective laser melting", following different manufacturing strategies. Int J Adv Manuf Technol 51(5):639–647
12. Laleh M, Hughes AE, Xu W et al (2019) Unexpected erosion-corrosion behaviour of 316L stainless steel produced by selective laser melting. Corros Sci 155:67–74
13. Wang D (2011) Study on the fabrication properties and process of stainless steel parts by selective laser melting. South China University of Technology
14. Zhang L, Wang J (2014) Effect of temperature and loading mode on environmentally assisted crack growth of a forged 316L SS in oxygenated high-temperature water. Corros Sci 87:278–287
15. Čapek J, Machova M, Fousova M et al (2016) Highly porous, low elastic modulus 316L stainless steel scaffold prepared by selective laser melting. Mater Sci Eng C 69:631–639
16. Kong DC, Ni XQ, Dong CF et al (2018) Bio-functional and anti-corrosive 3D printing 316L stainless steel fabricated by selective laser melting. Mater Design 152
17. Al-Tamimi AA, Al-Qahtani MS, Liu F et al (2021) A review on powder bed fusion additive manufacturing for metallic fixation implants. In: Virtual prototyping & bio manufacturing in medical applications. Springer, Cham, pp 235–256
18. He J, He FL, Li DW et al (2016) Advances in Fe-based biodegradable metallic materials. RSC Adv 6(114):112819–112838
19. Li Y, Jahr H, Lietaert K et al (2018) Additively manufactured biodegradable porous iron. Acta Biomater 77:380–393
20. Li Y, Jahr H, Pavanram P et al (2019) Additively manufactured functionally graded biodegradable porous iron. Acta Biomater 96:646–661
21. Shuai C, Yang W, Yang Y et al (2020) Selective laser melted Fe-Mn bone scaffold: microstructure, corrosion behavior and cell response. Mater Res Expr 7(1):015404
22. Brånemark R, Brånemark PI, Rydevik B et al (2001) Osseo integration in skeletal reconstruction and rehabilitation. J Rehabil Res Dev 38(2):1–4
23. Elias CN, Lima JHC, Valiev R et al (2008) Biomedical applications of titanium and its alloys. JOM 60(3):46–49

24. Zhang LC, Attar H (2016) Selective laser melting of titanium alloys and titanium matrix composites for biomedical applications: a review. Adv Eng Mater 18(4):463–475
25. Wang D, Liu L, Deng G et al (20211) Recent progress on additive manufacturing of multi-material structures with laser powder bed fusion. Virtual Phys Prototyping 1–37
26. Yang L, Han C, Wu H et al (2020) Insights into unit cell size effect on mechanical responses and energy absorption capability of titanium graded porous structures manufactured by laser powder bed fusion. J Mech Behav Biomed Mater 109:103843
27. Ahn YK, Kim HG, Park HK et al (2017) Mechanical and microstructural characteristics of commercial purity titanium implants fabricated by electron-beam additive manufacturing. Mater Lett 187:64–67
28. Attar H, Bermingham MJ, Ehtemam-Haghighi S et al (2019) Evaluation of the mechanical and wear properties of titanium produced by three different additive manufacturing methods for biomedical application. Mater Sci Eng A 760:339–345
29. Attar H, Calin M, Zhang LC et al (2014) Manufacture by selective laser melting and mechanical behavior of commercially pure titanium. Mater Sci Eng A 593:170–177
30. Han C, Li Y, Wang Q et al (2018) Continuous functionally graded porous titanium scaffolds manufactured by selective laser melting for bone implants. J Mech Behav Biomed Mater 80:119–127
31. Fan H, Fu J, Li X et al (2015) Implantation of customized 3-D printed titanium prosthesis in limb salvage surgery: a case series and review of the literature. World J Surg Oncol 13(1):1–10
32. Wang D, Wang Y, Wu S et al (2017) Customized a Ti6Al4V bone plate for complex pelvic fracture by selective laser melting. Materials 10(1):35
33. ASTM F (2013) Standard specification for wrought titanium-6aluminum-4vanadium ELI (extra low interstitial) alloy for surgical implant applications (UNS R56401). ASTM F136, 13
34. Rafi HK, Karthik NV, Gong H et al (2013) Microstructures and mechanical properties of Ti6Al4V parts fabricated by selective laser melting and electron beam melting. J Mater Eng Perform 22(12):3872–3883
35. Zhai Y, Galarraga H, Lados DA (2016) Microstructure, static properties, and fatigue crack growth mechanisms in Ti-6Al-4V fabricated by additive manufacturing: LENS and EBM. Eng Fail Anal 69:3–14
36. Van Hooreweder B, Moens D, Boonen R et al (2012) Analysis of fracture toughness and crack propagation of Ti6Al4V produced by selective laser melting. Adv Eng Mater 14(1–2):92–97
37. Ataee A, Li Y, Fraser D et al (2018) Anisotropic Ti-6Al-4V gyroid scaffolds manufactured by electron beam melting (EBM) for bone implant applications. Mater Des 137:345–354
38. Wang D, Wang H, Chen X et al (2022) Densification, Tailored Microstructure, and Mechanical Properties of Selective Laser Melted Ti–6Al–4V Alloy via Annealing Heat Treatment. Micromachines 13(2):331
39. Wong KC, Scheinemann P (2018) Additive manufactured metallic implants for orthopaedic applications. Sci China Mater 61(4):440–454
40. Wang H, Zhao B, Liu C et al (2016) A comparison of biocompatibility of a titanium alloy fabricated by electron beam melting and selective laser melting. PLoS ONE 11(7):e0158513
41. Sing SL, An J, Yeong WY et al (2016) Laser and electron-beam powder-bed additive manufacturing of metallic implants: a review on processes, materials and designs. J Orthop Res 34(3):369–385
42. Niinomi M, Nakai M (2011) Titanium-based biomaterials for preventing stress shielding between implant devices and bone. Int J Biomater 2011
43. Wu YC, Kuo CN, Shie MY et al (2018) Structural design and mechanical response of gradient porous Ti-6Al-4V fabricated by electron beam additive manufacturing. Mater Des 158:256–265
44. Chlebus E, Kuźnicka B, Kurzynowski T et al (2011) Microstructure and mechanical behaviour of Ti-6Al-7Nb alloy produced by selective laser melting. Mater Charact 62(5):488–495
45. Xu C, Sikan F, Atabay SE et al (2020) Microstructure and mechanical behavior of as-built and heat-treated Ti–6Al–7Nb produced by laser powder bed fusion. Mater Sci Eng A 793:139978

46. Rotaru H, Armencea G, Spîrchez D et al (2013) In vivo behavior of surface modified Ti6Al7Nb alloys used in selective laser melting for custom-made implants. A preliminary study. Rom J Morphol Embryol 54(3 suppl):791–796
47. Fischer M, Laheurte P, Acquier P et al (2017) Synthesis and characterization of Ti-27.5 Nb alloy made by CLAD® additive manufacturing process for biomedical applications. Mater Sci Eng C 75:341–348
48. Schulze C, Weinmann M, Schweigel C et al (2018) Mechanical properties of a newly additive manufactured implant material based on Ti-42Nb. Materials 11(1):124
49. Zhao D, Han C, Li J et al (2020) In situ fabrication of a titanium-niobium alloy with tailored microstructures, enhanced mechanical properties and biocompatibility by using selective laser melting. Mater Sci Eng C 111:110784
50. Liang H, Zhao D, Feng X et al (2020) 3D-printed porous titanium scaffolds incorporating niobium for high bone regeneration capacity. Mater Des 194:108890
51. Zhao D, Liang H, Han C et al (2021) 3D printing of a titanium-tantalum gyroid scaffold with superb elastic admissible strain, bioactivity and in-situ bone regeneration capability. Addit Manuf 47:102223
52. Zhou Y-L, Niinomi M (2008) Microstructures and mechanical properties of Ti–50 mass% Ta alloy for biomedical applications. Alloys Compd 466:535–542
53. Zhao D, Han C, Li Y et al (2019) Improvement on mechanical properties and corrosion resistance of titanium-tantalum alloys in-situ fabricated via selective laser melting. J Alloy Compd 804:288–298
54. Sing SL, Yeong WY, Wiria FE (2016) Selective laser melting of titanium alloy with 50 wt% tantalum: microstructure and mechanical properties. J Alloy Compd 461–470
55. Sing SL, Wiria FE, Yeong WY (2018) Selective laser melting of titanium alloy with 50 wt% tantalum: effect of laser process parameters on part quality. Int J Refract Metals Hard Mater 120–127
56. Soro N, Attar H, Brodie E et al (2019) Evaluation of the mechanical compatibility of additively manufactured porous Ti–25Ta alloy for load-bearing implant applications. J Mech Behav Biomed Mater 97:149–158
57. Attar H, Bönisch M, Calin M et al (2014) Comparative study of microstructures and mechanical properties of in situ Ti–TiB composites produced by selective laser melting, powder metallurgy, and casting technologies. J Mater Res 29(17):1941–1950
58. Han C, Babicheva R, Chua JDQ et al (2020) Microstructure and mechanical properties of (TiB+TiC)/Ti composites fabricated in situ via selective laser melting of Ti and B4C powders. Addit Manuf 36:101466
59. Han C, Wang Q, Song B et al (2017) Microstructure and property evolutions of titanium/nano-hydroxyapatite composites in-situ prepared by selective laser melting. J Mech Behav Biomed Mater 71:85–94
60. Han C, Li Y, Wang Q et al (2018) Titanium/hydroxyapatite (Ti/HA) gradient materials with quasi-continuous ratios fabricated by SLM: material interface and fracture toughness. Mater Des 141:256–266
61. Disegi JA, Kennedy RL, Pilliar R (1999) Cobalt-base alloys for biomedical applications. ASTM, Danvers
62. Mahapatro A (2012) Metals for biomedical applications and devices. J Biomater Tissue Eng 2(4):259–268
63. ASTM F75-12 (2012) Standard specification for chromium-6 molybdenum alloy castings and casting alloy for surgical implants (UNS R30075). ASTM Int
64. Zhang M, Yang Y, Song C et al (2018) An investigation into the aging behavior of CoCrMo alloys fabricated by selective laser melting. J Alloy Compd 750:878–886
65. Xiang DD, Wang P, Tan XP et al (2019) Anisotropic microstructure and mechanical properties of additively manufactured Co–Cr–Mo alloy using selective electron beam melting for orthopedic implants. Mater Sci Eng A 765:138270
66. Tan XP, Wang P, Kok Y et al (2018) Carbide precipitation characteristics in additive manufacturing of Co-Cr-Mo alloy via selective electron beam melting. Scripta Mater 143:117–121

67. Ghani SAC, Mohamed SR, Harun WSW et al (2017) Physical evaluations of Co-Cr-Mo parts processed using different additive manufacturing techniques. In: AIP Conference proceedings, vol 1901(1). AIP Publishing LLC, pp 100001
68. Hong JH, Yeoh FY (2020) Mechanical properties and corrosion resistance of cobalt-chrome alloy fabricated using additive manufacturing. Mater Today Proc 29:196–201
69. Song C, Zhang M, Yang Y et al (2018) Morphology and properties of CoCrMo parts fabricated by selective laser melting. Mater Sci Eng A 713:206–213
70. Zhang G, Yang Y, Song C et al (2018) Study on biocompatibility of CoCrMo alloy parts manufactured by selective laser melting. J Med Biol Eng 38(1):76–86
71. Hedberg YS, Qian B, Shen Z et al (2014) In vitro biocompatibility of CoCrMo dental alloys fabricated by selective laser melting. Dent Mater 30(5):525–534
72. Han C, Yao Y, Cheng X et al (2017) Electrophoretic deposition of gentamicin-loaded silk fibroin coatings on 3D-printed porous cobalt–chromium–molybdenum bone substitutes to prevent orthopedic implant infections. Biomacromolecules 18(11):3776–3787
73. Alvarez K, Nakajima H (2009) Metallic scaffolds for bone regeneration. Materials 2(3):790–832
74. Black J (1994) Biologic performance of tantalum. Clin Mater 16(3):167–173
75. Wauthle R, Van Der Stok J, Yavari SA et al (2015) Additively manufactured porous tantalum implants. Acta Biomater 14:217–225
76. Guo Y, Xie K, Jiang W et al (2018) In vitro and in vivo study of 3D-printed porous tantalum scaffolds for repairing bone defects. ACS Biomater Sci Eng 5(2):1123–1133
77. Bobyn JD, Poggie RA, Krygier JJ et al (2004) Clinical validation of a structural porous tantalum biomaterial for adult reconstruction. JBJS 86(suppl_2):123–129
78. Thijs L, Sistiaga MLM, Wauthle R et al (2013) Strong morphological and crystallographic texture and resulting yield strength anisotropy in selective laser melted tantalum. Acta Mater 61(12):4657–4668
79. Witte F, Ulrich H, Rudert M et al (2007) Biodegradable magnesium scaffolds. Part 1: Appropriate inflammatory response. J Biomed Mater Res Part A 81(3):748–756
80. Witte F, Reifenrath J, Müller PP et al (2006) Cartilage repair on magnesium scaffolds used as a subchondral bone replacement. Materialwiss Werkstofftech Entwicklung Fertigung Prüfung Eigenschaften Anwendungen Technischer Werkst 37(6):504–508
81. Li Y, Zhou J, Pavanram P et al (2018) Additively manufactured biodegradable porous magnesium. Acta Biomater 67:378–392
82. Xu R, Zhao MC, Zhao YC et al (2019) Improved biodegradation resistance by grain refinement of novel antibacterial ZK30-Cu alloys produced via selective laser melting. Mater Lett 237:253–257
83. Sezer N, Evis Z, Koc M (2021) Additive manufacturing of biodegradable magnesium implants and scaffolds: review of the recent advances and research trends. J Magnes Alloys 9(2):392–415
84. Ng CC, Savalani MM, Lau ML et al (2011) Microstructure and mechanical properties of selective laser melted magnesium. Appl Surf Sci 257(17):7447–7454
85. Wen P, Voshage M, Jauer L et al (2018) Laser additive manufacturing of Zn metal parts for biodegradable applications: processing, formation quality and mechanical properties. Mater Des 155:36–45
86. Hermawan H (2018) Updates on the research and development of absorbable metals for biomedical applications. Prog Biomater 7(2):93–110
87. Demir AG, Monguzzi L, Previtali B (2017) Selective laser melting of pure Zn with high density for biodegradable implant manufacturing. Addit Manuf 15:20–28
88. Wen P, Jauer L, Voshage M et al (2018) Densification behavior of pure Zn metal parts produced by selective laser melting for manufacturing biodegradable implants. J Mater Process Technol 258:128–137
89. Li Y, Pavanram P, Zhou J et al (2020) Additively manufactured biodegradable porous zinc. Acta Biomater 101:609–623

90. Qin Y, Wen P, Voshage M et al (2019) Additive manufacturing of biodegradable Zn-xWE43 porous scaffolds: formation quality, microstructure and mechanical properties. Mater Des 181:107937
91. Röttger A, Boes J, Theisen W et al (2020) Microstructure and mechanical properties of 316L austenitic stainless steel processed by different SLM devices. Int J Adv Manuf Technol 108:769–783
92. Vrancken B, Thijs L, Kruth JP et al (2012) Heat treatment of Ti6Al4V produced by selective laser melting: microstructure and mechanical properties. J Alloy Compd 541:177–185
93. Zhou L, Yuan T, Li R et al (2017) Selective laser melting of pure tantalum: densification, microstructure and mechanical behaviors. Mater Sci Eng A 707:443–451
94. Wei K, Gao M, Wang Z et al (2014) Effect of energy input on formability, microstructure and mechanical properties of selective laser melted AZ91D magnesium alloy. Mater Sci Eng A 611:212–222
95. Wang AM (2018) Study on personalized design method and verification of total knee prosthesis manufactured by 3D printing technology. South China University of Technology
96. Korkmaz IH, Kaymaz İ, Yildirim ÖS (2016) A new design for maximum conformity of total knee prosthesis to femur and tibia. J Long-Term Effects Med Implants 26(4)
97. Xiao DM (2013) Molding of porous structure of implants and direct manufacturing by selective laser melting. South China University of Technology
98. van Arkel RJ, Ghouse S, Milner PE et al (2018) Additive manufactured push-fit implant fixation with screw-strength pull out. J Orthop Res 36(5):1508–1518
99. Chen X, Wang D, Dou W et al (2020) Design and manufacture of bionic porous titanium alloy spinal implant based on selective laser melting (SLM). Comput Model Eng Sci 124(3):1099–1117
100. Bellia-Munzon G, Martinez J, Toselli L et al (2020) From bench to bedside: 3D reconstruction and printing as a valuable tool for the chest wall surgeon. J Pediatr Surg 55(12):2703–2709
101. Zhang H, Huang LJ et al (2020) Clinical application of 3D-printed titanium thoracic rib implants in chest wall reconstruction. Chin Clin J Thorac Cardiovasc Surg 27(03):268–273
102. Mobbs RJ, Coughlan M, Thompson R et al (2017) The utility of 3D printing for surgical planning and patient-specific implant design for complex spinal pathologies: case report. J Neurosurg Spine 26(4):513–518
103. Mai SZ (2016) Study on the forming processes and properties of customized CoCr alloy crowns and fixed bridges manufactured by selective laser melting. South China University of Technology
104. Sun TT (2010) Research on direct manufacturing process of customized lingual brackets by selective laser melting. South China University of Technology
105. Wang XL (2016) Study on process optimization and property of titanium alloy manufactured by selective laser melting. South China University of Technology
106. Yao WJ (2016) The first FDA-approved 3D-printed titanium cranial implant. Chin Titanium Ind 01:49
107. Xiao R (2018) Study on the lightweight design and manufacturing of customized porous mandibular prothesis based on selective laser melting. South China University of Technology
108. Nickels L (2012) World's first patient-specific jaw implant. Met Powder Rep 67(2):12–14
109. Liverani E, Fortunato A, Leardini A et al (2016) Fabrication of Co–Cr–Mo endoprosthetic ankle devices by means of selective laser melting (SLM). Mater Des 106:60–68
110. Williams JM, Adewunmi A, Schek RM et al (2005) Bone tissue engineering using polycaprolactone scaffolds fabricated via selective laser sintering. Biomaterials 26(23):4817–4827
111. Smith MH, Flanagan CL, Kemppainen JM et al (2007) Computed tomography-based tissue-engineered scaffolds in craniomaxillofacial surgery. Int J Med Rob Comput Assist Surg 3(3):207–216
112. Zhang DS, Wang JH, Wang D et al (2017) Preliminary application of individualized metal drill guide template in upper cervical spine surgery for young children. Chin Orthop J Clin Basic Res 9(01):11–16

113. Zhang GD, Yu Z, Chen X et al (2018) Accurate placement of cervical pedicle screws using 3D-printed navigational templates. Orthopade 47(5):428–436
114. He XR, Yang YQ, Wu WH et al (2010) research on direct forming of comminuted fracture surgery orienting model by selective laser melting. J Biomed Eng 27(03):519–523
115. Mai QG, Gu C, Lin XZ et al (2017) Treatment of complex acetabular fractures with personalized metal 3D printing of acetabular wing plate combined with lateral incision approach of rectus abdominis. Chin J Surg 55(03):172–178
116. Zheng P, Yao Q, Xu P et al (2017) Application of computer-aided design and 3D-printed navigation template in locking compression pediatric hip plate^{TM} placement for pediatric hip disease. Int J CARS 12(5):865–871
117. Yang YQ, He XR, Wu WH et al (2009) Personalized orthopedic surgery templates are directly molded by selective laser melting. Chin J Lasers 36(09):2460–2464
118. He XR (2010) Study on personalized surgical templates directly formed by selective laser melting. South China University of Technology

Chapter 6
Wire Arc Additive Manufacturing: Systems, Microstructure, Defects, Quality Control, and Modelling

Zhe Chen and Gim Song Soh

6.1 Background

Wire Arc Additive Manufacturing (WAAM) stems from conventional arc welding technologies. It is classified as a process of Direct Energy Deposition (DED) in the field of additive manufacturing. As its name indicates, the feedstock of WAAM is in terms of metallic wire, and the energy source is an electric arc by which the wire is fused and deposited at a designed pattern with a motion system. Once solidified, deposited metal forms a complete layer, and a near-net shape part can be printed by repeating this process layer by layer. Compared with powder-bed additive manufacturing methods, WAAM has many competitive edges in the equipment cost, buy-to-fly ratio, and processing efficiency with a deposition rate of 1–10 kg/h [1–3]. Additionally, WAAM does not suffer from volume confinement in an enclosed chamber or limited platform. Therefore, it is especially suitable for the fabrication of medium to large-scale metallic parts with simple to middle structural complexity. Although processing resolution and surface finishing of WAAM-processed parts are usually inferior to counterparts by laser-based methods, WAAM is still gaining momentum in both academia and multiple industries.

For example, marine propeller brackets were successfully manufactured by WAAM for large-scale power systems in ships, with excellent comprehensive mechanical performances [4]. WAAM not only saved mold cost and material loss from conventional subtractive methods but also avoided noticeable deformation and

Z. Chen
Agency for Science, Technology and Research (A*STAR), Singapore Institute of Manufacturing Technology, 2 Fusionopolis Way, Singapore 138634, Singapore
e-mail: chen_zhe@simtech.a-star.edu.sg

G. Song Soh (✉)
Engineering Product Development, Singapore University of Technology and Design, 8 Somapah Rd, Singapore 487372, Singapore
e-mail: sohgimsong@sutd.edu.sg

K. Zhou (ed.), *Additive Manufacturing*,
https://doi.org/10.1007/978-3-031-04721-3_6

shrinkage during processing. With proper tool path planning, different structures in this propeller bracket were printed sequentially, and the final dimension error was controlled within 0.8 mm. Apart from the marine application, WAAM is also employed in the aviation and aircraft manufacturing field. Take wing spars and landing gear assembly made of Ti6Al4V alloy as an instance, they were completely fabricated by WAAM and a material saving of 220 kg per part was achieved [3]. Wing ribs of an aluminum alloy with a length of 2.5 m were also processed by WAAM and a material saving of 500 kg per part was enabled. Moreover, WAAM is also capable of reducing the lead time of components significantly, sometimes from several months to a few days. This is very meaningful for product testing, such as wind tunnel testing for a wing structure. With the WAAM technique, the time from the release of structure design to testing data collection is greatly shortened, and this feature advantage is more and more attractive to many sectors, inclusive of the oil and gas fields, automotive industry, construction, and infrastructure.

For research work from institutes, the number of papers published on WAAM has been increasing steadily since 2012. The printability of most commonly employed metallic materials by WAAM is being studied, such as stainless steel, aluminum alloys, nickel-based superalloys, and titanium alloys, to find a possibility of fabricating parts more efficiently and economically, while maintaining comparable mechanical performances to conventional techniques, such as casting and forging. A number of positive results have been obtained on quality control, microstructure and mechanical properties improvement, and components manufacturing for real applications.

Based on these extensive works, this chapter is intended to provide a general literature review on the recent development of WAAM research, including an introduction of WAAM systems and fundamentals, prevailing materials employed in WAAM and related challenges, common defects, quality control methods, and bead modeling work of WAAM. A summary and outlook of WAAM research are also presented at the end of the chapter.

6.2 WAAM Systems and Fundamentals

A WAAM system can be considered as an integration of an off-the-shelf welding facility with a motion system, which can be a robotic arm, an automated gantry, or a computer numerical control machine. In light of processing flexibility and equipment cost, a robotic arm coupled with a welding torch is one of the most commonly employed WAAM systems, as is shown in Fig. 6.1. This representative WAAM system consists of a robotic arm and a respective controller, a programmable welding power source, a welding torch with a wire-feeding system, a gas supply for local inert atmosphere shielding, and a welding positioner where a substrate for the WAAM process will be located. In addition, there can be many other sensors employed in a WAAM system for various data collection and in-process monitoring, such as a laser

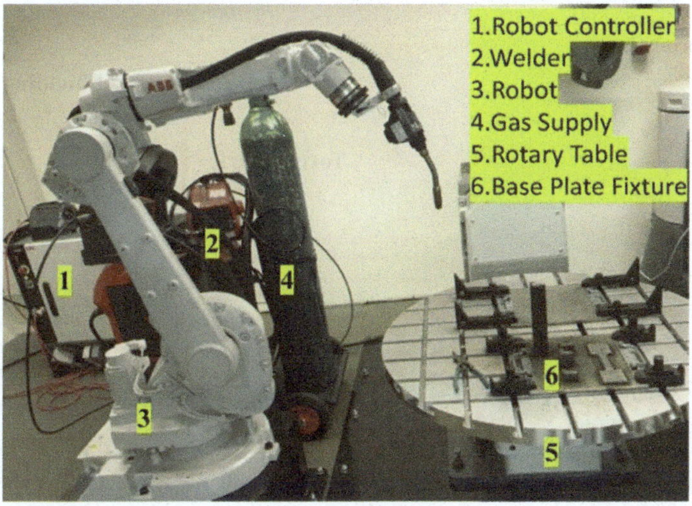

Fig. 6.1 Schematic of a representative WAAM system [5]

scanner, thermal camera, and sound recorder for defect analysis. All these contents will be elaborated on in the following sections.

Since WAAM is derived from arc welding technologies, the categories of WAAM systems are proposed based on different arc welding methods, mainly including Gas Metal Arc Welding (GMAW), Gas Tungsten Arc Welding (GTAW), and Plasma Arc Welding (PAW). Fundamentals related to each system will be introduced in this section.

6.2.1 GMAW-Based WAAM

GMAW is usually the most common choice in the WAAM system, because of its higher deposition rate and easier toolpath planning than other methods. As shown in Fig. 6.2, the welding wire acts as a consumable electrode that is coaxially located within the torch, while an electric arc is formed between wire and substrate. The arc arises from a high-intensity electric field exerted by a welding power source and consists of electrons, ionized inert gas and metal vapor, and excited atoms. Thermal energy released from the electric arc melts both wire and substrate, and metal droplets transfer from wire to molten pool under electromagnetic force, gravity, surface tension, and plasma pressure. The electric arc and molten pool are all locally shielded by inert gas fed through a welding torch, which keeps the stability of the arc and protects the molten pool from external contamination. However, it is worth mentioning that primary local inert gas shielding may not be sufficient for some highly reactive metals, such as titanium so a trailing shield device or even a chamber filled with highly pure Argon gas is necessary for a quality WAAM process.

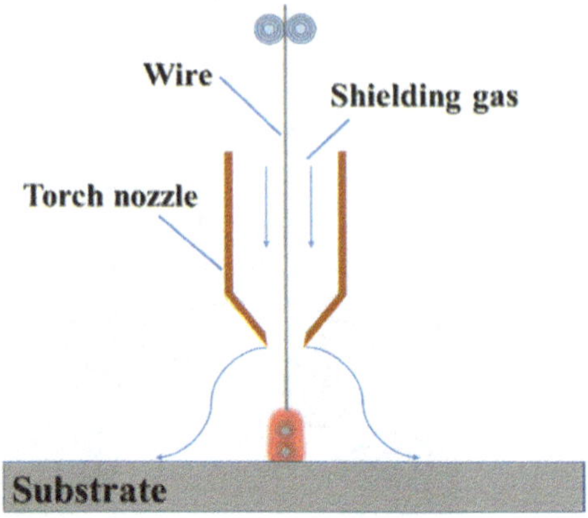

Fig. 6.2 Schematic of the GMAW process

For a GMAW-based WAAM system, the wire electrode is usually connected with positive polarity to achieve a faster melting rate of wire, due to heat concentration at the anode. Meanwhile, the substrate is connected with negative polarity so that a cleaning effect at the cathode can be expected, where oxide film on the previously printed layer is broken and removed under the bombardment of heavy cations in the arc. The processing parameters of GMAW mainly contain welding current, voltage, torch traveling speed, and wire feeding rate, which determine the heat input or energy density for printed 3D structures. GMAW often suffers from higher heat input that will adversely influence the resulting microstructure and thereby deteriorating mechanical properties of WAAM-processed parts. Hence, the Cold Metal Transfer (CMT) technique, which is a variant of GMAW and famous for much lower heat input, is very prevailing in the GMAW-based WAAM system. The key feature for CMT is the wire retraction motion when a short-circuit signal is detected between the wire electrode and molten pool, and the motion exactly facilitates stable metal droplet transfer nearly without spatters. Meanwhile, the heat input is also decreased due to the alternating arc-off time, compared with the conventional GMAW-based WAAM process.

6.2.2 GTAW-Based WAAM

Gas Tungsten Arc Welding employs a non-consumable tungsten electrode, and therefore a separate wire feeding device adjacent to the torch is required, as is illustrated in Fig. 6.3. The electric arc is generated between a tungsten electrode and a substrate. GTAW has an advantage over GMAW in terms of arc stability, due to the non-consumable tungsten electrode, and therefore quality deposited beads with consistent appearance can be achieved.

To increase arc penetration depth and protect the tungsten electrode from overburning, the tungsten electrode is usually connected with negative polarity and

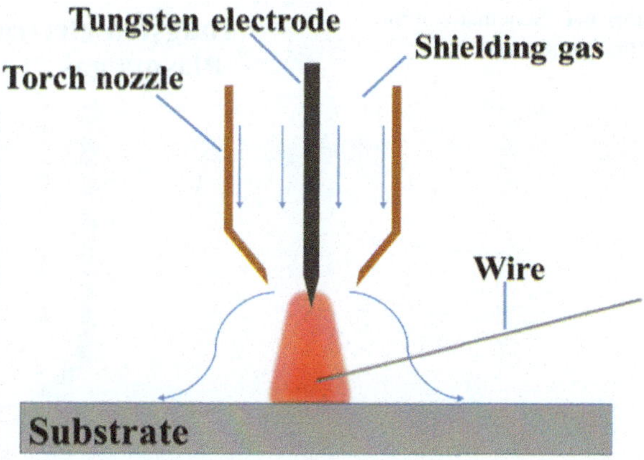

Fig. 6.3 Schematic of the GTAW process

coupled with a water-cooling system. The main processing parameters for GTAW-based WAAM include welding current, welding voltage, and torch traveling speed. Note that, a higher welding current will shorten the electrode life, hence the total arc power needs to be controlled appropriately. Otherwise, there may exist a unique flaw in GTAW-deposited metals, namely inclusions of tungsten, which arise from exceeding the current capability of electrode or accidental contact between the tungsten electrode and molten pool. Consequently, the torch traveling speed and respective deposition rate of GTAW-based WAAM (1–2 kg/h) is inferior to GMAW [6].

6.2.3 PAW-Based WAAM

Plasma Arc Welding is a modified GTAW technique where a non-consumable tungsten electrode is employed to generate an arc between the tip of an electrode and a substrate. Unlike the free and divergent arc formed in GTAW, the plasma arc in PAW is intensively confined by combinational effects from a water-cooled copper nozzle with a smaller orifice as well as gas cooling effects from a secondary coaxial inert gas feeding envelope, as shown in Fig. 6.4. Consequently, the electric arc is highly compressed radially and forms a plasma arc with an energy density 10–100 times higher than the conventional GTAW arc. Therefore, the processing efficiency of PAW-based WAAM is also increased with a deposition rate in the range of 2–4 kg/h [6].

6.3 Materials Used in WAAM

Since the fundamental process of WAAM is welding, it can be applied to a whole range of materials, as long as the corresponding materials are available as feedstock.

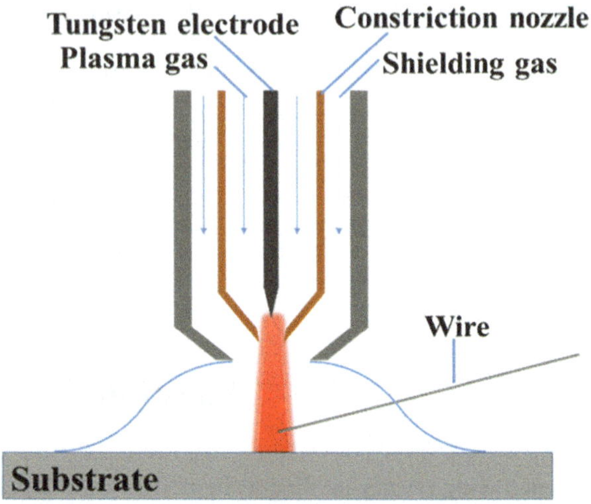

Fig. 6.4 Schematic of the PAW process

Usually, the mechanical performances of parts fabricated by WAAM may be different from that of welding joints using the same feedstock. This is due to different constrain conditions and complex thermal history experienced during the WAAM process. For different materials, the microstructure is usually featured with columnar grains along the built direction, because of the high-temperature gradient. Therefore, the as-fabricated parts are usually anisotropic in terms of mechanical properties. This section is concerned with a review of the most-commonly studied materials in the WAAM process, including steel, nickel alloys, aluminum alloys, titanium alloys, magnesium alloys, and intermetallic compounds. Characteristic microstructures and representative processing challenges for different alloys fabricated by WAAM will be elaborated.

6.3.1 Steel

Steel is the most employed metallic material in the world, because of its mass-abundance in the earth's crust, low cost, and satisfactory mechanical performances as structural components. Different kinds of steel are utilized in multiple industries, including but not limited to, infrastructure, vehicles, electrical appliances, and machining tools. With the WAAM process, the lead time and cost of steel parts can be further reduced, especially for those large-scale components without demanding requirements on fine structural features. There have been many research works performed on mild steel [7–10], carbon steel [11], and alloy steel [12–15], with stainless steel being especially attractive. This is because stainless steel possesses excellent corrosion resistance, heat resistance, and mechanical strength, which enables it to be used in many high-end situations, such as surgical instruments, space components, and marine/offshore components. Stainless steel is divided into many different categories, i.e., martensitic stainless steel, ferritic stainless steel, austenitic stainless steel, and duplex stainless steel. A number of systematic WAAM works can be found for different types of stainless steel [2, 16–23].

Debroy et al. [24] performed a systematic study on the printability of 316 stainless steel, which is one type of the most representative austenitic stainless steel. The printability of 316 stainless steel by WAAM was evaluated in terms of residual stress, distortion, composition change, and lack of fusion. It was found that the composition changes of 316 steel by WAAM are lower than that by laser-based techniques. This is mainly because the molten pool for the WAAM process is larger so that the ratio of free surface to volume of molten pool is lower, and therefore less material would vaporize during printing. In addition, the penetration depth for WAAM is usually larger owing to the impingement effects of metal droplets. Hence, the lack of fusion flaw does not frequently occur in WAAM, compared with 316 stainless steel printed by powder-bed methods. However, due to the low energy density of the electric arc, the overall heat input and heat accumulation is more significant for WAAM. Therefore, the residual stress and distortion for printed 316 stainless steel are also higher. Usually, a post-heat treatment is required to relieve thermal stress.

Similar to welding situations, the printed stainless steel parts in the WAAM process also experience repetitive rapid heating and cooling cycles. Hence, it is expected that phase transformation is in a non-equilibrium state. For example, in 316 stainless steel parts printed by WAAM, the microstructure consists of a major austenitic phase as well as remaining δ-ferrite due to insufficient transformation at a rapid cooling rate. Since microstructure will finally affect and determine the performance of parts, it will be helpful to establish a relationship between microstructure and resulting properties. Li et al. [25, 26] investigated the microstructure and mechanical properties of 316L steel processed by WAAM. The printed part was composed of coarse columnar austenitic grains in the middle and small dendrites towards the surface, as is shown in Fig. 6.5, which was a featured grain morphology along the temperature gradient. The printed 316 stainless steel parts presented were found to have comparable mechanical properties as their wrought counterparts and met the industry requirements. Furthermore, the microstructure of the austenitic matrix included the vermicular δ phase and

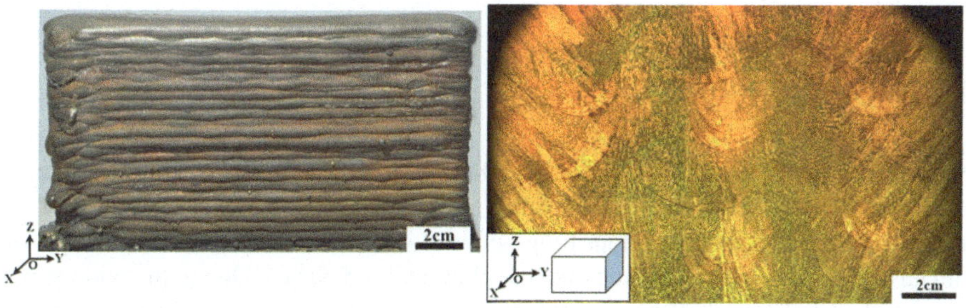

Fig. 6.5 Wall structure of 316L steel and grain morphology in cross-section [25]

σ phase, and the concentration of these phases can be adjusted by heat treatment. The mechanical testing results revealed that the σ phase jeopardized the ductility of printed parts, while the δ phase reduced the yielding and ultimate tensile strength. It was believed that the microcracks propagated from the σ phase to the austenitic matrix and led to the failure of samples during tensile tests.

In addition, elements partition and segregation will also occur during the melting and solidification process, which is closely related to the molten pool size and heat input. Zheng et al. [27] studied the influence of linear energy on corrosion resistance and microstructure of 316 stainless steel parts manufactured by WAAM. The results indicated that higher linear energy input resulted in a relatively lower cooling rate, which favored the growth of austenitic dendrite and led to larger dendrite spacing. Moreover, the slow cooling process allowed diffusion and segregation of alloying elements, such as Cr and Mo, to δ-ferrite. Consequently, the concentration of Cr/Mo was insufficient in the austenitic matrix to provide enough corrosion resistance.

Similar to austenitic steel processing, the residual phase effect is also a characteristic of WAAM of martensitic stainless steel. Lunde et al. [28] conducted a thin wall printing of 420 martensitic stainless steel by WAAM and found that the microstructure mainly consisted of the martensitic matrix with residual austenitic phase and δ-ferrite. The austenitic phase was mainly attributed to the heat accumulation of high interpass temperature during the printing process. The martensitic microstructure presented high strength and hardness, yet very poor ductility. Therefore, post-heat treatment is necessary to further improve overall mechanical performances. Nemani et al. [29] studied the effects of post heat treatment, including quenching and tempering, on microstructure and mechanical properties of WAAM-processed 420 martensitic stainless steel. It was found that austenitizing temperature of 1150 °C was optimum to fully eliminate δ-ferrite and carbides, and pure martensitic microstructure could be achieved after the quenching process. However, in order to further improve the properties of WAAM-processed parts, a tempering process was necessary to soften the matrix as well as to control the shape and size of carbide precipitations. The optimum tempering temperature in this research was identified as 400 °C, which rendered a well-tempered martensite matrix with uniformly distributed spherical chromium carbide. The resulting parts possessed both satisfactory strength (1142 MPa) and ductility (elongation of 11.8%).

6.3.2 Nickel Alloys

Nickel-based alloys are usually characterized by excellent mechanical performances and corrosion resistance even at elevated temperatures. Therefore, they usually act as key components serving under tough conditions in various industries, such as oil and gas, aerospace, aviation, nuclear, and chemical industries. Due to the wide applications of nickel-based alloys, intensive research has been done to process the alloys with WAAM, to increase production efficiency and achieve improved customization. This section will focus on the WAAM work of some typical nickel-based alloys [30–33].

Inconel 718 is one of the most representative precipitation-strengthening nickel alloys in which nano-scale intermetallic compounds are homogeneously distributed in face-centered-cubic γ matrix. The main precipitations include the γ' $Ni_3(Al, Ti, Nb)$ phase and γ'' Ni_3Nb phase, which are exactly coherent with γ matrix in terms of crystallographic structure. However, with certain heat processing, some detrimental phases may also occur in Inconel 718 and jeopardize ductility, strength, and fatigue performances of parts, such as Laves phase and δ phase which are incoherent with the γ matrix. Compared with laser-based printing techniques or conventional wrought processes, WAAM has a larger molten pool and therefore macro-segregation is usually observed in many reported works [34, 35]. In these works, island-like Laves phases along the dendritic region and grain boundaries (direct evidence of segregation) of key alloying elements, such as Nb and Ti, can be seen as shown in Fig. 6.6. Consequently, there is a deficiency of alloying elements in the matrix to precipitate strengthening phases γ' and γ''. This mainly accounts for the poor mechanical strength of as-printed Inconel 718 parts by WAAM. In addition, the grain size and morphology will also influence its mechanical properties. For WAAM-printed Inconel 718 parts, there are coarse columnar grains located near the substrate and growing along the building direction, the diameter of columnar grain can reach as much as 150 μm. In contrast, wrought counterparts, have finely equiaxed grains with an average size of 26.7 μm, which explains its higher mechanical strength.

In order to improve mechanical performances of WAAM-printed Inconel alloys, post heat treatment is necessary to remove the segregation and homogenize alloying elements in the matrix and contribute to precipitate forming. Usually, a standard post-heat treatment plan for Inconel alloys consists of 2 steps, solution-annealing and age-hardening. Lv et al. [36] studied the effects of post-heat treatment on microstructure and mechanical properties of Inconel superalloys printed by WAAM. It was found a solution annealing process at 980 °C effectively eliminated elements segregation

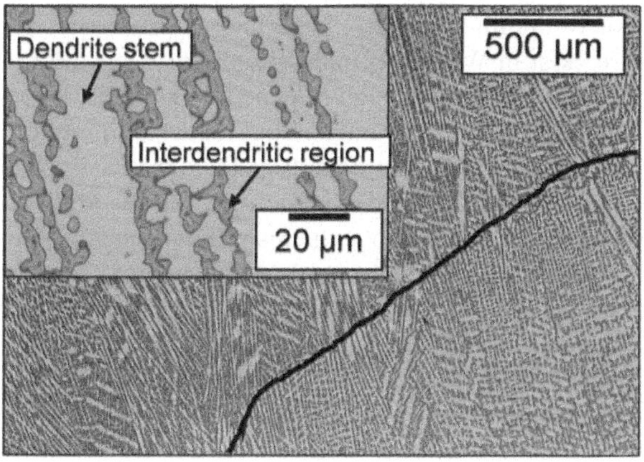

Fig. 6.6 Microstructure image of as-deposited WAAM Inconel 718 [34]

and dissolved Laves phase into the matrix, which otherwise appeared in between the dendrite region. Thereafter, an age-hardening process directly led to the formation of nanoscale strengthening phases, including γ' and γ'' discs, which raised the tensile strength from 872 MPa for as-fabricated samples to 1152 MPa for heat-treated samples. Such mechanical properties are comparable to wrought counterparts.

Inconel 625 is another representative nickel alloy that possesses excellent corrosion resistance, oxidization resistance, and thermal stability, acting as a remarkable candidate for seawater and chemical processing applications. Unlike Inconel 718, this kind of alloy is strengthened by solid solution hardening effects of alloying elements (Nb/Mo) in the Ni–Cr matrix. The microstructure of Inconel 625 produced by WAAM is comparable to Inconel 718, which features a primary cellular structure near the substrate, columnar dendrite structure with secondary dendritic arms in the middle of the part, and equiaxed grains along the top layers. Laves phase would also appear along the space between dendrites in as-printed parts. Similarly, a solution annealing process at 980 °C would dissolve the detrimental Laves phase and raise the ultimate tensile strength, while not changing the grain morphology [37–39]. However, with the increase in heat treatment time, an undesirable δ phase would also emerge and aggregate along grain boundaries. Thus, an optimum set of heat treatment parameters need to be determined based on the mechanical requirements for the resulting application, case by case.

6.3.3 Aluminum Alloys

There is a great mass abundance of aluminum elements in the earth, and aluminum can be further alloyed with different elements to form various systems of alloys, such as Al–Mg, Al–Mn, Al–Si, Al–Cu, and Al–Zn. Usage of aluminum alloys is very common in the aerospace, construction, and automotive industries, due to their lightweight, outstanding corrosion resistance, and low costs. It will be therefore very attractive to print aluminum alloys by WAAM for higher production efficiency and on more complex structures [40, 41]. However, due to the unique characteristic of aluminum alloys, material property issues that occurs in a welding process also exist in WAAM-printed aluminum alloys. For example, aluminum has higher thermal conductivity, lower density, and denser surface oxide than other metallic materials. All these characteristics render aluminum to have a huge porosity tendency in welding, which significantly reduces the loading area and deteriorates its mechanical performance. To overcome this, various ideas have been proposed to decrease the porosity and improve the resulting mechanical properties.

One of the straightforward strategies is to exert external force to close the microscale pores in WAAM-printed aluminum alloys. Gu et al. [42] studied the impacts of layer-wise cold rolling and post heat treatment on reducing porosity in the WAAM process of 2319 and 5087 aluminum alloys. It was found that pores consistently occurred in as-fabricated samples for both alloys. Figure 6.7 presents the pores

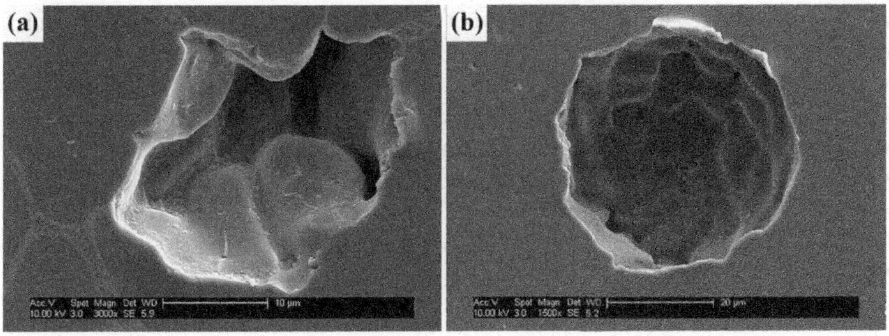

Fig. 6.7 Morphology of pores in the as-deposited **a** 2319 alloy and **b** 5087 alloy [42]

morphology comparison for these alloys. Due to the difference in chemical compositions and random eutectic phase of Al_2Cu, the morphology of pores in 2319 aluminum alloy was more irregular and smaller, while the pores in 5087 were more spherical and larger because of free coalescence. When exerting a compressional force of 15 kN by an interlayer rolling process, the porosity density in both aluminum alloys was effectively reduced from 0.176% to 0.029% for 2319, and from 0.232% to 0.061% for 5087. Such porosity closure effects from plastic deformation were even more significant if raising the force to 45 kN, where pores larger than 5 μm were all eliminated. However, the results also revealed that post-heat treatment would recover the flattened pores, possibly due to vacancies diffusion and release/coalescence of hydrogen which were originally trapped in dislocations.

Moreover, the hot wire technique can be also integrated into a WAAM process to reduce the porosity in WAAM-printed aluminum parts [43]. It has been reported that contaminants attached to the surface of the wire, such as water and oil, can be a source of hydrogen during welding and finally dissolve into the molten pool to generate hydrogen pores. With the hot wire technique, heat generated along the wire can effectively vaporize or decompose the contaminants, and therefore cut off one source of hydrogen during printing. Research indicates that the density of aluminum parts processed with hot-wire WAAM increases from 96.8% to 99.6%, as compared with conventional WAAM. The resulting mechanical properties are also improved, with tensile strength increasing from 350 to 399 MPa and elongation increasing from 5.2% to 12.0%.

One of the main reasons for porosity in parts printed by the WAAM process is the sharp discrepancy of hydrogen solubility in the liquid and solid-state of aluminum, where hydrogen precipitates and coalesces during metal cooling and solidification. Therefore, it will be favorable to remove the hydrogen porosity, if the WAAM condition allows small pores to escape from the molten pool. Conventional CMT welding mode usually results in a finger-shaped welding bead, where the floating distance for the pore is too long to escape before metal solidification. In contrast, CMT pulse mode generates beads with a shorter floating distance and larger surface area, and therefore the porosity is decreased in aluminum samples. CMT pulse advanced (CMT-PADV) mode is also adopted to achieve WAAM of aluminum parts, including but not limited to 2000 series and 5000 series [44–46]. It indicates that CMT-PADV performs very well in porosity control for aluminum parts printing, because of favorable conditions for pores to escape, such as molten pool oscillation induced by variable polarity, and proper escaping distance along the building direction. Moreover, there is a negative polar cleaning effect for welding wires in CMT-PADV to remove contaminants. The resulting porosity rate is reduced from 0.347% to 0.06% by modifying CMT as CMT-PADV mode, and mechanical strength of as-fabricated parts (Al 5556) increases from 266.4 to 282.2 MPa [47].

6.3.4 Titanium Alloys

Titanium alloys is a class of the most attractive metallic materials that possess comprehensively exceptional properties, inclusive of high specific strength, great toughness, robust thermal stability, excellent corrosion resistance, and biocompatibility. Therefore, titanium alloys are very popular in various high-end industries, such as aerospace and aviation, marine offshore, and bio-implants fabrication. The major drawback which prohibits the wide application of titanium alloys might be their high costs, which mainly arises from a large extraction expense from ore and costly processing procedures. To fabricate titanium alloy structures more economically and efficiently, WAAM technology is also explored, especially for large-sized components, due to its high deposition rate and low buy-to-fly ratio. As such, numerous works have been conducted recently for improving the microstructure and properties of titanium printed parts by the WAAM process.

The Ti6Al4V alloy is one of the most representative titanium alloys for many applications, which consists of hpc-structured α and bcc-structured β phases at room temperature due to its stabilizing elements. Several thermal features occur in the WAAM process, such as directional heat dissipation and layer-by-layer thermal cycles, and it is widely reported that millimeter-scale prior β phase grows epitaxially within the Ti6Al4V part. Additionally, a high cooling rate at certain layers would lead to phase transformation from β to Widmanstätten and/or martensite α. Since certain layers in the WAAM process will be further thermally affected by following deposited layers, meta stable martensite will decompose and transform into acicular α phase and β phase. The WAAM-printed Ti6Al4V alloy would finally be composed of the coarse columnar β phase, α phase along the grain boundary of the prior β phase, fine acicular α, and post precipitated β phase within the prior β phase [48–51]. Although the acicular α phase assists in the strengthening of the titanium alloy, the grain boundary α, which delineates prior β grains, severely compromises the ductility of printed Ti6Al4V parts. Therefore, it is necessary to refine the grain size and eliminate the grain boundary α phase for performance improvement.

Bermingham et al. [48] found that doping 0.13 wt.% boron to Ti6Al4V wire could effectively remove the α phase along the grain boundary and also confine the growth of the prior β phase. The induced constitutional supercooling by boron

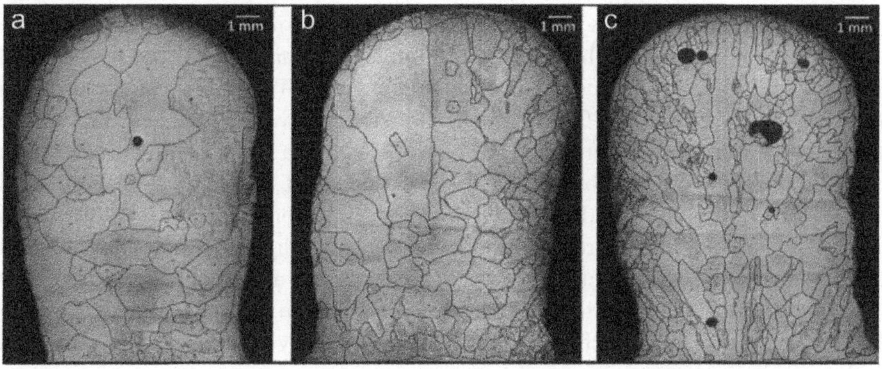

Fig. 6.8 Microstructure of as-deposited Ti with **a** 0.04% Si, **b** 0.19% Si, and **c** 0.75% Si [52]

also transformed the α phase to an equiaxed form. With all these microstructure modification effects, the plasticity of as-printed Ti6Al4V was improved without much loss of strength. Such an effect was more significant after a post-heat treatment, where mechanical strength for boron-doped Ti6Al4V was increased by 10%, and ductility was raised by 40%. Similarly, doping Si to pure titanium welding wire can also lead to a grain refinement effect, as is shown in Fig. 6.8. It was found that the width of the prior columnar β phase was noticeably decreased when 0.75 wt.% of Si was doped into pure titanium feedstock, and the density of equiaxed grains was also improved due to constitutional supercooling [52].

To achieve grain refinement, the hot wire technique is also employed to decrease the heat input, and therefore the cooling rate of the molten pool becomes higher and transformation to equiaxed grains can be accomplished. Lu et al. [50] studied the effects of the hot wire technique on grain morphology and mechanical performances of the WAAM-printed TiAlMoZr alloy by the tungsten inert gas welding method. The resistance heat in the feeding wire effectively reduced the overall heat input from the welding arc, and dimensions of the originally coarse columnar β phase were sharply narrowed with the occurrence of equiaxed grains. Moreover, the grain size of the α phase within the prior β phase also tended to be thinner, with the average width changing from 1.1 to 0.89 μm. Consequently, compared with conventional WAAM of TiAlMoZr, the yield strength and ultimate tensile strength of as-fabricated parts all rose when hot wire current reached 100 A, and the parts also became more isotropic in terms of strength and ductility along the transverse and longitudinal orientation.

In addition, systematic research on post heat treatment is also conducted to adjust the microstructure and improve mechanical performances of WAAM-printed titanium alloy. For example, Lin et al. [53] investigated the effects of different heat treatment plans on WAAM-printed Ti6Al4V alloy. The microstructure of the as-fabricated Ti6Al4V part by tungsten inert gas welding consisted of continuous α phase along grain boundaries of prior β phase, and as well as basket-weave structured α within columnar prior β grains. When an annealing heat treatment at 600 °C was conducted, no remarkable change of microstructure was appreciated. However, a slight increase in mechanical strength was observed due to the stress-relieving effect. When the annealing temperature reached 850 °C, the intergranular α grains got coarser and therefore deteriorated the mechanical performances of Ti6Al4V printed parts. If a solution treatment at 930 °C plus a water quenching were conducted, martensite would form along with intergranular α phase. Usually, a further aging process was required to refine the microstructure. It was found that aging the solution-treated Ti6Al4V at 550 °C converted partial martensites into secondary α and β phases. Such a process can lead to an increased in mechanical strength, but if not controlled well can lead to a decrease in ductility. It was reported that an appropriate aging temperature was 800 °C, where martensite was all transformed to fine grains of secondary α and β phases, and both mechanical strength and ductility were increased compared with as-fabricated samples.

6.3.5 Magnesium Alloys

As one of the lightest structural and functional materials, magnesium alloys possess a remarkable strength-to-weight ratio, good machinability, and excellent electromagnetic shielding performances. With these merit features, magnesium alloys are widely applied in automobile manufacturing, electronic industry, and the energy-saving field, where it is processed by forging and casting conventionally. However, it is hard to achieve complex shaped components by cold working because of its poor ductility, and hot working may induce oxidization issues and affect processing efficiency. For the casting of magnesium alloy, slag inclusion tends to occur because of oxidization, which will deteriorate the resulting mechanical properties of parts. With additive manufacturing techniques gaining momentum, various AM methods have been explored to fabricate magnesium alloy parts. Although laser-based AM methods usually possess a higher processing resolution, the reflection of the laser by magnesium alloy is very strong so effective processing is hard to achieve. In addition, magnesium powders used for the powder-bed AM method have a huge oxidization tendency, which not only increases contamination concentration but also poses a threat to operational safety. Considering these issues, WAAM of magnesium alloy is a promising way to remove the barriers, and lots of research works have been done for mechanical and metallurgical studies.

For example, a 50-layer single wall structure of AZ80M magnesium alloy was manufactured by GTAW-based WAAM, where microstructure consisted of Mg matrix (the α phase) and hardening precipitations of $Mg_{17}Al_{12}$ (the β phase) [54]. It was found that thermal history during the WAAM process played an essential part in shape forming and microstructure distribution. Specifically, heat accumulation during the metal deposition process affected heat dissipation and molten pool size, and interpass temperature gradually increased and stabilized at a certain level. As a consequence, the deposited bead width raised at first several layers and became more uniform in the intermediate region. Additionally, microstructure along the building direction was also influenced by the thermal history of WAAM. Top layers were featured with equiaxed grains and significant segregation of β phase in the dendritic region. Such segregation effect was weakened by heat from following deposited layers, where dendritic precipitations partially dissolved. When interpass temperature was stable after several layers of deposition, the dissolved phases precipitated again along grain boundaries in terms of divorced eutectic strip structures, as is shown in Fig. 6.9. As for the general microstructure in the middle part of the printed wall, it was characterized as equiaxed grains. However, coarse grains and micro defects tended to form at boundaries between layers, which finally resulted in anisotropic mechanical properties with mechanical strength along its horizontal orientation 30% higher than the building orientation, although strength along both orientations outperformed casted counterparts.

Thin walls of the AZ31 Mg alloy were also fabricated via GMAW-WAAM [55]. Different from grain structures of AZ80M mentioned above, columnar predominated in the middle region of walls, instead of equiaxed grains. Specifically, the

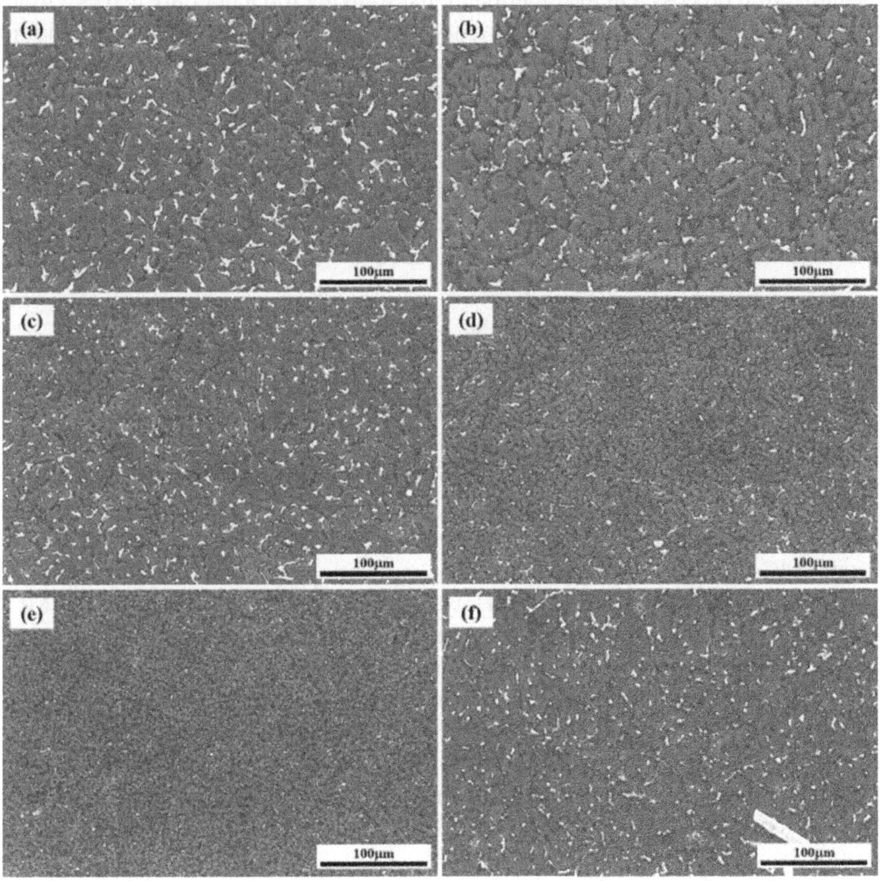

Fig. 6.9 Microstructures at different layers: **a–e** the 50th–46th layers, **f** the 25th layer [54]

microstructure also consisted of α matrix and strip-like eutectic of α Mg and β precipitations in inter-dendritic space. The columnar grains grew along the building direction due to thermal gradients. Therefore, the printed AZ31 alloy also presented significant anisotropic mechanical properties, but samples along the building direction possessed higher strength compared with transverse samples. This is because vertical dendrites in columnar grains increased crack resistance, especially for those transgranular cracks. While for transverse samples, cracks can easily propagate along interdendritic space owing to the properties difference of Mg matrix and eutectic phase.

These anisotropic properties of printed Mg alloy parts will pose a potential threat to real applications, and therefore hinder the extension of WAAM-printed Mg alloy to engineering fields. To solve this problem, a weaved CMT method was proposed to achieve reliable WAAM of AZ91 Mg alloy [56]. On the one hand, the CMT technique is famous for lower heat input compared with other conventional arc welding technologies. Therefore, CMT-based WAAM will lead to rapid solidification and crystallization process. On the other hand, the weaving torch not only leaves a more uniform temperature field around the molten pool but also induces oscillation in liquid metals during solidification, which is favorable to dendrites. With all the synergetic effects, the printed AZ91 alloy was featured with uniform equiaxed grains, instead of coarse columnar dendrites, as is shown in Fig. 6.10. Consequently, the resulting mechanical properties were also very consistent in different orientations, such as ultimate tensile strength and elongation. The anisotropy level was only 2%, which was much lower than other WAAM-printed Mg alloy parts, ranging from 23 to 28%.

6.3.6 Intermetallic Compounds

Apart from the solid-solution-based alloys as mentioned above, intermetallic compounds (IMCs) are also gaining popularity due to their distinguished mechanical properties, especially at elevated temperatures, such as TiAl and NiAl alloys. These IMCs are promising alternatives to conventional Ni-based superalloys, but with lower density, higher oxidization resistance, and excellent strength. Therefore, certain IMCs can make great candidates for applications in tough service conditions, such as gas turbines and jet engines. However, one of the main drawbacks of IMCs is significant brittleness arising from their crystal structures, which makes it challenging to fabricate IMCs parts by conventional processing methods, e.g., rolling, casting, and forging. In contrast, additive manufacturing provides an avenue to near-net-shape IMCs parts due to its unique advantages. While selective laser melting or electron beam melting methods may suffer from costly powder feedstock, complex vacuum conditions, and expensive equipment, a novel method of IMCs printing via WAAM is proposed by feeding twin wires into a molten pool and achieving in situ reaction between dissimilar metals.

For example, Pan et al. [57] accomplished different TiAl IMCs phase control in the printed parts via dual-wire WAAM technology. While fixing welding current, travel

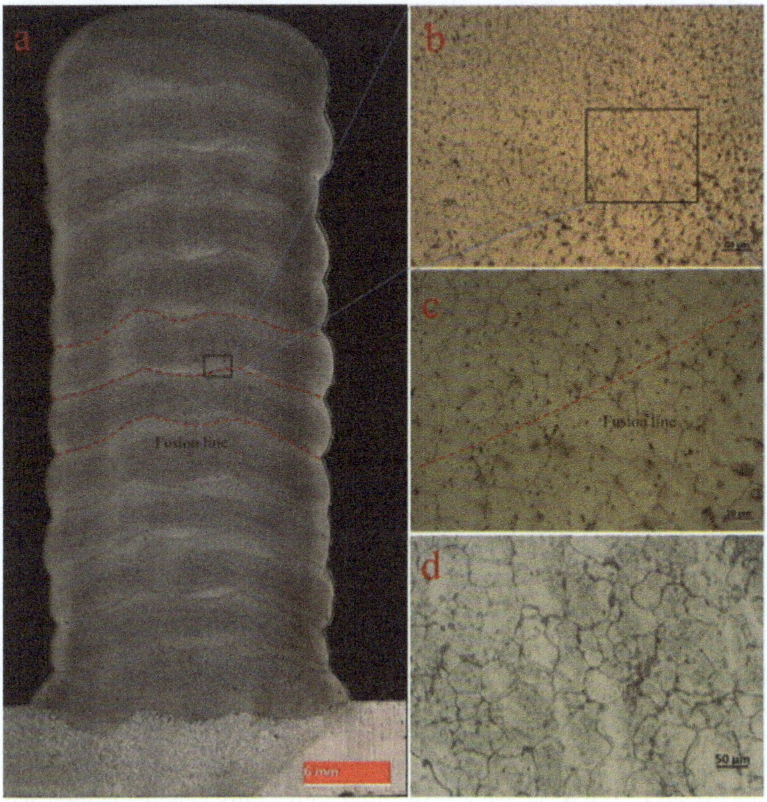

Fig. 6.10 **a** The overall macrostructure, **b** low and **c** high magnification images of the middle region, **d** as-cast substrate microstructure [56]

speed, and interpass temperature, the wire feed rates of pure Al wire and pure Ti wire were precisely regulated and thereby controlling the composition of deposited IMC parts. Specifically, when the atomic concentration of Al in deposited parts was 35%, a single phase of Ti_3Al could be achieved. While the concentration of Al was further increased to 45%, a mixture of Ti_3Al and TiAl occurred. When the concentration of Al reached 55%, the deposited part was composed of TiAl single-phase, instead. As can be expected, different phase compositions would result in different mechanical and chemical properties. It was found the Ti_3Al phase would increase the overall hardness and tensile strength of printed parts, while the TiAl phase tended to raise the ductility of printed IMC. The oxidization resistance could be improved by raising Al concentrations. The proper combination of different phases can be applied for situations with different property requirements.

However, it is reported that GTAM-based WAAM results in stronger heat input to printed parts so that the residual stress is in a high level. Additionally, coarse grains and inhomogeneous microstructure of IMCs will form under intense heat input. To

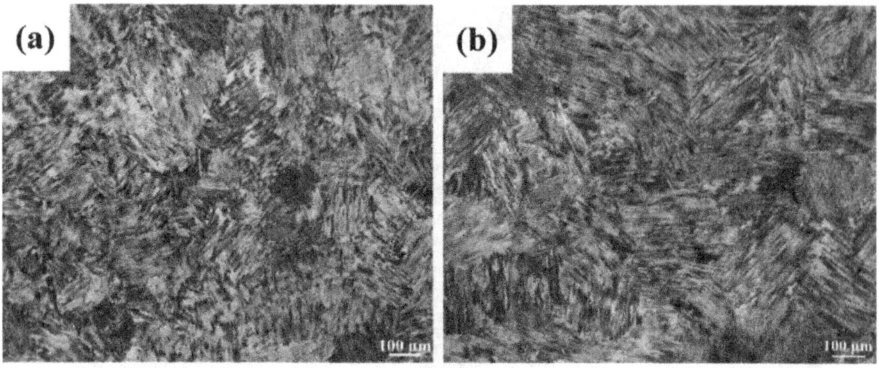

Fig. 6.11 L microstructure of the lamellar colony in as-deposited TiAl alloy at (**a**) the upper part, **b** the middle part [58]

alleviate this issue, plasma arc welding (PAW) is proposed to be employed in a dual-wire WAAM system. The high energy density of PAW does not only decrease the total heat input but also allows sufficient mixing of alloying elements in the molten pool, which is conducive to avoiding macrosegregation in the final parts. For example, Hua et al. [58] fabricated TiAl IMCs via dual-wire PAW-based WAAM technology. It revealed that uniform distribution of Ti/Al elements was achieved under intense plasma arc effects. Moreover, homogeneous lamellar structures of TiAl and Ti_3Al phases were detected in different locations of printed walls as shown in Fig. 6.11. Specifically, each layer consisted of two different microstructure regions due to reheating effects of WAAM. One was featured with a fully lamellar colony, while the other was characterized as lamellar structures with γ-TiAl between dendrites space.

Apart from TiAl IMCs, NiAl IMCs have also been studied and fabricated via the WAAM method [59]. Similarly, a dual-wire WAAM system is employed, and final parts compositions can be exactly adjusted by regulating wire feeding rates. However, instead of using pure Ni or Al wires, alloying elements, such as Cr, Mn, and Fe, are indispensable in feedstock, due to the significant brittleness of pure NiAl IMCs. So far, multiple research works have been conducted to investigate the microstructure, compositions, and phases formed in NiAl printed parts. Reported works indicated that various Ni–Al intermetallic compounds with different morphology emerged with varying ratios of Ni to Al. By raising the Ni concentration in the molten pool during the WAAM process, the microstructures transformed from M-NiAl/Ni_3Al to bulky Ni_3Al with Ni matrix or γ'-Ni_3Al. The M-NiAl IMCs presented needle and lath shapes, while the Ni_3Al phase was in the shape of bulk, granulate and laminar. The hardness curve regarding Ni content was approximately parabola shape. This is because the M-NiAl phase possessed higher hardness compared with Ni_3Al, while its concentration gradually decreased with increasing Ni content. Therefore, the overall hardness declined in the first stage. Thereafter, with precipitation of hard γ'-Ni_3Al, the average hardness enhanced again.

6.4 Defects and Quality Control of WAAM

For a WAAM process, there are many inherent challenges and defects to overcome, which are closely related to its unique material processing features, inclusive of but not limited to higher heat input, strong heat accumulation, larger molten pool size, slower cooling rate than laser-based methods, and in situ heat treatment effects from the following layers.

6.4.1 Challenges and Defects in WAAM

One of the most critical challenges for WAAM is residual stress and related deformation. Residual stress is mainly derived from non-uniform plastic deformation in the metal deposition process. Specifically, the thermal expansion and shrinkage of different metallic layers, under repetitive heating and cooling cycles, cannot be aligned or synced in a continuous printing process. This means previously deposited metal layers will restrict the shrinkage of subsequent solidifying layers. Consequently, a pair of residual tensile and compressional stress will remain within a printed part, even when all the external clamping force is released. Research indicates that residual stress level and deformation of parts in a WAAM process are especially more significant than laser-based printing technologies, due to their stronger heat accumulation [24]. For some scenarios where the residual stress is higher than the yield stress, permanent plastic deformation would occur, especially for thin-wall parts that has a geometrically asymmetrical structure, affecting dimension accuracy or part tolerances. On the other hand, the residual stress will also superpose with external loading force when the part is under service. Once the overall stress level exceeds the ultimate tensile stress of respective materials, cracks or fractures will occur and propagate more easily in regions dominated with residual tensile stress and leading to a failure of printed parts. It is therefore of great importance to remove or relieve thermal stress during/after the WAAM process to maintain dimension accuracy and expected mechanical performances. Detailed methods for thermal stress relieving will be elaborated in the following sections.

Apart from residual stress, segregation is also a long-existing problem in a WAAM process, where enrichment of alloying elements at the dendritic region and grain boundaries can be observed. Compared with laser-based AM techniques, the molten pool size for WAAM is larger and the solidification rate is also lower, hence the segregation phenomenon is more remarkable. This phenomenon results in a more heterogeneous microstructure and deteriorates the mechanical performances of printed parts. For example, segregation can deprive the metal matrix of solute atoms which otherwise are supposed to be uniformly distributed within crystal lattices of base materials and achieve solution hardening effects. Consequently, both hardness and mechanical strength will be far away from expectation. Similarly, for some precipitation strengthening alloy, like Inconel 718, the key precipitation-forming elements

(Nb and Ti) will also segregate as Laves intermetallic compounds at the dendritic region, thereby inhibiting the generation of nanoscale precipitates in the matrix. Testing results indicate that both mechanical strength and corrosion resistance of as-fabricated Inconel 718 are inferior to its counterparts without segregation effects. Additionally, the combination of residual stress and segregation will also result in solidification cracks along the welding beads [60]. In a WAAM process, segregation of certain alloying elements may lead to the eutectic formation with a lower melting temperature at the dendritic region, such as the laves phase in nickel-based alloys. At the final stage of solidification, the quantity of eutectic liquid may not be enough to fill in the gaps caused by grain shrinkage from residual stress. Therefore, solidification cracks will occur along dendritic boundaries, especially when preheating or interpass temperature is too high. Such cracks will usually act as the source of fracture under external loading. In contrast, for high-strength low-alloy steel, such as 4340 steel, high temperature preheating is necessary for suppressing cracks propagation. This is because the hardenability tendency of such steels is very noticeable, so that brittle martensite or bainite can easily emerge in the fusion zone and heat-affected zone of WAAM-printed parts. Under this circumstance, the tensile residual stress is at a high level due to the lack of plastic deformation, and its brittle microstructure facilitates the formation and propagation of the crack.

Porosity is considered as another undesirable issue often found in WAAM-printed parts. Based on the formation mechanism of porosity, there are generally two categories of pore defects in parts. If the pore size is small and exists in a more spherical shape, it could be caused by trapped gas in the molten pool. This occurs due to the difference in gas solubility between solid and liquid metals. Gas atoms tend to precipitate and escape from a solidifying molten pool in a cooling process. Therefore, thermal conductivity, buoyancy on pores, and the surface tension of liquid metal will all affect the formation tendency of porosity in metal parts. The porosity rate for the WAAM process is higher because the feedstock for WAAM is more likely to be contaminated by moisture or oil hidden in surface notches, as compared with well-preserved powder feedstock for other AM methods. On the other hand, if the pores are large, more irregular, and observed along beads boundaries or edges. This is more likely to be attributed to a lack of fusion voids or gaps. This stems from inappropriate tool path planning or poor combinations of processing parameters. For example, in a CMT-based WAAM process, the arc penetration depth will be very limited if the welding current is low, so that the adjacent layers cannot be bonded reliably, and voids will occur in certain positions at the boundaries. In addition, the cross-section of a welding bead is a hump shape, of which the top region is far less flat than powder bed printing methods. Therefore, an appropriate overlapping distance between beads should be well studied to form a smooth layer for quality deposition of the following layers.

One more representative issue in microstructures of parts printed by WAAM is coarse columnar grains along the building direction. It has been well reported that the mechanical and chemical properties of as-fabricated components by WAAM are very anisotropic, and coarsely aligned columnar grains can mainly account for this phenomenon. Compared with laser-based printing methods, the solidification rate of the molten pool for WAAM is slower, and the heat accumulation effect is stronger. Additionally, there is a quite large temperature gradient between the substrate and molten pool. Consequently, the overall conditions are more favorable to the epitaxial growth of columnar grains along the direction of the maximum temperature gradient. Prior printed layers will be further affected by subsequently deposited metals and therefore enlargement of columnar grains could be detected due to repetitive thermal effects. These coarse columnar grains will not only reduce the mechanical strength and toughness of parts but also weaken the corrosion resistance. Hence, it has been a popular research topic to realize the transition from coarse columnar grains to fine equiaxed grains in WAAM-printed parts.

In light of these challenges and defects encountered in a WAAM process, many quality control methods are also proposed to adjust the microstructure and further improve the properties of as-fabricated parts. Some commonly employed strategies and advanced techniques will be presented in the following sections.

6.4.2 WAAM Process Control

WAAM of certain materials or structure features can be very sensitive to processing parameters, and slight disturbance from the environment or machine instability can still cause defects, even though the parameters may have already been optimized. Therefore, it is necessary to develop an in-process monitoring system to collect necessary sensor data, for quality control, toolpath adjustment, and automation improvement. During a WAAM process, there are many kinds of signals emitted from the welding pool, such as visual signals, acoustical signals, and thermal radiation signals. It will be meaningful if these signals can be collected appropriately for the quality control of the WAAM process. For example, different metal transfer modes or solidification processes will emit distinctive signals. With continuous in-process signals collection and analysis, a defect formed in WAAM can be detected early for intervention, and the toolpath can therefore be adjusted to remove the current defect and modify subsequent steps to avoid similar defects from occurring. Otherwise, it would cause a huge waste of both material and time, if a certain flaw region was not detected until a full part was completed and ended up being discarded. In this section, we will introduce the research and developments of some of the representative real-time monitoring systems for WAAM.

6.4.2.1 Vision Sensing

Vision sensing is the most straightforward monitoring technique, which can provide real-time morphology of molten pools, electric arcs, feeding wires, transfer of metal droplets, and information of defects if any. For example, in a PAW-based WAAM process, a separate wire feeding device is required besides the torch, and the orientation of the feed wire will affect the resulting quality and dimension of deposited beads. Therefore, it will be favorable if a wire feeder position can be monitored and corrected in real-time to stabilize the beads' dimension. Zhan et al. [61] developed an online wire position monitoring system with a color welding camera. Based on the collected pictures and respective image processing techniques, wire edges and wire-feeder nozzle edges could be identified, and then the deflection angle of the feeding wire was automatically calculated. Next, with this angle feedback, the orientation of the feeding wire could be thus adjusted to an expected position accordingly for better metal deposition quality.

Similarly, Zhang et al. [62] built up a vision sensing system with a CCD camera and a series of image processing technology for monitoring and controlling a GMAW-based WAAM system. Geometrical feature data of deposited metal beads were extracted, analyzed, and employed on an online control system. With the collected real-time information of bead dimensions, the heat input of the WAAM process was adjusted accordingly by modifying the welding parameters via a closed-loop control. In this way, the bead width in a thin wall structure was much more consistent than its counterpart, which is printed using an open-loop control strategy. See Fig. 6.12. Moreover, both energy and materials costs were also conserved to some level, considering the reduced heat input as well as the need for less post-machining.

In addition to monitoring torch/wire position and beads dimensions, vision sensing is also employed in the real-time detection of defects, which helps to locate the flaw region automatically. For instance, Tang et al. [63] developed an online welding defects detection system for a WAAM process, based on image data collected by a CMOS camera and a deep learning algorithm. The key geometrical features of deposited beads were extracted and processed by a Convolutional Neural Networks (CNN), and then a Support Vector Machine (SVM) was further established to achieve

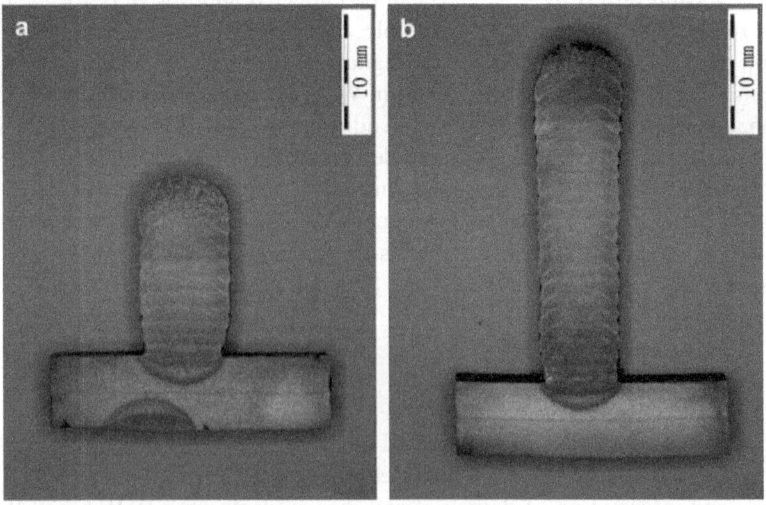

Fig. 6.12 A thin wall structure printed with **a** open-loop control and **b** closed-loop control [62]

Fig. 6.13 A vision sensing system based on laser profilometers [64]

the classification of beads region, based on the features extracted by CNN. A sufficient dataset was required to train the integrated model offline. Thereafter, such a reliable model was established to accomplish real-time defects detection, such as pore and undercut, with an identification accuracy of 95.29%.

Apart from industrial cameras, vision sensing can also be realized by Laser profilometers. It was found that a pair of laser profilometers was enough to collect complete geometrical data of each printed layer for simple structures, as illustrated in Fig. 6.13 [64]. Cameras are sometimes limited by only offering 2D images of printed parts from the top or side view. In contrast, with complete data collected by a laser profilometer set, a 3D model of printed parts could be established which assists in multi-angle dimensions inspection and toolpath planning for necessary post-machining.

6.4.2.2 Acoustic Sensing

In an arc welding process, various distinctive acoustic signals are emitted simultaneously with fluctuation of molten pool mode [65], change of metal transfer mode [66], phase transformation [67], and variation of plasma arc [68]. Hence, these acoustic signals can provide insightful information related to the quality of deposited metal layers. It has been found that typical welding defects generate distinctive acoustic signals, and an online acoustic signal sensing and the analyzing system can assist to detect defects early. For example, Polajnar et al. [69] established a sound pressure monitoring system for the GMAW process. It revealed that the sound signal was even more sensitive to disturbance in arc welding conditions than current signals, such as the scenario of geometrical defects in welding bead and abnormal high-frequency oscillation of plasma arc.

Furthermore, machine learning algorithms are also integrated with acoustic sensing techniques to analyze collected data and achieve classification between normally deposited metal and different defects cases [70]. For example, a random forest classification algorithm was employed for defect identification in an arc welding process. The acoustic signals during welding were collected and analyzed by data mining software to extract features. A dataset of these features was used to train the random forest model to distinguish quality weld bead, lack of fusion, and burn through the flaw. The results indicated that the accuracy of 88.69% could be achieved for defect identification [71].

6.4.2.3 Thermal Sensing

For materials fabricated by WAAM, temperature control is required during the printing process to prevent the deposited parts from significant heat accumulation and hence poor mechanical properties. For example, the hot cracking tendency for high alloyed materials will be increased if the inter-pass temperature is not well controlled. On the other hand, for some types of low-alloy high-strength steel, elevated-temperature preheating is often necessary to lower the cooling rate, reduce the thermal stress level, and avoid cracks due to phase transformation in a rapid cooling process. In both scenarios, thermal sensing is indispensable to maintain print quality and a wide-range pyrometer will be enough to achieve inter-pass temperature control [72].

Furthermore, thermal sensing is also capable of recording real-time thermographs where molten pool morphology, key solidification parameters, and possible defects information can be inferred as well. For example, Zhang et al. [73] employed a thermography method to investigate a proper cooling strategy for a GMAW-based WAAM process. Real-time thermal images of thin-wall structures were recorded by an infrared camera, as is shown in Fig. 6.14. Based on the collected images, core thermal parameters for each deposited layer could be obtained, including molten pool size, mean temperature, temperature gradient, and cooling rate. An optimized alternate-cooling-time strategy was thus proposed based on thermal analysis to print a structure more efficiently with high quality.

Additionally, thermography was also employed for real-time defects identification and dimension measurements in a welding process [74, 75]. Take GTAW as an example, various defects including lack of fusion, lack of penetration, cracks, and tungsten inclusions could all be identified with features extracted from thermal images collected by an IR camera. Moreover, dimensional characterizations of longitudinal cracks along welding beads could also be achieved by thermography, which otherwise was very challenging to be detected by conventional visual inspections.

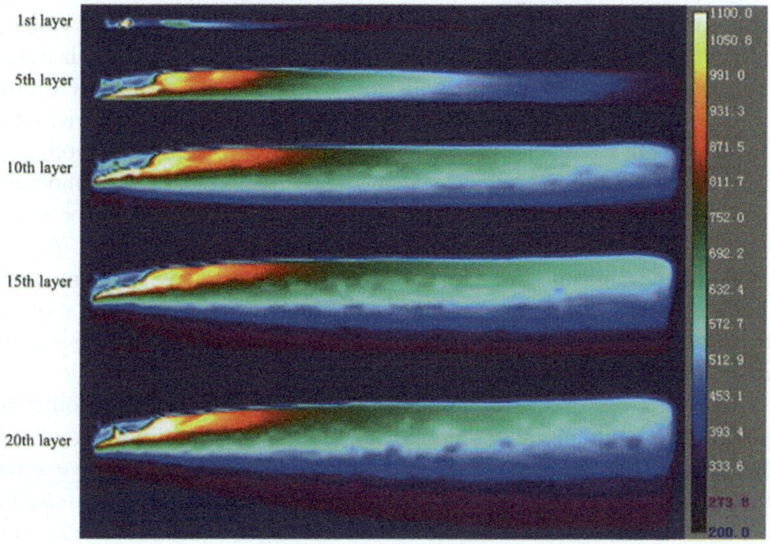

Fig. 6.14 Real-time thermal graph collected by an infrared (IR) camera [73]

6.4.3 Microstructure Control

6.4.3.1 Oscillation Techniques

As mentioned in Sect. 6.4.1, the transformation from coarse columnar grains to fine equiaxed grains has been very attractive to improve the overall mechanical performances and achieve isotropic properties for WAAM-printed parts. It has been reported that molten pool oscillation during arc welding is effective in refining grains and lower defects rate. This is mainly because emerged dendrites will be broken into segments in a turbulent flow of oscillated molten pool, and even freshly formed grains on the wall can get detached if the oscillation is strong. These particulates play a key role in the heterogeneous nucleation of liquid metal in the molten pool, thereby enhancing the nucleation rate and refining grain size. With the increase of grain boundary region, segregation effects in WAAM-printed parts can be also relieved to a lower level. Additionally, oscillation of molten pool is also favorable for the pore to escape and reduce the porosity rate in as-printed parts. Therefore, much research has been conducted to integrate oscillation processes to WAAM for quality improvement.

Modulation of welding power source is usually the first choice to realize molten pool oscillation because no additional devices are required and it is less complex. For example, pulse mode welding current was employed in GTAW-based WAAM of the AZ31 magnesium alloy [76]. The authors studied the influences of pulse welding current on microstructure and mechanical performances and concluded that pulses indeed led to molten pool oscillation. They found that material performance was further maximized when the pulse frequency was adjusted in the range of 5–10 Hz, owing to the resonance effects on the molten pool. Also, with modulated pulse current, fine equiaxed grains with an average size of 21 μm were observed in the microstructure of the as-printed magnesium alloy, and higher yield strength and tensile strength were achieved. In addition to pulse mode modulation, a current waveform with variable polarity was also found effective at stirring molten pools using electromagnetic and plasma force. Take a CMT-based WAAM process as an instance [77], it was found that variable polarity mode significantly transformed coarse columnar grains into fine equiaxed grains for aluminum alloy, compared with other conventional welding modes. The strong stirring effects in the molten pool caused columnar dendrites to split as equiaxed dendrites, which was the main reason for grain refinements and a higher tensile strength value (333 MPa).

Apart from arc power supply modulation, an ultrasonic wave is another type of energy source to facilitate molten pool oscillation during welding. For example, in the CMT process of aluminum alloys, an ultrasonic device was directly exerted on a welding substrate to provide ultrasonic oscillation. It was found that equiaxed grains were obtained in beads with ultrasonic vibration, compared with columnar grains for a conventional CMT process. Cavitation effects and forced streaming in the oscillated molten pool were believed to be the main reasons for fostering favorable conditions for homogeneous equiaxed grains, such as an increase in undercooling and fracture of dendrites. The resulting hardness and corrosion resistance were also highly improved [78]. Similar effects were also observed in GMAW-based WAAM of bronze alloys [79], where the role of an ultrasonic peening device coupled with the substrate was investigated. Microstructure images indicated that columnar dendrites were broken when ultrasonic vibration was applied and was further transformed to fine cellular structures with twinning and dislocations. Tensile properties and nano hardness of respective samples were all improved for the ultrasonic-wave-facilitated WAAM process. It is worth mentioning that ultrasonic vibration can consume much energy if the printed parts are large, which might inhibit the wide application of this method. As an alternative, the mechanical vibration of a welding torch can also be employed to achieve molten pool oscillation and exert similar effects on microstructure and mechanical properties improvement [80].

An external magnetic field has also been explored to be coupled with a WAAM system, and many positive effects on forming, microstructure, and mechanical properties are reported. It indicated that magnetic field contributed to molten pool oscillation and thereby breaking prime dendrite arms and resulting in a dendrites refinement and alignment phenomenon [81]. The dendrite average size could be decreased by 60%, and the orientation of dendrites was also more uniform. Additionally, the magnetic field-induced Lorentz force resulted in stirring effects on the molten pool, which enhanced the convection of liquid metal and alleviated macrosegregation in WAAM. Consequently, even the distribution of alloying elements could be expected. With all these positive improvements of microstructure under magnetic field, the resulting mechanical properties also raised compared with counterparts by conventional WAAM, including microhardness, yield strength, ultimate tensile strength, and even ductility.

Different from employing a controlled electric or magnetic field, weaving of welding torch is also able to exert oscillating effects on the molten pool. For example, a weaving CMT method was employed to achieve WAAM of magnesium alloy wall structures [56]. It was found that the strong electromagnetic force from plasma could effectively stir the molten pool during the weaving process. Such oscillation effects from weaving torch not only broke dendrites, refined grain size but also contributed to the transformation from columnar grains to equiaxed grains. As a consequence, the microstructure of the printed wall was much more uniform than the counterparts with the conventional WAAM method. Even the isotropic properties of the resulting parts were highly improved since the difference among grains orientation was narrowed down.

6.4.3.2 Integrated Cold Work

In-process cold work, such as peening and rolling, has been long integrated into a welding system to mitigate residual stress, control thermal deformation, and improve microstructure by inducing the necessary driving force for recrystallization refinement. Therefore, it can be expected that integrated cold work can also be used for quality control of WAAM-printed parts, and many positive research works have been reported for various materials processed by WAAM.

As elaborated in Sect. 6.3.4, the prior β phase with coarse columnar grains is a challenging problem for the WAAM of titanium alloys. Donoghue et al. [82] found that the β grains could be effectively refined by exerting an interlayer rolling process, with a high rolling force, on each deposited Ti6Al4V layer by GTAW-based WAAM. Moreover, the texture of both α and β phases was mitigated due to the recrystallization induced by heat effects from the subsequent layers. The induced strain energy and heterogeneous deformation from rolling facilitated recrystallization refinement and centimeter-scale columnar grains were transformed to fine equiaxed grains with an average grain size of smaller than 100 μm, as shown in Fig. 6.15.

Interlayer rolling was also utilized in aluminum alloy printing with the GMAW-based WAAM process. It is known that the residual stress level in AM-printed aluminum alloys can be even higher than its yield stress, hence significant deformation usually occurs after the printing process. However, with the introduction of the interlayer rolling process, it was found that the deformation of as-printed aluminum parts could be nearly eliminated and the residual stress was well-controlled [83]. This is because auxiliary interlayer rolling fosters strengthening precipitates which played a key role in enhancing the strength and hardness of printed parts. Similar beneficial effects in microstructure and mechanical performances from cold rolling were also observed in tantalum and nickel alloy printed by WAAM [84, 85].

Apart from interlayer rolling, ultrasonic peening is another commonly employed cold work technique. For example, Wang et al. [86] studied the WAAM printing of the Ti6Al4V alloy augmented with ultrasonic peening, where the peening force was exerted to each deposited layer from three directions. They found that there was no appreciable strain or deformation occurring after the peening process, a feature

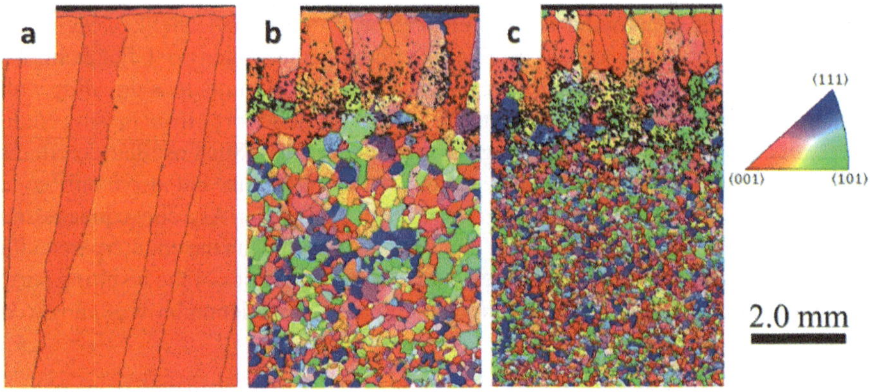

Fig. 6.15 Inverse pole figure (IIPF) electron backscatter diffraction EBSD maps of **a** as-printed wall and rolled WAAM wall under **b** a rolling force of 50 kN, **c** a rolling force of 75 kN [82]

that is very different from cold rolling. However, the microstructure changed due to the refinement of the prior β phase and secondary α phase. They also concluded that ultrasonic vibration fostered grains refinement thus increasing substructures and grains deformation, raising the ultimate tensile strength of the printed part, and mitigating its anisotropy.

6.4.3.3 Post Heat Treatment

Post heat treatment is one of the most effective and commonly employed techniques to relieve residual stress, eliminate detrimental phases, adjust microstructure, and therefore improve mechanical and chemical performances of WAAM-printed parts. As elaborated in previous sections, metallic parts fabricated by WAAM are formed in a high non-equilibrium state with repetitive rapid heating and cooling cycles. Since most structural alloys are strengthened either by solution hardening, precipitation strengthening, or a hybrid mechanism, printing alone is far from sufficient to allow for complete diffusion of alloying elements or precipitation of strengthening phases. However, heat treatment will promote favorable conditions for microstructure in equilibrium. This section will present some representative research works about the effects of heat treatment on as-printed parts by WAAM.

Heat treatment is a universally employed stress relief process that is conducted at an appropriate range of dwelling temperature with a very slow cooling rate, where the materials get softened under no phase transformation. The residual stress in a WAAM-printed part is therefore relieved with local plastic deformation accommodated. For example, stress-relief heat treatment for as-printed Ti6Al4V by WAAM was carried out at 753 K for 2 h [87]. The results indicated that both mechanical strength and ductility of samples were enhanced, compared with an as-printed sample.

In contrast, solution-annealing is typically conducted at a much higher range of temperature to achieve chemical homogeneity and eliminate segregation of alloying elements. The higher temperature of solution annealing provides more driving force to dissolve detrimental phases. However, it is worth mentioning that grain growth may arise in the solution annealing process if the dwelling temperature is elevated with a long holding time. This will degrade mechanical strength and hardness. Usually, a solution annealing treatment will be followed by an age-hardening treatment for uniform precipitation of strengthening phases. Take the Inconel 718 alloy as an example, a solution annealing at 980–1050 °C is performed to dissolve brittle Laves phase and homogenize the distribution of Nb, Al, and Ti elements in the Ni–Cr matrix. Age-hardening steps at 720 and 620 °C will be carried out subsequently to generate nanoscale γ' an γ'' phases, which considerably enhance mechanical performances and corrosion resistance of WAAM-printed parts with 91% increase in hardness [88]. Similar advantageous effects are also observed in the WAAM of the AZ80M magnesium alloy. It was reported that the eutectic structure was fully dissolved and finer continuous precipitates were evenly distributed along the matrix after a solution annealing treatment was performed at 410 °C, and followed by an age-hardening process at 180 °C for 25 h [89]. Such heat treatment improved the microstructure accounted for the simultaneously rising mechanical strength and ductility of WAAM-printed magnesium alloy and also led to an increase in isotropy.

6.4.4 Processing Accuracy Control

It is challenging to achieve excellent control of dimension accuracy and geometry integrity of parts printed using a WAAM system alone, because of its layer-by-layer process which can lead to error accumulation along the built direction. This problem is especially significant when it comes to overhanging structures, where any deviation from the desired overhanging angle may occur if the layer thickness was not well controlled [90]. Fluctuation of beads dimensions may also take place at different height positions of a printed part, due to the difference in heat accumulation, cooling rate, and noise disturbance from the welding power supply. These variations in beads geometries will affect the overlapping of beads as well as the quality of subsequent layers, where porosity and lack of fusion defects may arise. In addition, for some reactive metals, the surface oxide will emerge on the surface of a freshly deposited layer, which may accumulate throughout the print process and stabilize as oxide islands, leading to discontinuity flaws.

In light of these inherent features and potential problems for WAAM, an intelligent hybrid-WAAM (H-WAAM) technique is proposed to remove these barriers with a sensing and machining system integrated with WAAM [91]. Specifically, a laser-based sensing system was proposed to capture the geometrical data of each deposited layer, and the data were compared with the pre-designed pattern to make decisions on the print adjustment of subsequent layers. A milling tool is installed into a motion system and employed to machine the surface or side of the deposited layer according to the data analysis. This approach not only removes possible defects, such as pores, oxide layers, and uneven surfaces but also renders the dimension of each layer highly consistent. With such a system, the geometrical integrity and mechanical performances of printed parts can be well controlled. In addition, some research also indicated that H-WAAM could adjust residual stress levels by mitigating high-level tensile stress with induced compressive stress from the milling process, and the tensile elongation anisotropy could be controlled to be within 5% if the milled thickness was smaller than 0.4 mm [92]. An optimization of H-WAAM processing parameters was also studied to increase materials utilization and processing efficiency while keeping to satisfactory print quality [93]. It also indicated that a high torch speed with a wire feed rate corresponding to fixed bead width was supposed to be coupled with a higher spindle speed of the milling tool and a moderate tool feeding rate for a sustainable fabrication process.

6.5 Modeling of Print Geometry

In WAAM, parts are built systematically by depositing overlapping weld beads in both the horizontal and vertical directions. Hence, each deposited layer acts as the substrate for the printing of the next layer. Thus, it is important that the printed layer is of sufficient flatness in order to provide proper substrate support for the next layer's deposition. Irregularities among the supporting layer's surface lead to error accumulation, resulting in undesirable concave/convex surfaces as the print progresses vertically, leading to print failure. The uniformity of the deposited layer depends largely on the knowledge of the printed bead geometry and appropriate overlapping distance. They are tightly coupled and various analytical formulations have been proposed to model each bead profile and suitable stepover distance.

6.5.1 Single-Bead Model

On single-bead modeling, the majority of the research has been focused on developing mathematical models to relate welding parameters such as the welding current, welding voltage, torch speed, torch angle, wire feed-rate to its resulting bead width [94], height [95], penetration [96], etc. This is typically achieved using machine learning techniques such as regression [97] or neural networks [98]. For those who did not, parametric models such as parabola, cosine, and arc functions have been proposed to model bead profiles accurately. They are subsequently built upon to determine adjacent bead's overlapping distance for optimal printing. For instance, Xiong et al. [99] determined the bead cross-section profile based on various parametric models and compared the actual area of the bead section with that as predicted by the parabola, cosine, and arc functions. He concluded that the ratio of wire feed rate to welding speed greatly influences the shape of the bead profile, and provided

Table 6.1 Existing bead models used to predict single-bead profiles

Model description	Model function	Bead height, h	Bead width, w
Parabola model	$z = ax^2 + c$	c	$2\sqrt{-\frac{c}{a}}$
Cosine model	$z = a\cos bx$	a	$\frac{\pi}{b}$
Arc model	$z = 2\sqrt{a^2 - x^2} - b$	$a - b$	$2\sqrt{a^2 - b^2}$

a guide on which of these models to use based on the input parameters. Ding et al. [100] subsequently extend this result to determine the optimal overlapping distance so that flat multi-bead deposits can be made. Table 6.1 summarises the existing bead models that have been used to approximate cross-sectional profiles of single beads.

Despite these pioneering works, the proposed models are still limited to capturing the entire spectrum of bead profiles, especially for those with significant tail ends curves. A recent approach that uses planar quaternion features with a non-linear neural network has been shown to retain the curvature characteristics of the bead most accurately [101].

6.5.2 *Multi-Bead Model*

A layer of WAAM-printed part is formed by depositing multiple single beads side-by-side. Since the single bead has a curved cross-sectional area, the resulting surface formed by two adjacent overlapping beads is unlikely to be completely flat. They will form two possible valley geometries in-between: (a) A concave valley, or (b) A convex valley as shown in Fig. 6.16.

To achieve a flat valley, the optimal stepover ratios for achieving smooth multi-beads need to be determined and this has been widely investigated in recent literature [99, 100, 102, 103]. Most of this research assumed the single bead to take a parabolic profile and proposed a variety of multi-bead overlapping models such as

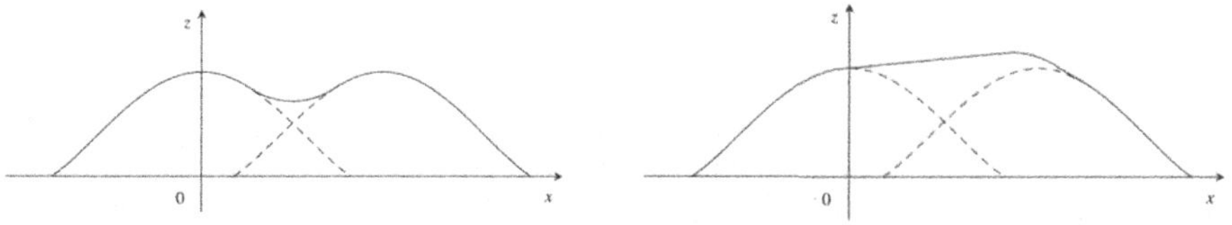

(a) *A concave valley formed between two overlapping beads*

(b) *A convex valley formed between two overlapping beads*

Fig. 6.16 Possible geometries formed by two overlapping beads

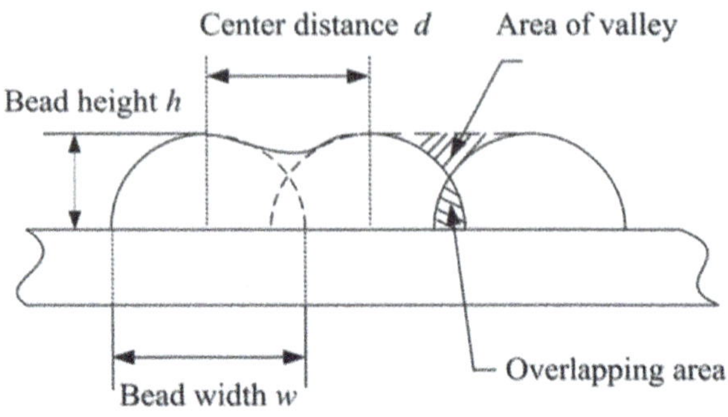

Fig. 6.17 Sketch of the flat-top overlapping model

Flat-Top, Tangent, Varying Ratio, etc. with corresponding optimal stepover ratios. These models are summarised as follows.

6.5.2.1 Flat-Top Overlapping Model (FOM)

The FOM is based on the theory that the overlapping area of two single beads is equal to the area of the valley that is formed between them, which forms an ideal flat-top surface between the peak of both beads [99, 102]. See Fig. 6.17. For the parabolic single-bead model, the optimal stepover distance is found to be $d = \frac{2}{3}w$.

6.5.2.2 Tangent Overlapping Model (TOM)

The TOM is observed from experiments with mild steel. It was found that the material formed between the valley of two single beads does not form a flat, horizontal surface, but instead, they form a line (BC) that is tangential to the second bead [100], see Fig. 6.18. For the parabolic single-bead model, the optimal stepover distance is found to be $d = 0.738w$.

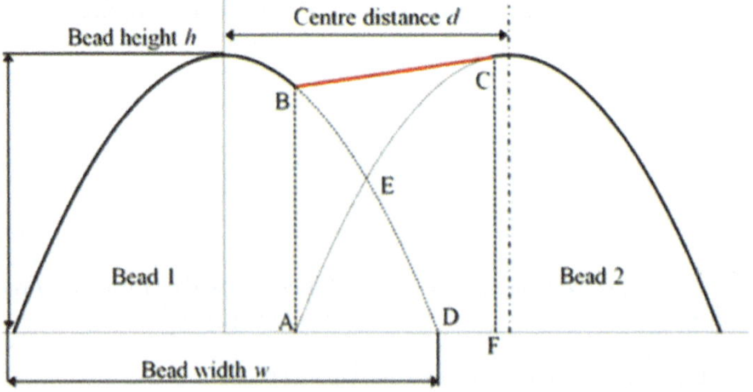

Fig. 6.18 Sketch of the tangent overlapping model

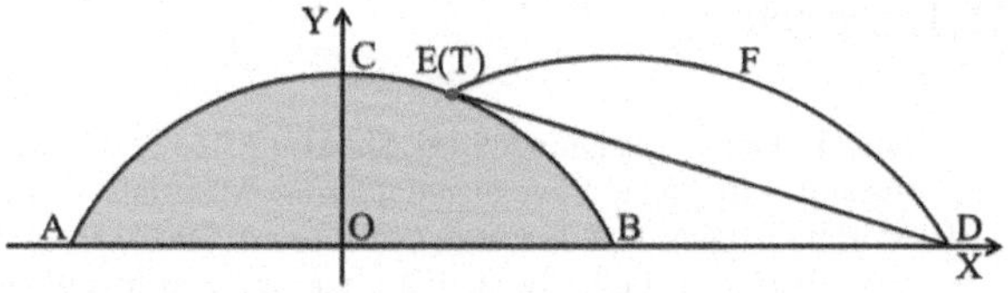

Fig. 6.19 Sketch of varying-ratio overlapping model

6.5.2.3 Varying-Ratio Overlapping Model

The VOM models the two single beads slightly differently: The first bead is represented by a parabola function, while the second bead profile was approximated by another rotated parabola function. The model assumes the chord length (DE) of the second bead profile to be equal to the width of the first bead (AB), and the chord (line DE) of the second bead to be tangent to the first bead's curvature (curve ACB), see Fig. 6.19. For parabolic single beads, the optimal stepover distance was found to vary with bead geometry. It was found that they reduced with an increase of the width to height ratio (WHR) and lay within the region of $d = [0.6317, 0.7732]w$ [103].

6.6 Summary and Outlook

Despite significant efforts in studying the various aspects of the WAAM process, it was found that many areas of WAAM still need significant development and innovation. They can range from process control for residual stress and segregation reduction, defect identification and correction techniques, and bead modeling and simulation (digital twining). One of the biggest hurdles in achieving print accuracy seems to lie in multi-bead modeling. They can help to optimize the selection of process parameters for a given part design, and determine the optimal stepover distances for overlapping beads such that the printed layer can achieve the required feature and yet retain a sufficiently flat surface for subsequent layer support. However, due to difficulties in capturing the single bead's curvature and the material behavior during bead overlapping, challenges in optimizing process parameters and stepover distances arise. It becomes hard to identify the ideal input parameters for the print when the final geometry corresponding to the print parameters remains unknown. More research is required on this front to understand how different materials flow under varying thermal factors to establish a more accurate print simulation. This is just one of the many gaps to address. However, to make it truly viable for commercial application, a multi-disciplinary approach is still required to solve this problem from a system perspective, to achieve an integrated AM workflow that allows for the design to the final part, leveraging on material and process understanding, adaptive process control and artificial intelligence for defects identification and correction.

References

1. Ding D, Pan Z, Van Duin S, Li H, Shen C (2016) Fabricating superior NiAl bronze components through wire arc additive manufacturing. Materials 9(8):652
2. Martina F, Ding J, Williams S, Caballero A, Pardal G, Quintino L (2019) Tandem metal inert gas process for high productivity wire arc additive manufacturing in stainless steel. Addit Manuf 25:545–550
3. Williams SW, Martina F, Addison AC, Ding J, Pardal G, Colegrove P (2016) Wire+arc additive manufacturing. Mater Sci Technol 32(7):641–647
4. He T, Yu S, Shi Y, Huang A (2020) Forming and mechanical properties of wire arc additive manufacture for marine propeller bracket. J Manuf Process 52:96–105
5. Xiong Y, Park S-I, Padmanathan S, Dharmawan AG, Foong S, Rosen DW et al (2019) Process planning for adaptive contour parallel toolpath in additive manufacturing with variable bead width. Int J Adv Manuf Technol 105(10):4159–4170
6. Wu B, Pan Z, Ding D, Cuiuri D, Li H, Xu J et al (2018) A review of the wire arc additive manufacturing of metals: properties, defects and quality improvement. J Manuf Process 35:127–139
7. Colegrove PA, Coules HE, Fairman J, Martina F, Kashoob T, Mamash H et al (2013) Microstructure and residual stress improvement in wire and arc additively manufactured parts through high-pressure rolling. J Mater Process Technol 213(10):1782–1791
8. Ding D, Pan ZS, Cuiuri D, Li HJ (2014) A tool-path generation strategy for wire and arc additive manufacturing. Int J Adv Manuf Technol 73(1–4):173–183
9. Kapil S, Legesse F, Kulkarni P, Joshi P, Desai A, Karunakaran KJ (2016) Hybrid-layered manufacturing using tungsten inert gas cladding. Prog Addi Manuf 1(1):79–91
10. Nilsiam Y, Sanders P, Pearce JM (2017) Slicer and process improvements for open-source GMAW-based metal 3-D printing. Addi Manuf 18:110–120
11. Venturini G, Montevecchi F, Scippa A, Campatelli G (2016) Optimization of WAAM deposition patterns for T-crossing features. Procedia CIRP 55:95–100
12. Caballero A, Ding J, Ganguly S, Williams SJ (2019) Wire+Arc additive manufacture of 17–4 PH stainless steel: effect of different processing conditions on microstructure, hardness, and tensile strength. J Mater Process Technol 268:54–62
13. Feng Y, Zhan B, He J, Wang KJ (2018) The double-wire feed and plasma arc additive manufacturing process for deposition in Cr–Ni stainless steel. J Mater Process Technol 259:206–215
14. Xiong J, Lei Y, Chen H, Zhang GJ (2017) Fabrication of inclined thin-walled parts in multi-layer single-pass GMAW-based additive manufacturing with flat position deposition. J Mater Process Technol 240:397–403
15. Yang D, Wang G, Zhang GJ (2017) A comparative study of GMAW-and DE-GMAW-based additive manufacturing techniques: thermal behavior of the deposition process for thin-walled parts. Int J Adv Manuf Technol 91(5):2175–2184
16. Wu W, Xue J, Wang L, Zhang Z, Hu Y, Dong C (2019) Forming process, microstructure, and mechanical properties of thin-walled 316L stainless steel using speed-cold-welding additive manufacturing. Metals 9(1):109
17. Wang L, Xue J, Wang QJ (2019) Correlation between arc mode, microstructure, and mechanical properties during wire arc additive manufacturing of 316L stainless steel. Mater Sci Eng: A 751:183–190
18. Rodriguez N, Vázquez L, Huarte I, Arruti E, Tabernero I, Alvarez (2018) Wire and arc additive manufacturing: a comparison between CMT and TopTIG processes applied to stainless steel. Weld World 62(5):1083–1096
19. Stützer J, Totzauer T, Wittig B, Zinke M, Jüttner S (2019) GMAW cold wire technology for adjusting the ferrite–austenite ratio of wire and arc additive manufactured duplex stainless steel components. Metals 9(5):564

20. Hosseini VA, Högström M, Hurtig K, Bermejo MAV, Stridh L-E, Karlsson LJ (2019) Wire-arc additive manufacturing of a duplex stainless steel: thermal cycle analysis and microstructure characterization. Weld World 63(4):975–987
21. Karpagaraj A, Baskaran S, Arunnellaiappan T, Kumar NR (2020) A review on the suitability of wire arc additive manufacturing (WAAM) for stainless steel 316. In: Advances in mechanical design, materials and manufacture: proceeding of the second international conference on design, materials and manufacture (ICDEM 2019)
22. Jin W, Zhang C, Jin S, Tian Y, Wellmann D, Liu W (2020) Wire arc additive manufacturing of stainless steels: a review. Appl Sci 10(5)
23. Rodrigues TA, Duarte V, Miranda RM, Santos TG, Oliveira JP (2019) Current status and perspectives on wire and arc additive manufacturing (WAAM). Materials (Basel) 12(7)
24. Mukherjee T, DebRoy T (2019) Printability of 316 stainless steel. Sci Technol Weld Joining 24(5):412–419
25. Chen X, Li J, Cheng X, He B, Wang H, Huang Z et al (2017) Microstructure and mechanical properties of the austenitic stainless steel 316L fabricated by gas metal arc additive manufacturing. Mater Sci Eng, A 703:567–577
26. Chen X, Li J, Cheng X, Wang H, Huang ZJ (2018) Effect of heat treatment on microstructure, mechanical and corrosion properties of austenitic stainless steel 316L using arc additive manufacturing. Mater Sci Eng: A 715:307–314
27. Wen D, Long P, Li J, Huang L, Zheng Z (2020) Effects of linear heat input on microstructure and corrosion behavior of an austenitic stainless steel processed by wire arc additive manufacturing. Vacuum 173:109131
28. Lunde J, Kazemipour M, Salahi S, Nasiri A (eds) (2020) Microstructure and mechanical properties of AISI 420 stainless steel produced by wire arc additive manufacturing. In: TMS 2020 149th annual meeting and exhibition supplemental proceedings. Springer
29. Nemani AV, Ghaffari M, Salahi S, Nasiri AJ (2021) Effects of post-printing heat treatment on the microstructure and mechanical properties of a wire arc additive manufactured 420 martensitic stainless steel part. Mater Sci Eng: A 813:141167
30. Attallah MM, Jennings R, Wang X, Carter LN (2016) Additive manufacturing of Ni-based superalloys: the outstanding issues. MRS Bull 41(10):758–764
31. Dhinakaran V, Ajith J, Fathima Yasin Fahmidha A, Jagadeesha T, Sathish T, Stalin B (2020) Wire arc additive manufacturing (WAAM) process of nickel based superalloys—a review. Mater Today: Proc 21:920–925
32. Hosseini E, Popovich VA (2019) A review of mechanical properties of additively manufactured Inconel 718. Addit Manuf 30
33. Zadi-Maad A, Basuki A (2018) The development of additive manufacturing technique for nickel-base alloys: a review. In: AIP conference proceedings
34. Seow CE, Coules HE, Wu G, Khan RHU, Xu X, Williams S (2019) Wire + arc additively manufactured Inconel 718: effect of post-deposition heat treatments on microstructure and tensile properties. Mater Des 183
35. Xu X, Ding J, Ganguly S, Williams S (2019) Investigation of process factors affecting mechanical properties of INCONEL 718 superalloy in wire + arc additive manufacture process. J Mater Process Technol 265:201–209
36. Wang K, Liu Y, Sun Z, Lin J, Lv Y, Xu B (2020) Microstructural evolution and mechanical properties of Inconel 718 superalloy thin wall fabricated by pulsed plasma arc additive manufacturing. J Alloy Compd 819
37. Tanvir A, Ahsan MR, Ji C, Hawkins W, Bates B, Kim DB (2019) Heat treatment effects on Inconel 625 components fabricated by wire+ arc additive manufacturing (WAAM)—part 1: microstructural characterization. Int J Adv Manuf Technol 103(9):3785–3798
38. Tanvir A, Ahsan MR, Seo G, Kim J-d, Ji C, Bates B et al (2020) Heat treatment effects on Inconel 625 components fabricated by wire+ arc additively manufacturing (WAAM)—part 2: mechanical properties. Int J Adv Manuf Technol 110(7):1709–1721
39. Yangfan W, Xizhang C, Chuanchu SJS, Technology C (2019) Microstructure and mechanical properties of Inconel 625 fabricated by wire-arc additive manufacturing. Surf Coat Technol 374:116–123

40. Derekar KS (2018) A review of wire arc additive manufacturing and advances in wire arc additive manufacturing of aluminium. Mater Sci Technol 34(8):895–916
41. Köhler M, Fiebig S, Hensel J, Dilger K (2019) Wire and arc additive manufacturing of aluminum components. Metals 9(5):608
42. Gu J, Ding J, Williams SW, Gu H, Ma P, Zhai Y (2016) The effect of inter-layer cold working and post-deposition heat treatment on porosity in additively manufactured aluminum alloys. J Mater Process Technol 230:26–34
43. Fu R, Tang S, Lu J, Cui Y, Li Z, Zhang H et al (2021) Hot-wire arc additive manufacturing of aluminum alloy with reduced porosity and high deposition rate. Mater Des 199:109370
44. Aldalur E, Suárez A, Veiga F (2021) Metal transfer modes for wire arc additive manufacturing Al–Mg alloys: influence of heat input in microstructure and porosity. J Mater Process Technol 117271
45. Cong B, Ding J, Williams S (2015) Effect of arc mode in cold metal transfer process on porosity of additively manufactured Al-6.3% Cu alloy. Int J Adv Manuf Technol 76(9–12):1593–1606
46. Zhang C, Gao M, Zeng XJ (2019) Workpiece vibration augmented wire arc additive manufacturing of high strength aluminum alloy. J Mater Process Technol 271:85–92
47. Gierth M, Henckell P, Ali Y, Scholl J, Bergmann JPJ (2020) Wire arc additive manufacturing (WAAM) of aluminum alloy AlMg5Mn with energy-reduced gas metal arc welding (GMAW). Materials 13(12):2671
48. Bermingham M, Kent D, Zhan H, StJohn D, Dargusch M (2015) Controlling the microstructure and properties of wire arc additive manufactured Ti–6Al–4V with trace boron additions. Acta Mater 91:289–303
49. Wu B, Pan Z, Ding D, Cuiuri D, Li H (2018) Effects of heat accumulation on microstructure and mechanical properties of Ti6Al4V alloy deposited by wire arc additive manufacturing. Addit Manuf 23:151–160
50. Lu T, Liu C, Li Z, Wu Q, Wang J, Xu T et al (2020) Hot-wire arc additive manufacturing Ti–6.5 Al–2Zr–1Mo–1V titanium alloy: pore characterization, microstructural evolution, and mechanical properties. J Alloy Compd 817:153334
51. Veiga F, Gil Del Val A, Suárez A, Alonso U (2020) Analysis of the machining process of titanium Ti6Al-4V parts manufactured by wire arc additive manufacturing (WAAM). Materials 13(3):766
52. Mereddy S, Bermingham MJ, StJohn DH, Dargusch MS (2017) Grain refinement of wire arc additively manufactured titanium by the addition of silicon. J Alloy Compd 695:2097–2103
53. Wang J, Lin X, Wang M, Li J, Wang C, Huang W (2020) Effects of subtransus heat treatments on microstructure features and mechanical properties of wire and arc additive manufactured Ti–6Al–4V alloy. Mater Sci Eng, A 776:139020
54. Guo Y, Quan G, Jiang Y, Ren L, Fan L, Pan H (2021) Formability, microstructure evolution and mechanical properties of wire arc additively manufactured AZ80M magnesium alloy using gas tungsten arc welding. J Magnes Alloy 9(1):192–201
55. Yang X, Liu J, Wang Z, Lin X, Liu F, Huang W et al (2020) Microstructure and mechanical properties of wire and arc additive manufactured AZ31 magnesium alloy using cold metal transfer process. Mater Sci Eng, A 774:138942
56. Bi J, Shen J, Hu S, Zhen Y, Yin F, Bu X (2020) Microstructure and mechanical properties of AZ91 Mg alloy fabricated by cold metal transfer additive manufacturing. Mater Lett 276:128185
57. Wang J, Pan Z, Cuiuri D, Li H (2019) Phase constituent control and correlated properties of titanium aluminide intermetallic alloys through dual-wire arc additive manufacturing. Mater Lett 242:111–114
58. Wang L, Zhang Y, Hua X, Shen C, Li F, Huang Y et al (2021) Fabrication of γ-TiAl intermetallic alloy using the twin-wire plasma arc additive manufacturing process: Microstructure evolution and mechanical properties. Mater Sci Eng, A 812:141056
59. Meng Y, Li J, Gao M, Zeng X (2021) Microstructure characteristics of wire arc additive manufactured NiAl intermetallic compounds. J Manuf Process 68:932–939

60. Ye X, Hua X, Wang M, Lou S (2015) Controlling hot cracking in Ni-based Inconel-718 superalloy cast sheets during tungsten inert gas welding. J Mater Process Technol 222:381–390
61. Zhan Q, Liang Y, Ding J, Williams S (2017) A wire deflection detection method based on image processing in wire+ arc additive manufacturing. Int J Adv Manuf Technol 89(1–4):755–763
62. Xiong J, Zhang G, Qiu Z, Li Y (2013) Vision-sensing and bead width control of a single-bead multi-layer part: material and energy savings in GMAW-based rapid manufacturing. J Clean Prod 41:82–88
63. Tang S, Wang G, Zhang H, Wang R (eds) (2017) An online surface defects detection system for AWAM based on deep learning. In: Solid freeform fabrication 2017: proceedings of the 28th annual international solid freeform fabrication symposium–an additive manufacturing conference
64. Xu F, Dhokia V, Colegrove P, McAndrew A, Williams S, Henstridge A et al (2018) Realisation of a multi-sensor framework for process monitoring of the wire arc additive manufacturing in producing Ti-6Al-4V parts. Int J Comput Integr Manuf 31(8):785–798
65. Saad E, Wang H, Kovacevic R (2006) Classification of molten pool modes in variable polarity plasma arc welding based on acoustic signature. J Mater Process Technol 174(1–3):127–136
66. Pal K, Bhattacharya S, Pal SK (2010) Investigation on arc sound and metal transfer modes for on-line monitoring in pulsed gas metal arc welding. J Mater Process Technol 210(10):1397–1410
67. Van Bohemen S, Hermans M, Den Ouden G (2001) Monitoring of martensite formation during welding by means of acoustic emission. J Phys D Appl Phys 34(22):3312
68. Chen B, Chen S (2011) A study on applications of multi-sensor information fusion in pulsed-GTAW. In: Robotic welding, intelligence and automation. Springer, pp 245–252
69. Polajnar I, Bergant Z, Grum J (eds) (2013) ARC welding process monitoring by audible sound. In: 12th international conference of the Slovenian society for non-destructive testing: application of contemporary non-destructive testing in engineering, ICNDT—conference proceedings
70. Asif K, Zhang L, Derrible S, Indacochea JE, Ozevin D, Ziebart B (2020) Machine learning model to predict welding quality using air-coupled acoustic emission and weld inputs. J Intell Manuf 1–15
71. Sumesh A, Rameshkumar K, Mohandas K, Babu RS (2015) Use of machine learning algorithms for weld quality monitoring using acoustic signature. Procedia Comput Sci 50:316–322
72. Artaza T, Suárez A, Veiga F, Braceras I, Tabernero I, Larrañaga O et al (2020) Wire arc additive manufacturing Ti6Al4V aeronautical parts using plasma arc welding: analysis of heat-treatment processes in different atmospheres. J Market Res 9(6):15454–15466
73. Yang D, Wang G, Zhang G (2017) Thermal analysis for single-pass multi-layer GMAW based additive manufacturing using infrared thermography. J Mater Process Technol 244:215–224
74. Rodríguez-Martín M, Lagüela S, González-Aguilera D, Martínez J (2016) Thermographic test for the geometric characterization of cracks in welding using IR image rectification. Autom Constr 61:58–65
75. Vasudevan M, Chandrasekhar N, Maduraimuthu V, Bhaduri A, Raj B (2011) Real-time monitoring of weld pool during GTAW using infra-red thermography and analysis of infra-red thermal images. Weld World 55(7):83–89
76. Guo J, Zhou Y, Liu C, Wu Q, Chen X, Lu J (2016) Wire arc additive manufacturing of AZ31 magnesium alloy: grain refinement by adjusting pulse frequency. Materials 9(10):823
77. Zhang C, Li Y, Gao M, Zeng X (2018) Wire arc additive manufacturing of Al-6Mg alloy using variable polarity cold metal transfer arc as power source. Mater Sci Eng, A 711:415–423
78. Tian Y, Shen J, Hu S, Wang Z, Gou J (2018) Effects of ultrasonic vibration in the CMT process on welded joints of Al alloy. J Mater Process Technol 259:282–291
79. Chen W, Chen Y, Zhang T, Wen T, Yin Z, Feng X (2020) Effect of ultrasonic vibration and interpass temperature on microstructure and mechanical properties of Cu-8Al-2Ni-2Fe-2Mn alloy fabricated by wire arc additive manufacturing. Metals 10(2)

80. Biradar N, Raman R (2012) Grain refinement in Al-Mg-Si alloy TIG welds using transverse mechanical arc oscillation. J Mater Eng Perform 21(11):2495–2502
81. Wang Y, Chen X, Shen Q, Su C, Zhang Y, Jayalakshmi S et al (2021) Effect of magnetic field on the microstructure and mechanical properties of inconel 625 superalloy fabricated by wire arc additive manufacturing. J Manuf Process 64:10–19
82. Donoghue J, Antonysamy AA, Martina F, Colegrove PA, Williams SW, Prangnell P (2016) The effectiveness of combining rolling deformation with wire-arc additive manufacture on β-grain refinement and texture modification in Ti–6Al–4V. Mater Charact 114:103–114
83. Hönnige J, Colegrove PA, Ganguly S, Eimer E, Kabra S, Williams S (2018) Control of residual stress and distortion in aluminium wire+ arc additive manufacture with rolling. Addit Manuf 22:775–783
84. Marinelli G, Martina F, Ganguly S, Williams S (2020) Grain refinement in an unalloyed tantalum structure by combining wire+ arc additive manufacturing and vertical cold rolling. Addit Manuf 32:101009
85. Xu X, Ganguly S, Ding J, Seow CE, Williams S (2018) Enhancing mechanical properties of wire + arc additively manufactured INCONEL 718 superalloy through in-process thermomechanical processing. Mater Des 160:1042–1051
86. Gou J, Wang Z, Hu S, Shen J, Tian Y, Zhao G et al (2020) Effects of ultrasonic peening treatment in three directions on grain refinement and anisotropy of cold metal transfer additive manufactured Ti-6Al-4V thin wall structure. J Manuf Process 54:148–157
87. Bermingham M, Nicastro L, Kent D, Chen Y, Dargusch M (2018) Optimising the mechanical properties of Ti-6Al-4V components produced by wire+ arc additive manufacturing with post-process heat treatments. J Alloy Compd 753:247–255
88. Kindermann RM, Roy M, Morana R, Prangnell PB (2020) Process response of Inconel 718 to wire+ arc additive manufacturing with cold metal transfer. Mater Des 195:109031
89. Guo Y, Quan G, Celikin M, Ren L, Zhan Y, Fan L et al (2021) Effect of heat treatment on the microstructure and mechanical properties of AZ80M magnesium alloy fabricated by wire arc additive manufacturing. J Magnes Alloy
90. Lam TF, Xiong Y, Dharmawan AG, Foong S, Soh GS (2020) Adaptive process control implementation of wire arc additive manufacturing for thin-walled components with overhang features. Int J Adv Manuf Technol 108(4):1061–1071
91. Dharmawan AG, Xiong Y, Foong S, Soh GS (eds) (2019) Development of an automated and adaptive system for robotic hybrid-wire arc additive manufacturing (h-waam). In: IFToMM international symposium on robotics and mechatronics. Springer
92. Zhang S, Gong M, Zeng X, Gao M (2021) Residual stress and tensile anisotropy of hybrid wire arc additive-milling subtractive manufacturing. J Mater Process Technol 293:117077
93. Li F, Chen S, Shi J, Tian H, Zhao Y (2017) Evaluation and optimization of a hybrid manufacturing process combining wire arc additive manufacturing with milling for the fabrication of stiffened panels. Appl Sci 7(12):1233
94. Kim I-S, Son J-S, Yarlagadda PKDV (2003) A study on the quality improvement of robotic GMA welding process. Robot Comput-Integr Manuf 19(6):567–572
95. Li P, Fang MTC, Lucas J (1997) Modelling of submerged arc weld beads using self-adaptive offset neutral networks. J Mater Process Technol 71(2):288–298
96. Palani PK, Murugan N (2007) Optimization of weld bead geometry for stainless steel claddings deposited by FCAW. J Mater Process Technol 190(1):291–299
97. Kannan T, Murugan N (2006) Effect of flux cored arc welding process parameters on duplex stainless steel clad quality. J Mater Process Technol 176(1):230–239
98. Kim I-S, Son J-S, Lee S-H, Yarlagadda PKDV (2004) Optimal design of neural networks for control in robotic arc welding. Robot Comput-Integr Manuf 20(1):57–63
99. Xiong J, Zhang G, Gao H, Wu L (2013) Modeling of bead section profile and overlapping beads with experimental validation for robotic GMAW-based rapid manufacturing. Robot Comput-Integr Manuf 29(2):417–423
100. Ding D, Pan Z, Cuiuri D, Li H (2015) A multi-bead overlapping model for robotic wire and arc additive manufacturing (WAAM). Robot Comput-Integr Manuf 31:101–110

101. Oh XY, Soh GS (2020) A study on the machine learning framework for the geometric modelling of wire arc bead profile. In: Volume 6: 25th design for manufacturing and the life cycle conference (DFMLC)
102. Suryakumar S, Karunakaran KP, Bernard A, Chandrasekhar U, Raghavender N, Sharma D (2011) Weld bead modeling and process optimization in hybrid layered manufacturing. Comput Aided Des 43(4):331–344
103. Hu Z, Qin X, Li Y, Yuan J, Wu Q (2020) Multi-bead overlapping model with varying cross-section profile for robotic GMAW-based additive manufacturing. J Intell Manuf 31(5):1133–1147

Chapter 7
Additive Manufacturing of Ceramics: Materials, Characterization and Applications

Jiaming Bai, Jinxing Sun, and Jon Binner

Abbreviations

AM	Additive manufacturing
CMC	Ceramic matrix composites
PDCs	Polymer derived ceramics
HPCs	High performance ceramics
3DP	Three dimensional printing
CAD	Computer aided design
CAM	Computer aided manufacturing
CIM	Ceramic injection molding
VP	Vat-polymerization
SLA	Stereolithography
DLP	Digital light processing
TPP	Two-photon polymerization
DIW	Direct ink writing
RC	Robocasting
BJ	Binder jetting
IJP	Inkjet printing
SLS	Selective laser sintering
SLM	Selective laser melting

J. Bai (✉) · J. Sun
Department of Mechanical and Energy Engineering, Southern University of Science and Technology, Shenzhen 518055, China
e-mail: baijm@sustech.edu.cn

J. Sun
e-mail: sjx476526525@163.com

J. Sun · J. Binner
School of Metallurgy and Materials, University of Birmingham, Birmingham B15 2TT, UK
e-mail: J.Binner@bham.ac.uk

K. Zhou (ed.), *Additive Manufacturing*,
https://doi.org/10.1007/978-3-031-04721-3_7

LENS	Laser engineered net shaping
DMD	Digital mirror device
PZT	Lead zirconate titanate
CVI	Chemical vapor infiltration
PIP	Precursor infiltration and pyrolysis
RMI	Reactive melt infiltration
CIP	Cold isostatic pressing
DOD	Drop on demand
DED	Direct energy deposition
H-AM	Hybrid additive manufacturing
LTCC	Low temperature co-fired ceramic
P-BJ	Powder-based binder jetting
S-BJ	Slurry-based binder jetting
P-SLS	Powder-based selective laser sintering
S-SLS	Slurry-based selective laser sintering
4PB/4P	Four-point bending test
3PB/3P	Three-point bending test
In	Indentation fracture toughness
SEVNB	Single edge V-notched beam
TZP	Tetragonal zirconia polycrystal
TBC	Thermal barrier coating
LTD	Low temperature degradation
HIP	Hot isostatic pressing
UHTCs	Ultra-high-temperature ceramics
CSD	Calcium sulfate dehydrate
β-TCP	β-Tricalcium phosphate
HA	Hydroxyapatite
PVOH	Polyvinyl alcohol
C_f	Carbon fiber
SiC_w	SiC whisker
Si_3N_{4w}	Si_3N_4 whisker
CNTs	Carbon nanotubes
SWCNTs	Single-wall carbon nanotubes
MWCNTs	Multi-wall carbon nanotubes
GO	Graphene oxide
3Y-TZP	3 Mol.% yttria stabilized tetragonal zirconia polycrystal
BCZT	$Ba_{0.85}Ca_{0.15}Zr_{0.1}Ti_{0.9}O_3$
KNN	$(K_{0.5}Na_{0.5})NbO_3$
BTO	Barium titanate
ZTA	Zirconia toughening alumina
ATZ	Alumina toughening zirconia
ZrO_{2p}	Zirconia particles
Al_2O_{3p}	Alumina particles
LMP	Ceramics with low melting point
HLA	Ceramics with high laser absorption

LRI	Ceramics with low refractive index
All	A wide range of ceramic materials
LAUV	Ceramics with low absorption at UV wavelength
F	Functional component
S	Structural components
Bio	Bioceramics
CM	Casting mold
C	Cellular parts
E	Extremely high precision parts
R	Component repair and coat

7.1 Background

Advanced ceramics, also commonly known as high-performance ceramics (HPCs), can be classified into four distinct material categories: structural, functional, bioceramics, and ceramic matrix composites. They offer a wide range of useful properties, involving high mechanical strength and hardness, good chemical and thermal resistance and biocompatibility, as well as some very specific electronic, magnetic, and nuclear properties, with demanding applications in the aerospace, automotive, defense, energy, environment, electronics, chemical and biomedical engineering industries [1, 2].

Due to the inherent brittleness of ceramic materials, conventional manufacturing methods are limited, and it is complicated to shape such materials when compared to other material types, such as metals and polymers. Conventional processing techniques, such as slip casting, injection molding and tape casting involve long processing times and are expensive. Furthermore, ceramic objects with highly complex geometries and interconnected channels are impossible to produce through those methods as molds are usually involved in conventional techniques. Moreover, post machining operations, such as milling and drilling, are almost always required. Such operations are expensive and can lead to damage that deteriorates properties. Hence, a new method that allows the formation of near net complex shapes is needed.

Additive manufacturing (AM), also known as three-dimensional (3D) printing, is a process of making three-dimensional components by adding materials layer-upon-layer guided by 3D computer aided design (CAD) model data, instead of removing material. 3D printing enables the creation of parts with customized and highly complex architectures with high precision, whilst also minimizing the amount of expensive machining required. As such, the drawbacks of conventional manufacturing methods can be effectively mitigated.

Compared to metallic and polymeric materials, ceramic AM is in its technological infancy. The introduction of AM to produce ceramic components offers tremendous hope in terms of addressing the limitations of conventional methods [3]. A comparison of the conventional method of ceramic injection molding (CIM) and AM process

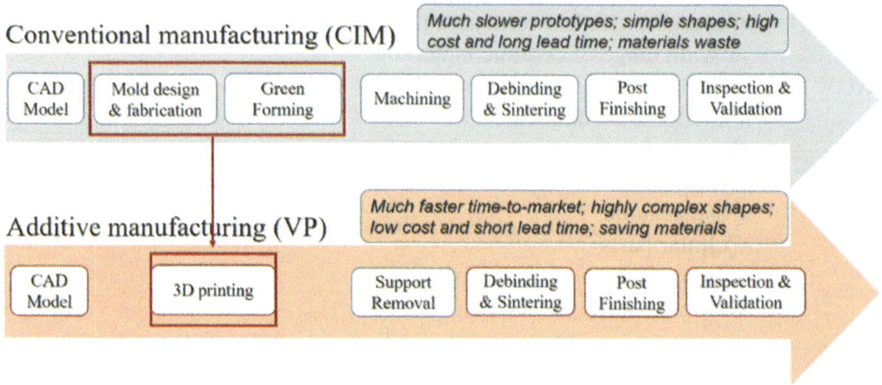

Fig. 7.1 Comparison of the conventional and AM process chains

chains for the fabrication of advanced ceramics, shown in Fig. 7.1, is presented to comprehensively illustrate the benefits of AM. It is clear that AM provides an alternative green forming method and therefore much of the ancillary processing required to achieve a final ceramic component is almost identical to conventional processing [4]. The advantages of AM methods have an impact on the fabrication of ceramic parts in general. This is especially true when compared to CIM, which has a long leading time and costly post machining due to the costly and laborious fabrication of the required molds [5]. In this case, AM enables a significantly faster time-to-market, in combination with a shorter product life cycle, without the involvement of a mold. In addition, while CIM can create a range of more complex features compared to other conventional ceramic processing routes, it is limited by the tooling geometry and the process of extraction from the tool. This is not the case with AM; the latter allows the efficient fabrication of highly complex shapes with detailed feature resolution directly from a CAD model owing to the freedom of design. In this context, AM can be the key technology to enable a transition from manufacturing-oriented design towards design-oriented manufacturing, which can be a huge advantage for the construction of advanced ceramic components such as customized cellular scaffolds or catalyst carriers, which cannot be created via conventional methods. Finally, AM generates significantly less waste than conventional ceramic processing, which can help to significantly reduce material costs. In summary, ceramic AM technology has the potential to reduce production costs and lead times through the reduction of materials loss and the number of fabrication steps, as well as removing the limitation of the geometrical complexity of parts..

AM has gained significant attention for processing advanced ceramic components over the past two decades and can be described in the following sequence in terms of frequency of use: vat-polymerization; direct ink writing/robocasting (DIW/RC);

binder jetting/three-dimensional printing (BJ/3DP); inkjet printing (IJP); selective laser melting/sintering (SLM/SLS) and laser engineered net shaping (LENS) as well as emerging hybrid AM processes. Each ceramic AM route requires different process parameters, and feedstocks and offers both potential capabilities and limitations. When selecting a suitable AM technology to fabricate ceramic pasts, both the advantages and limitations must be considered.

This chapter comprehensively reviews recent progress in the AM of technical ceramics. An in-depth analysis of the advantages, limitations, opportunities and issues arising with each AM route is provided, with the emphasis being in the areas of materials, properties and their applications, as shown in Fig. 7.2. Section 7.2 describes popular AM technologies that have been already used to fabricate advanced ceramics. The most commonly used advanced ceramics and representative specimens produced by AM are provided in Sects. 7.3 and 7.4, with more details on achievements of final properties in terms of mechanical, electrical and biological properties, their applications, and a comparison with the conventional process. Section 7.5 provides a summary and outlook, as well as an understanding of the advantages and limitations of AM, emphasizing the challenges to the AM of advanced ceramics. This chapter is specifically aimed at ceramists, AM specialists and product designers who may consider applying their individual knowledge to explore the potential of ceramic AM.

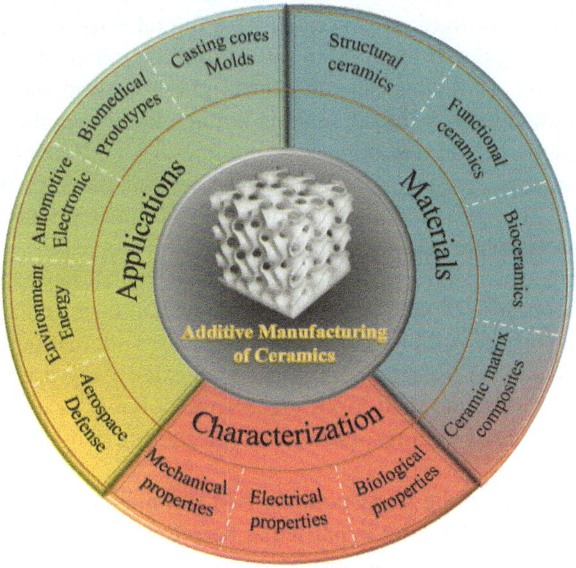

Fig. 7.2 Schematic illustration of the main topics covered in this chapter

7.2 Overview of Ceramic AM Technology

7.2.1 Vat-polymerization

Vat-polymerization (VP), the first commercially available AM technology, has been gaining the most interest among AM technologies for processing ceramics. VP uses a vat of photo reactive liquid ceramic slurry or paste, out of which the ceramic green part is built layer-by-layer based on a CAD file. Sometimes, a wiper blade is needed to level the fresh ceramic slurry after printing each layer. Three variants of the VP have been developed: stereolithography (SLA), digital light processing (DLP), and two-photon polymerization (TPP), with the latter being capable of creating ceramic parts with extremely high accuracy and resolution, as compared to SLA and DLP. SLA and DLP remain the most widely adopted technologies in the fabrication of ceramics today. Table 7.1 shows a direct comparison among SLA, DLP and TPP. Indeed, they are based on distinct light patterning techniques such as vector scanning, mask projection and two-photon, respectively, to induce the photopolymerization process to form ceramic green parts [6]. The SLA was the first proposed rapid solid freeform fabrication technique, created by Charles Hull in 1986. Since 1994, it has been the most carried out technique to shape advanced ceramics [7]. In SLA, a certain wavelength light beam is used to selectively solidify a ceramic suspension to produce parts. While the DLP approach uses a light mask to cure the entire layer at once. The light mask is generated by a digital micromirror device (DMD), which has a rectangular array of up to several million mirrors that can be rotated individually to an on or off state. The ultra-fast light projection allows the DLP printing time to be dramatically reduced, making it much faster than the SLA point-line-layer scanning mode [8]. In addition, most commercial DLP printers on the market offer higher accuracy for the manufacturing of ceramic parts [9, 10].

Two-photon polymerization (TPP) is an emerging AM technology which is used to fabricate ceramic components. TPP involves the polymerization being activated by a near-infrared femtosecond laser, which has great advantages for fabricating arbitrary 3D micro/nano structures with ultra-resolution and accuracy. However, this technique is limited to printing small area microstructures as a result of a short travel stage [11]. Recently, TPP has been extensively used to manufacture ceramic parts by curing the mixture of preceramic polymer or ceramic slurry. In 2019, Cui et al. fabricated high resolution PZT piezoelectric micro-lattices using the TPP method [12].

Among the many ceramic AM techniques, VP-based ceramic 3D printing technologies stand out in particular because of their outstanding level of accuracy, good surface quality, anisotropic microstructure and exceptional mechanical properties [13]. To obtain the high-quality ceramic parts, several key requirements along with challenges need to be realized and are discussed below.

Table 7.1 Comparison of three kinds of vat polymerization process

VP	DLP	SLA	TPP
Schematic diagram	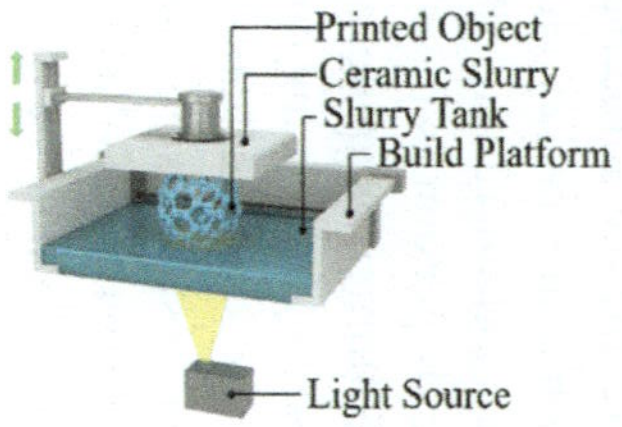	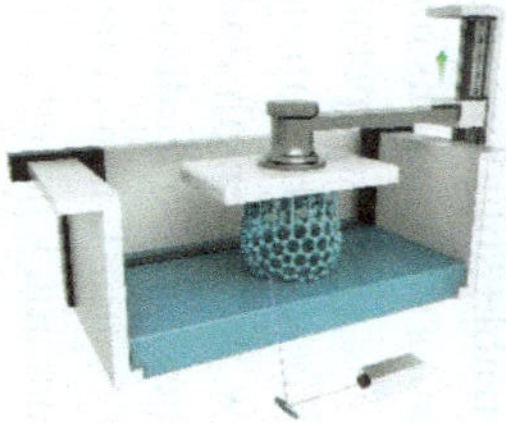	
Setup	Bottom-up/top-down	Bottom-up/top-down	Bottom-up/top-down
Resolution	High	High	Extremely high
Speed	High	Medium	Low
Build volume	Large	Large	Small
Materials	Material with low refractive index and low absorption at laser source		

7.2.1.1 Ceramic Suspensions

The preparation of a suitable ceramic suspension is an essential step in the VP process. In general, ceramic suspensions can be categorized according to the slurry constituents into ceramic powder-based slurry and preceramic polymer-based slurry; they should satisfy a few requirements to be used for the production of ceramic parts: (a) have long-term stability and have a satisfactory viscosity to make sure the scraper can recoat the suspensions homogeneously, or be self-levelling without a scraper; (b) have a high ceramic loading to minimize the debinding process, a low dimensional shrinkage, a high level density and superior mechanical properties of sintered parts [14–16].

Ceramic powder-based slurry—It is typically comprised of ceramic powders, monomer, oligomer, diluent (or solvent), photoinitiator and other additives, such as dispersant, light absorber and defoamer [3, 17]. In general, it can also be subdivided according to the photopolymerizable solutions used into aqueous-based and non-aqueous-based suspensions. Aqueous suspensions have low viscosity compared to non-aqueous based suspensions at the same temperature and shear rate and are much more environmentally friendly. However, the limiting factors of aqueous-based suspension mainly resolve around low part strength and part accuracy. Moreover, a large amount of water present in the slurry introduces several issues, including cracking caused by drying process. In contrast, non-aqueous based ceramic suspension offering higher part strength and accuracy have been extensively investigated in recent works [18].

The requirement of the ceramic powder-based slurry with low viscosity and high solid loading is usually mutually exclusive due to the strong attractive force between particles. It is therefore critical to overcome the attractive inter-particle interaction to obtain suitable suspensions. Dispersants are the key element of ceramic powder-based slurries as they can prevent the agglomerations of ceramic particles in the slurry. The reasonable selection of the dispersant is based on the ceramic material type, particle size, and the liquid medium; it is quite effective to disperse most ceramic particles in a non-aqueous based system using commercial dispersant. Sun et al. comprehensively studied the dispersing mechanism of various commercial dispersant, and evaluated the dispersed results using varied methods [15]. Mario et al. reported using the commercial dispersant Solspderse 41,000 to modify the alumina powder ($D_{50} = 0.1$–$0.3\ \mu m$) based on steric or electro-steric mechanisms, with the final suspensions showing homogeneous and shear thinning behavior [19].

Preceramic polymer-based slurry—It is formulated by mixing preceramic polymer, photoreactive polymer, photoinitiator, and other additives, such as ceramic filler [20]. A variety of advanced ceramics SiOC [21], SiCN [22], SiBCN [23] and SiC can be obtained from the pyrolysis of preceramic polymer. This kind of slurry typically shows extremely low viscosity, as compared to ceramic powder-based slurry. However, as-sintered ceramic can exhibit large shrinkage after pyrolysis and often leads to the formation of cracks with an improper pyrolysis profile [24].

7.2.1.2 Photocuring Behaviors

The ceramic particles suspended in photosensitive resins significantly disrupt the photopolymerization behavior of conventional resins due to the presence of light scattering, which leads to light attenuation in vertical directions, resulting in a decrease in cure depth and an increase in cure width. It is notable that cure depth must be slightly greater than the layer thickness to ensure a good interlayer bonding strength; an insufficient cure depth is subject to delamination of green body, further decreasing the mechanical properties of the final parts. It is highly necessary to predict the cure depth, in most cases by applying the Beer-Lambert law (Jacobs Equation) [25] and Griffith's approximation Equation [26].

$$C_d = \frac{2}{3}\frac{d}{Q}\frac{1}{f}\left(\frac{n_0}{n - n_0}\right)^2 \ln\left(\frac{E}{E_C}\right) \tag{7.1}$$

where d is the ceramic particle size, Q is the scattering efficiency term, f is the solid loading, n_0 and n are the refractive index of the photopolymer and ceramic powder, respectively. These parameters are suspension-dependent variables. Moreover, E is the incident exposure energy and E_c is the critical exposure energy. In the case of certain suspensions, the cure depth is directly affected by the exposure energy, which depends on the laser intensity, scan speed and laser beam diameter in the SLA process, whilst the exposure intensity and exposure time attribute to the cure depth in the DLP system. Ceramic suspension-dependent factors, such as ceramic particle size, the difference of refractive index and ceramic loading also influence the cure depth. By increasing the particle size and lowering the ceramic loading a high cure depth is enabled, owing to the reduced scattering center, but this is not beneficial for the properties of the final parts. Also, a lower difference of the refractive index yields a higher cure depth.

Light scattering can degrade the printing accuracy and surface quality due to the excess cure width. The curing profiles of ceramic suspension are shown in Fig. 7.3. It can be seen a mushroom shape can be observed. Gerald et al. studied the influence of cure strategies on the scattering of the DLP process, defining the overgrowth resulting from the scattering as originally exposed length subtracted from the actually measured length [27]. In comparison, when a laser beam with an ideal Gaussian lateral intensity distribution is used in the SLA route, the cured lines can be widened and cure depth can be decreased. Gentry et al. proposed a scattering model using a series of experiments to prove that small refractive index differences are helpful to improve cure depth [28]. Indeed, scattering issues are inevitable due to the refractive index contrast; the scattering model opens the door for a comprehensive understanding of light scattering in ceramic suspensions, further helping researchers to eliminate light scattering.

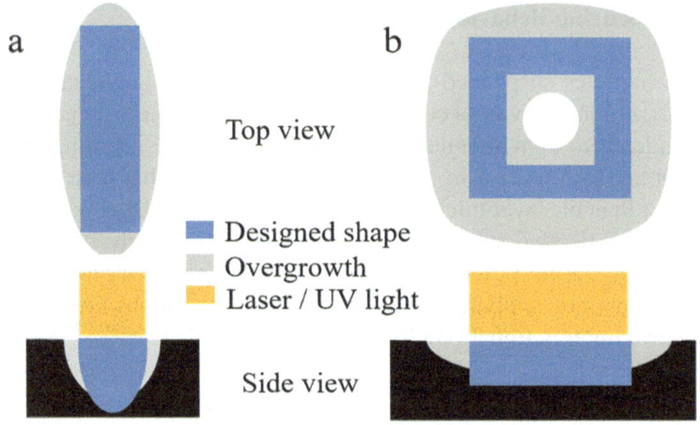

Fig. 7.3 Schematic illustration of curing profiles of suspension in the process of **a** SLA; **b** DLP

7.2.1.3 Post-processing

After obtaining the green parts, the post-processing steps: cleaning, debinding (or pyrolysis), along with sintering, are required to obtain the final parts.

Cleaning—The residual slurry in the green body should be firstly removed using cleaning solutions prior to the debinding process. The green part is immersed in the low viscosity liquid for a few hours and then follows an ultrasonic cleaning and compressed air treatment to remove the uncured materials. The selection of a suitable cleaning solvent to remove un-cured slurry from the green parts produced is important to avoid swelling, deformation and layer delamination. For example, researchers at Lithoz developed their commercial cleaning solution (LithaSol 20) to be used together with compressed air to completely remove uncured suspensions [29]. Hamada et al. successfully fabricated wollastonite-diopside glass–ceramics using DLP technology; the cleaning work involved the immersion of the green body in isopropanol, followed by careful drying with compressed air [30].

Debinding or pyrolysis—After cleaning, the binder needs to be thermally removed prior to sintering. Proper care should be taken in the careful optimization of the debinding profile, it includes the heating rate, dwelling temperature and duration. An improper debinding cycle is expected to introduce cracks, deformation and blistering, thus the optimization of the debinding process plays a determining role to produce defect-free bodies [31]. Extensive work has been done to investigate and analyze the binder decomposition process. For example, Sun et al. studied the debinding process of the ceramic VP route; an optimized approach was designed and developed based on the TGA result, with the optimized debinding cycle yielding a crack-free body and reduced debinding time [32].

Sintering—Sintering is a densification process and a relatively mature process in comparison to the debinding step. Sintering involves the green body experiencing shrinkage to remove porosity caused by binder removal. As a result, the density of as-sintered parts increases but parts shrink volumetrically.

7.2.2 *Direct Ink Writing*

Direct ink writing (DIW), originally developed in Sandia National Laboratories in 1996 and also known as robocasting, is one of the most versatile and most used ceramic AM technologies [33, 34]. A schematic diagram of the DIW process is shown in Fig. 7.4.

In the approach, a ceramic or preceramic polymer ink is extruded through nozzles (typical nozzle size is a few hundred microns) and built up into the desired ceramic object on a substrate [35, 36]. DIW offers great potential to fabricate porous ceramics with space filling and spanning structures. However, the achievable and quality resolution in the DIW process is much lower than that achieved by SLA or DLP. Moreover, robocasting is not only a cheaper and faster manufacturing process, but also benefits from relatively easy access in the complex shape of ceramic matrix composites through extruding mixture paste using one or two nozzles or a UV lamp [37]. Giorgia et al. presented a preceramic polymer ink containing chopped carbon fibers, enabling the fabrication of ceramic matrix composite structure with complex geometry [38]. Fiocco et al. reported a DIW technique to print silica-bonded calcite scaffolds, the paste was comprised of a silicone polymer, calcite powders and fillers [39]. Ideally, it can also be used to produce the overhang or spanning geometries by introducing expendable support material, such as concentrated water-soluble [40, 41] or combustible colloidal inks [42]. More importantly, almost all ceramic materials can be fabricated using DIW, typically including the fabrication of black ceramics such as SiC [43] and Si_3N_4 [44]. Considering its advancement, robocasting has attracted

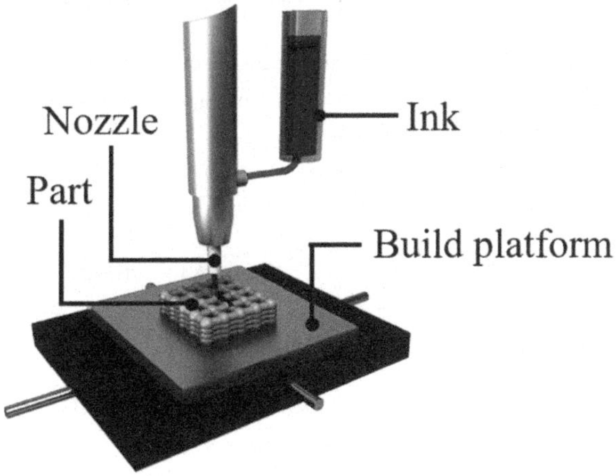

Fig. 7.4 Schematic diagram of direct ink writing process

tremendous attention in recent years, however fabricating dense ceramic components by DIW is still a challenge.

7.2.2.1 Ceramic Pastes Requirement

The DIW process requires precise control over the rheological behavior of ceramic pastes; they must easily flow through the nozzle during extrusion, but also have sufficient strength and stiffness to resist deformation caused by the force of gravity. Aqueous-based suspensions are preferred for DIW as compared to the suspensions for the VP process as the latter are typically characterized by high ceramic loading and low environmental contamination. Basically, aqueous ceramic ink is comprised of ceramic powders, deionized water, dispersant, binder and other additives, such as pH adjuster, coagulation agents and plasticizer [45].

Printability—Ceramic paste having pseudoplastic properties and shear thinning flow behavior plays an important role for successful printing. During the DIW, a desired well-dispersed paste facilitates deposition through the nozzle when the shear stress exceeds the sufficient yield stress, while the viscosity decreases dramatically as the shear rate increases, as described by the Herschel-Bulkley equation [46]. After extrusion, the paste undergoes quick gelation to support layer stacking. Amin et al. studied the relationship between rheology and printability for the fabrication of dense ceramics; they initially prepared a viscoelastic ink and determined the printability criterion considering the capillary forces and gravitational slumping [47]. In their work, the paste exhibited a linear viscoelastic response below the intersection ($\gamma^* = 10\%$). After this point, the loss modulus was greater than the storage modulus; the gel structure would be destroyed with enough shear stress, showing shear thinning flow behavior. Therefore, the main target of most research is to develop a desirable DIW paste with optimized ink ingredients. Ceramic ink composition in the literature is summarized in Table 7.2, which is a valuable reference for DIW researchers.

Table 7.2 Composition of ceramic inks used in the DIW process

No	Ceramic material	Solid loading	Dispersant	Binder	Other additive	Reference
1	Al_2O_3	73.3 wt.%	Acetic acid	PVP	Octanol	[41]
2	ZrO_2	60 vol.%	DARVAN C-N	methylcellulose	–	[48]
3	BCZT	41.6 vol.%	Dispex A40	HPMC	PEI	[49]
4	Ti_2AlC	81 wt.%	–	PEG, PVA, MC, CMC	–	[50]
5	Si_3N_4	61.8 wt.%	–	HPMC	PEI	[51]

7.2.3 *Binder Jetting*

Binder jetting (BJ), also known as 3D printing, was originally developed and applied to shape ceramic objects by Michal et al. at the Massachusetts Institute of Technology in the early 1990s [52, 53]. Figure 7.5 shows a schematic diagram of the BJ process. In this process, the liquid binder droplets are selectively deposited into a powder bed based on part geometry and finally form ceramic green bodies. Subsequently, following post-processing, the final parts are obtained [54]. Table 7.3 summarizes several advantages and drawbacks of binder jetting. The latter is capable of fabricating parts with highly porous structure and desired gradient properties suitable for practical applications. Moreover, a broad range of ceramic materials can be shaped using this process. However, BJ does not enable the fabrication of dense components. The improvement of part quality is strongly dependent on materials selection, printing parameters optimization and post-processing methods. Mohsen et al. presented a comprehensive binder jetting review based on its challenges [55]. The remaining challenges, from an aspect of powder selection and post-processing, are discussed below.

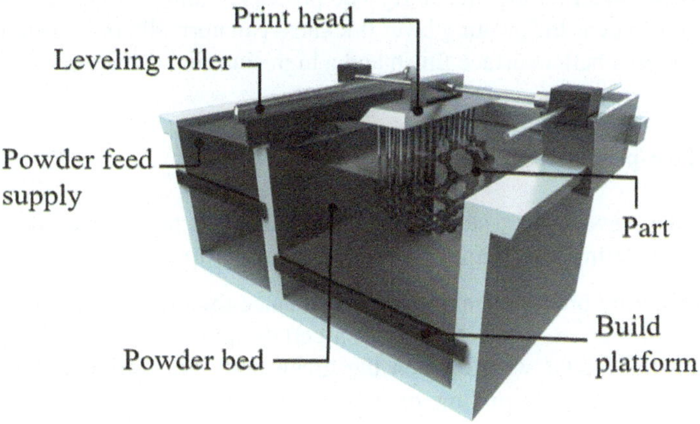

Fig. 7.5 Schematic diagram of binder jetting process

Table 7.3 Advantages and limitations of binder jetting

Advantages	Limitations
No support structure needed	Low green and sintered strength
Broad material available Multi-material/multi-colored features	Required more extra post processing steps
Cost efficient	Low surface finishing and resolution
Large build volume	–

7.2.3.1 Powder Selection

Generally, the powder size, powder shape and binder materials are critical physics that determine the green part quality. There are two binder jetting bed types: powder-based bed and slurry-based bed [56], with different powder size and shape requirements. The particle size has a significant influence on powder flowability and sinterability. In the case of powder-based bed binder jetting, coarse (typically in a range of 30–200 μm) [57] and spherical powders are usually preferred owing to their excellent flowability. However, the printed green body requires a high sintering temperature due to the low activity of coarse powder and the final density obtained rarely exceeds ~50%, resulting in low mechanical properties [58]. In order to improve the packing and sintered density, bimodal powder mixture with an optimized ratio is applied in the powder-based bed; the coarse powder facilitates the flowability, whilst the fine powder fills the gaps, aiming to increase the packing density and subsequently the mechanical properties of the final objects [56, 59, 60]. A slurry-based bed was developed in an effort to avoid the issue caused by the powder-based bed process, with fine powder being used to guarantee the sinterability. The slurry typically consists of fine powders mixing within a liquid to form a homogeneous slurry [61]. Compared to powder-bed binder jetting, the slurry-bed process requires a long printing time to build the same part as the printing layer thickness can normally be adjusted to a finer level, leading to a better surface finish and a high sintering density [62].

7.2.3.2 Post-processing

After printing, several post steps should be conducted to achieve ceramic parts: curing, de-powdering, debinding, sintering and infiltration or other steps.

Curing—A curing process is employed to enhance the mechanical strength of the green body through normally heating at a certain temperature (~185–200 °C), at which point the binder will toughen the green body via further polymerization, solvent evaporation, cross-linking, or some other mechanism [63].

De-powdering—After curing, excess powders may be removed from the cured green parts. Two mechanical methods can be applied to remove powder: vibration [64] or vacuuming [65]. Sometimes, samples are de-powdered by ultrasonic and microwave-induced boiling when the binder is not soluble in the liquid.

Debinding or drying—Binder burnout or drying is required before furnace sintering. Similar to the debinding process in VP, the precise control of the debinding temperature and time yields defect-free parts. In the powder-bed binder jetting process, the debinding process typically requires less time due to the high porosity. Alternatively, the liquid binder can be evaporated at a lower temperature.

Infiltration—It is a densification process using a secondary material (typically solutions containing nanofillers) via capillary forces to increase green density and mechanical strength, reducing the shrinkage. It is notable that the use of excess infiltrant materials can affect dimensional accuracy. Conversely, insufficient penetration of infiltrant materials is not beneficial for the improvement of mechanical properties. Pragnya et al. prepared densifiers of 15 wt.% suspensions with nanoparticles to penetrate the printed alumina body; the results showed the addition of nanoparticle densifiers to ceramic body can improve sintered density, increasing from 75 to 87%, and a 740% improvement in compressive strength [56]. Chemical vapor infiltration (CVI), Precursor infiltration and pyrolysis (PIP), and reactive melt infiltration (RMI) are near-net size densification processes that have been employed in the production of ceramic matrix composites [66]. Cold isostatic pressing (CIP) is an effective method as it is beneficial in increasing the density and mechanical properties. Carrijo et al. applied this technology to reduce the porosity from 48 to 3.6%, furthering the improvement of the mechanical properties [67]. Additionally, Carlos et al. reported on the fabrication of aluminum nitride components using a binder jetting machine; they carried out hot isostatic pressing (HIP) to reach a promising final density, increasing from 1.07 to 1.96 g/cm^3 [54].

Sintering—Sintering is the final step involving consolidation of ceramic particles. The sintering condition may include sintering temperature, atmosphere and dwell time; the precise conditions depend on the ceramic materials. For ceramic sintering, the effect of dwell is not as important as the sintering temperature.

7.2.4 Inkjet Printing

Inkjet printing (IJP), also known as materials jetting, was first reported to shape ceramics in 1995 [68]. A typical IJP setup is schematically depicted in Fig. 7.6. In inkjet printing, well-dispersed ceramic ink is selectively deposited onto a build substrate, and then the part was formed through a solidification process that can

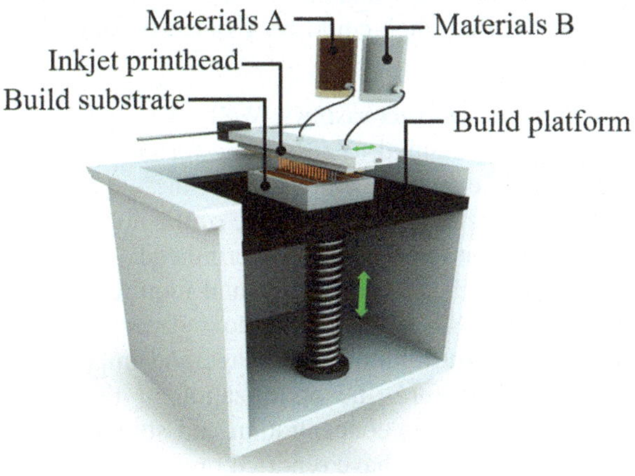

Fig. 7.6 Schematic diagram of inkjet printing process

occur through temperature-induced gelation or chemical reaction and solvent evaporation. Derby from the University of Manchester highlighted its great benefits in the fabrication of ceramic components [69]. Similar to binder jetting and robocasting, it can fabricate a wide range of ceramics, such as Al_2O_3 [70], ZrO_2 [71], PZT [72], SiC [73] and Si_3N_4 [74]. Inkjet printing is considered as a powerful microfabrication tool; the printed parts exhibit high resolution due to the size of individual droplets [74]. Furthermore, it is typically suitable for fabricating thin ceramic layers [7], and introduces the ability of part production with different materials, something which was not previously achievable using binder jetting [75].

Ceramic inkjet printing requires fewer post-processing steps than binder jetting. Based on these advantages, a commercial inkjet printer, XJet from Israel, was recently built to process the near net shape of ceramic parts with inks containing nano-sized ceramic particles. Prior to sintering; only a solvent evaporation process is needed [76]. The drawback of inkjet printing is that it is not easy to ensure consistency and reliability in the printing process due to serious nozzle clogging resulting from the agglomerates of ceramic particles in the ceramic ink. Hence, the preparation of stable and homogeneous ceramic slurry is essential to ensure a successful process [77].

7.2.4.1 Ceramic Ink Requirement

Ceramic inks are likely to be very similar to those needed for other slurry-based AM technologies, which can also be divided into two systems: solvent-based and wax-based according to the solution type. Solvent-based inks, which contain water or organic solvent as a liquid medium, have to evaporate before sintering. However, wax-based inks are solidified through phase changes to form a complete 3D part. There are several key requirements, as described in detail below.

Printability—A dimensionless number, Z, was proposed to define the printability of as-prepared inks [78, 79]. Derby et al. focused on the effect of ink formulation on the Z value and the "coffee ring" as a result of inhomogeneous deposition, suggesting that this value of a printable ceramic ink should be limited within a range of 1–10, and the "coffee ring" could be avoided by optimizing the formulation of the inks [7, 69].

Rheological characteristics—Well-dispersed ceramic inks with low enough viscosity are required to ensure the formation of repeatable and uniform drops at an appropriate drop-generation rate. Otherwise, existing agglomeration in inks tends to cause clogging and/or blockage of the nozzle, further resulting in failed printing and a high cost for replacing the printhead.

Generally, there is a fundamental requirement for ink viscosity in the range of 20–30 mPa s [80, 81]. However, low viscosity limits the ceramic loading, which is not expected to minimize shrinkage and improve the mechanical properties after sintering. Therefore, the optimization of the ink formulation, concluding solid loading, dispersants concentration and solutions, is a crucial step before printing. Ramakrishnan et al. have reported on the preparation of alumina and zirconia inks with an optimized concentration of dispersant (oleic acid); the highest ceramic loaded suspension of 5 vol.% was achieved, establishing the optimum level of apparent viscosity [82]. Anja et al. of RWTH Aachen University, successfully fabricated complex-shaped structures with high accuracy using 27 vol.% zirconia suspensions with the addition of 4 vol.% dispersant via inkjet printing. The investigations of the mechanical properties on printed zirconia specimens yielded a flexural strength of 1366 MPa and fracture toughness of 8.9 MPa $m^{1/2}$, respectively [83].

7.2.5 *Selective Laser Sintering/Melting*

Selective laser sintering/melting (SLS/SLM), proposed by Deckard and his academic advisor, Beaman, of the University of Texas in the mid-1980s, falls under the category of powder bed fusion technologies [84, 85]. Indeed, the processing schematic diagram (see Fig. 7.7) of SLS/SLM is very similar to that of binder jetting except for its use of thermal energy to bind the powder instead of a binding agent. In 1999, its variant, selective laser melting (SLM), was developed by Fockele and Schwarze at the Fraunhofer Institute for Laser Technology in Germany [86, 87]. In SLS/SLM processes, a high-energy laser beam (e.g., CO_2-based) was used to completely sinter or fuse ceramic powder. With post-processing, final parts could be achieved [88]. Table 7.4 shows a careful comparison between binder jetting and SLS/SLM process.

Although the SLM has been regarded as a direct ceramic AM technology without the need for separate debinding and sintering, the mechanical properties of the parts are not comparable to those obtained by indirect AM technologies since parts are subject to cracking because of ceramic materials having high melting points, low

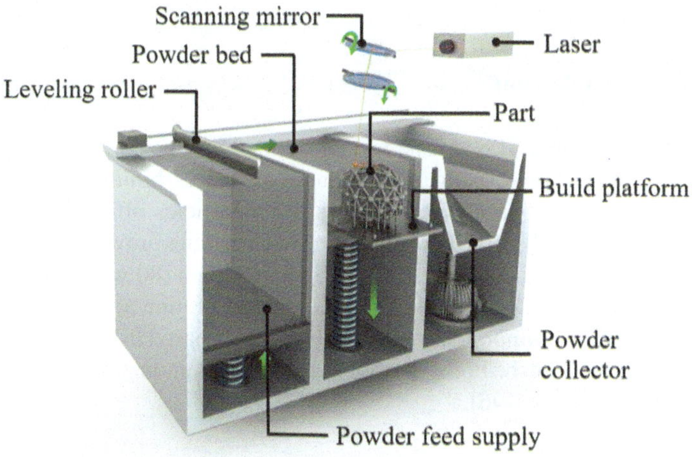

Fig. 7.7 Schematic diagram of selective laser sintering/melting processes

Table 7.4 Comparison between SLS/SLM and binder jetting [89–91]

	SLM	SLS	BJ
Power source	Laser beam	Laser beam	Binder
Preheating	Yes	No	No
Materials	Laser-materials interactions need to be considered		A large range of materials
Process types	Direct	Direct or indirect	Indirect
Feedstock	Powder-based	Slurry- or powder-based	Slurry- or powder-based
Supports	Yes	No	No
Post-processing	No	Indirect process requires de-powdering, drying, debinding, sintering, infiltration, CIP, HIP, etc	

thermal shock resistance and low thermal conductivity [5]. To eliminate the aforementioned issues, the use of preheating can avoid the crack formation resulting from thermal stresses. Liu et al. reported preheating at high temperature of the zirconia powder could not only effectively reduce cracks, but could also improve the sintered density, increasing from 84% to ~91% [92]. The existing literature has demonstrated success; however, SLM-produced ceramics are characterized by low strength, low surface quality and high porosity. Thus, an indirect AM method, SLS, has drawn significant attention from the scientific community; SLS involves a sacrificial binder with a low melting point being blended into the ceramic powder to obtain polymer-ceramic composite powders [90, 93, 94]. Previous ceramic SLS reviews have comprehensively investigated the recent achievements, powder preparation, sintering mechanisms, post-processing and advantages along with limitations [7, 88, 95, 96]. However, two major concerns, inherent characteristics of ceramics powder and the optimization of printing parameters, are less reported as yet.

7.2.5.1 Powder Selection

Similar to binder jetting, SLS can be further divided into powder-based and slurry-based systems. In the case of using powder-based processes, bimodal ceramic powder is preferred in a powder-based SLS process. Fine particles offer the high packing density and better laser absorption, while the coarse particles provide good flowability [97]. A mixture of two differently sized ceramic powders (80 wt.% of 22.5–45 μm and 20 wt.% 9–22.5 μm) is helpful to avoid crack generation and obtaining a dense part [92]. The ceramic components produced by slurry-based exhibited high density, strength and surface finish. For example, Tang et al. demonstrated the benefits of slurry-based SLS process [96].

7.2.5.2 Printing Parameters

The densification level and resultant mechanical properties of parts are influenced by printing parameters, such as laser powder, laser spot diameters, scanning speed, scanning spacing, layer thickness and preheat temperature [98]. Therefore, in order to fabricate parts with desirable properties, the printing parameters should be carefully optimized. Table 7.5 shows the SLS/SLM processing parameters reported in the literature. It can be seen that the different ceramic materials have their own optimized printing parameters. Typically, the laser energy density e (J/cm^2) can be calculated using Eq. 7.2:

$$e = \frac{P}{Hv} \tag{7.2}$$

where P is the laser power (W), H denotes the scanning spacing (mm) and v is the scanning speed (mm/s). Ergo, the laser energy density is directly proportional to the laser power and inversely proportional to the scanning spacing and speed. A proper laser energy density may be able to effectively sinter or melt binder or ceramic power, further leading to an improved strength of final parts.

7.2.6 *Laser Engineered Net Shaping*

Laser engineered net shaping (LENS), also named as direct energy deposition (DED) or 3D laser cladding [99], was developed by Sandia National Laboratories [100]. Unlike metal, it has received less attention in the fabrication of ceramic parts due to the low printing accuracy. Figure 7.8 shows how a combination of two different ceramic materials can be used to manufacture ceramic structure.

In the LENS process, ceramic powder is completely melted by a high-energy density laser [108]. When compared with the direct SLM route, the primary advantages for the LENS process concern its higher fabrication efficiency and the ability to repair or coat parts as well as composition grading by means of using two or more feeders [109, 110]. In addition, it has the striking advantage over other indirect AM processes of being a single-step process, meaning that it does not involve post-processing such as debinding and sintering. However, the disadvantages of this route are low dimensional tolerances and a considerable amount of porosity, as well as the limited geometrical complexity that can be achieved. Moreover, so far only a very limited range of ceramics materials can be fabricated using LENS [111]. To obtain a LENS produced ceramic component, the laser parameters should be optimized by combining with other assistive technology, such as ultrasonic or mechanical vibration and pre/post heating, etc. [112–114].

Table 7.5 The optimal processing parameters used in SLS/SLM process

	Ceramic	Preheating temperature/°C	Laser power/W	Laser beam diameter/μm	Scanning speed/mm s^{-1}	Scanning spacing/μm	Layer thickness/μm	References
SLS	Al_2O_3		21	200	1600	100	150	[101]
	Glass	1200	100	1100	250	550	250	[93]
	Bioglass	30–35	21	–	1800	0.1	80	[102]
	Bioglass	–	3	–	0.24	100	100	[103]
	SiC	–	18	300	3500	–	100	[104]
SLM	ZrO_2	2000	60	34	400	20	150	[92]
	ZrO_2	1860	60	200	200	50	50	[105]
	Al_2O_3	–	30	100	300	100	70	[106]
	SiC	–	350	70	2500	50	50	[107]

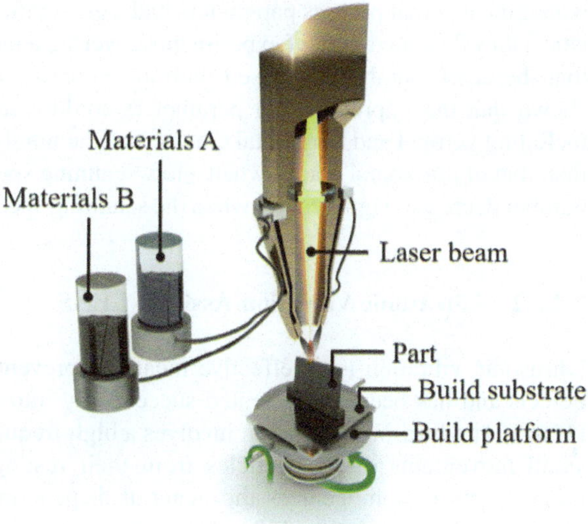

Fig. 7.8 Schematic diagram of laser engineered net shaping process

7.2.6.1 Process Parameters

Cracks are easily generated during the melting process when processing ceramic materials via LENS. Therefore, optimal parameters need to be methodically developed to make sure that the final part is crack-free, further yielding a good mechanical property. Table 7.6 shows some attempts at optimizing LENS process parameters on ceramics forming, as reported in the literature. One example comes from Niu et al.,

Table 7.6 The optimal printing parameters for the LENS process

Ceramics materials	Feedingrate/g min^{-1}	Feeding proportion	Laser power/W	Scanning speed/mm s^{-1}	References
ZrO_2–Al_2O_3	2	–	325–400	10	[112]
Al_2O_3	1.38	–	326	5	[116]
ZrO_2–Al_2O_3	2.09	58.5 wt.% Al_2O_3 41.5 wt.% ZrO_2	530	6.67	[117]
TCP	9/13	–	400/500	10/15	[118]
Al_2O_3–Y_2O_3–ZrO_2	8	54 wt.% Al_2O_3 27 wt.% Y_2O_3 19 wt.% ZrO_2	220	9.3	[119]

who indicated that process parameters had a great influence on the crack's characteristics; they designed a set of experiments covering a large process scope, concluding that the crack number decreased with an increase in scanning speed. The results shown that the improper laser parameters could result in the formation of cracks including vertical and horizontal cracks, and the number of vertical cracks is greater than that of horizontal cracks when a low scanning speed is used, although the crack number decreases significantly when the scanning speed exceeds 700 mm/min [115].

7.2.6.2 Ultrasonic Vibration-Assisted LENS

Ultrasonic vibration is an effective means to prevent the formation of fabrication defects and has been implemented successfully into metal solidification processes [120, 121]. Ultrasonic vibration involves a high frequency being applied to provoke small movements of the particles from their resting position, which refines the microstructure, homogenizes the material dispersion, smooths out thermal gradients and reduces porosity [122, 123]. In the past few years, ultrasonic vibration has been applied to suppress crack formation [115]. Hu et al. experimentally reported the effect of ultrasonic vibration on crack suppression, microstructure and the mechanical properties of LENS-manufactured parts, parts produced combining ultrasonic vibration showed smaller grain size and more homogeneous dispersion. In addition, the mechanical properties were relatively higher than those of parts made without ultrasonic vibration [112]. Yan et al. focused on the effect of ultrasonic power on the quality of LENS produced parts, reporting that the number of cracks decreased in an obvious manner with an increase of power. Furthermore, the porosity of parts decreased initially and then increased after the power exceeded 120 W [124].

7.2.7 Hybrid Additive Manufacturing Processes

Hybrid additive manufacturing (H-AM) technology, typically defined as the integration of two or more AM technologies, has been used to fabricate multimaterial, multistructural and multifunctional components. H-AM unlocks new opportunities and applications for fabricating various components with intricate features and enables them to become customized and multi-functional; features which could not be economically viable by processes on their own [125, 126]. Chong et al. extended the H-AM scope, covering either the combination of single additive manufacturing with subtractive manufacturing or two or more AM processes [127]. In the Collaborative Research Center, the researchers combined powder bed laser melting with sheet metal forming processes to produce large-scale parts with a short lead time [128]. Li et al. introduced a novel hybrid AM to integrate materials jetting and materials extrusion, creating a pyramid embedded with LEND blinking circuitry [129].

To date, most hybrid AM systems that have been successfully investigated have been in the fabrication of metal/metal [130–133], metal/polymer [134–136], and

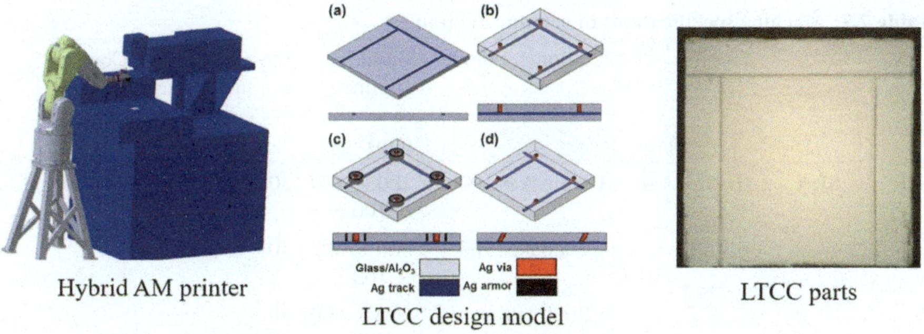

Fig. 7.9 Hybrid AM process chain: LTCC design and fabrication process [139]

polymer/polymer [137] multi-materials parts. Hybrid AM to shape ceramic components is still in an early stages of development [138]. More recently, the French company 3D CERAM had started working on H-AM techniques, developing a hybrid 3D printer [139]. Their H-AM system is based on the existing laser stereolithography 3D printer and adds a hybrid scraper tool. This hybrid printer is equipped with a dispensing system, enabling the integration of other materials into the ceramic substrate as it is built. The company has successfully 3D printed a multi-layer, low temperature, co-fired ceramic (LTCC) with embedded silver tracks. As detailed in Fig. 7.9, initially the alumina paste is spread using the scraper and then cured using a UV laser. After this layer is polymerized, an air jet is used to remove uncured paste from the area where another material is needed. The robocasting systems then fills the gap created by air blowing with Ag ink. Finally, the LTCC parts is formed.

7.2.8 Summary

The recent developments of several AM technologies used for ceramic fabrication over the past few decades demonstrate tremendous potential in the advantages of mass customization, short lead time and low cost for small runs. According to a strong annual growth of revenue generated by ceramic additive manufacturing, it is clear that there has been an increase in the number of commercial ceramic 3D printers and their different AM processes (and related information) are listed in Table 7.7. Each ceramic AM technology has been attracting more manufacturers to update 3D printers with high resolution, large build volume and low machine cost. Moreover, ceramists can select a suitable printer depending on the relevant favorable process and property requirements of components.

Several AM technologies have been proven to be successful in the fabrication of advanced ceramic materials. However, there is not a "best overall" ceramic AM technique, but rather the most suitable AM to be used to fabricate a particular ceramic component. Therefore, it is suggested that the selection of the AM process for certain

Table 7.7 Machine specifications of ceramic 3D printers

AM		Supplier	Equipment mode	Build volume ($l \times w \times h$ mm)	Layer resolution/μm	Country
VP	SLA	*3D CERAM*	Ceramaker 100	100 × 100 × 100	30	France
			Ceramaker 900	300 × 300 × 100	30	
			C3600 Ultimate	600 × 600 × 300	30	
		DWS	XFAB3500 PD	140 × 140 × 180	–	Italy
		Porimy	CSL 200	200 × 200 × 100	25–40	China
			CSL 100	100 × 100 × 50		
		Sunshine Machinery& equipment	CeraBuilder 100	100 × 100 × 100	30	China
			CeraBuilder 100 Pro	100 × 150 × 150		
			CeraBuilder2500	250 × 250 × 150		
		ZRapid	iAMC150	150 × 150 × 100	–	China
		Kings	Kings JS6000-C	600 × 600 × 400	–	China
	DLP	*lithoz*	CeraFab S25	64 × 40 × 320	25	Austria
			CeraFab S65	102 × 64 × 320	40	
			CeraFab S230	192 × 120 × 320	75	
		Admatec	Admaflex130	96 × 54 × 110	50	Netherlands
			Admaflex300	200 × 200 × 300	80	
		Prodways	Promaker V6000	120 × 500 × 150	32	France
		Ten dimensions	Autocera-M	96 × 54 × 200	50	China
		Changlang	Ceraform60/100	96 × 54 × 300	–	China

(continued)

Table 7.7 (continued)

AM		Supplier	Equipment mode	Build volume ($l \times w \times h$ mm)	Layer resolution/μm	Country
		Soonsolid	CeraRay CR-1	144 × 81 × 200	50	China
	TPP	*Nanoscribe*	Photonic professional GT2	100 × 100 × 8	0.4	Germany
DIW		*WASP*	DeltaWASP	200 × 200 × 400	–	Italy
		3D Potter	PotterBot micro	240 × 260 × 250	3000	USA
		Stoneflower	Stoneflower	–	300	Germany
		3D inks, LLC	Robocasting system	–	410	USA
		NB: Most of the printers used are self-developed based on a 3-axis platform				
BJ		*3D systems*	CeraJet	–	100	USA
			CubeJet		–	
		MicroJet	ComeTrue M10	100 × 100 × 60	–	Taiwan
			ComeTrue T10	200 × 160 × 150	556dpi	
		Kwambio	Ceram one	235 × 330 × 280	20	Ukraine
			Ceram Zero Max	150 × 150 × 150	300dpi	
			Ceram Two	400 × 350 × 380	600dpi	
		Wuhan Easy	Easy 3DP300	300 × 300 × 300	200	China
			Easy 3DP300	550 × 550 × 400		
		ExOne	Innovent +	160 × 65 × 65	–	USA
			X1 25Pro	400 × 250 × 250	–	
			X1 160Pro™	800 × 500 × 400	–	
		Voxeljet	VX 200	300 × 200 × 150	300dpi	Germany
			VX 500	500 × 400 × 300	600dpi	

(continued)

Table 7.7 (continued)

AM	Supplier	Equipment mode	Build volume ($l \times w \times h$ mm)	Layer resolution/μm	Country
		VX 1000	1000 × 600 × 500	80	
		VX 2000	2000 × 1000 × 1000	300dpi	
IJP	CeraDrop	CeraPrinter F series	305 × 305 × 10	–	France
	Xjet	Cermel 1400C	500 × 280 × 200	–	Israel
		Cermel 700C	500 × 140 × 200	–	
	Fujifilm	DMP-2850	210 × 315 × 0.5	–	USA
	Domino	Macrojet2	–	–	UK
	SUSS	PixDro LP50	–	–	Netherlands
SLS/SLM	3D systems	Phenix PXM	140 × 140 × 100	–	USA
	Wuhan Huake	HK C250	250 × 250 × 250	200	China
	EOS GmbH	P396	340 × 340 × 600	–	Germany
		P700	700 × 380 × 580	–	
	Wuhan Binhu	HRPS-IV	500 × 500 × 400	–	China
	SLM solution	SLM125	125 × 125 × 125	–	Germany
LENS	Optomec	LENS CS800	800 × 600 × 600	670	USA
		LENS CS1500	900 × 1500 × 900	–	

NB Most of the LENS printers used are self-developed based on a CNC platform

materials and end-use applications should be carefully considered, based on the significant advantages and, limitations, as well as various aspects of concern, such as costs, build volume, tolerance, etc., which are provided in Table 7.8. It is important to understand that each ceramic AM approach is characterized by its own capabilities.

Table 7.8 Comprehensive comparison of the different ceramic additive manufacturing technologies

AM		Feedstock	Cost			Build size	Tolerance	Structure Freedom	Support needed	Available ceramic materials	Preferred or potential applications
			Equipment	Materials	Process						
VP	SLA	Slurry	Medium	High	Medium	10 μm–300 mm	μm	High	Yes	LRI, LAUV	S, F, Bio, CMCs
	DLP	Slurry	Medium	High	Medium	10 μm–300 mm	μm	High	Yes		
	TPP	Slurry	High	High	high	1 μm–1 mm	nm–μm	High	Yes		E
DIW		Slurry	Low	Low	Low	1 mm–500 mm	μm–mm	Medium	Yes	All	S, F, Bio, CMCs
BJ	P-BJ[a]	Powder	Medium	Medium	Medium	10 mm–1 m	μm–mm	High	No	All	Bio, CM
	S-BJ[b]	Slurry	Medium	High	High	10 mm–1 m	μm	High	Yes	All	S
IJP		Slurry	High	Medium	Low	100 μm – 200 mm	μm	Medium	Yes	All	F, S, CMC
SLS	P-SLS[c]	Powder	High	Low	High	10 mm – 600 mm	μm–mm	High	No	All	C, CM
	S-SLS[d]	Slurry	High	High	High	10 mm – 600 mm	μm	High	Yes	All	F, S and CM
SLM		Powder	High	Medium	Low	10 mm – 100 mm	μm–mm	Medium	Yes	LMP, HLA	C, CM
LENS		Powder	Medium	Low	High	10 mm–1 m	mm	Low	Yes	LMP, HLA	S, CMC, R

[a]Powder-based binder jetting; [b]Slurry-based binder jetting; [c]Powder-based selective laser sintering; [d]Slurry-based selective laser sintering
LMP Ceramics with low melting point; *HLA* Ceramics with high laser absorption; *LRI* Ceramics with low refractive index; *All* A wide range of ceramic materials
LAUV Ceramics with low absorption at UV wavelength; *F* Functional component; *S* Structural components; *Bio* Bioceramics; *CM* Casting mould; *C* Cellular parts; *E* Extremely high precision parts; *R* Component repair and coat

For example, vat-polymerization, such as SLA and DLP, is a powerful manufacturing tool that can be used to shape bulk and cellular ceramic components, as well as ceramic matrix composites with excellent surface quality, good mechanical properties and very high precision. On the other hand, it is not suitable for ceramics having a high refractive index and absorption at a certain wavelength. As a result, its use can lead to insufficient cured depths. In addition, the fabrication of non-oxide ceramics may also be hindered by SLM and LENS, which would be mainly attributed to their very high melting points and poor thermal shock resistance. Direct sintering via SLM and LENS does not require expensive and time-consuming post-processing procedures, yet the extreme temperature gradients prevent the fabrication of highly dense parts. Other AM, such as DIW, BJ and IJP techniques are relatively independent from the physical properties of the materials and can be used to process all types of advanced ceramics. DIW and powder-based BJ are recognized as powerful tools for the cost-effective fabrication of porous ceramic scaffolds for biomedical, catalyst carrier and filter applications in which the requirement of mechanical properties is less important. Therefore, for specified ceramic materials and applications, ceramists must keep in mind that the selection of the best AM candidates should be based on the comprehensive investigation of all AM technologies.

7.3 Additive Manufacturing of Advanced Monolithic Ceramics

The most commonly studied advanced monolithic ceramics for AM are structural, functional and bioceramics. This section provides the current state of advanced monolithic ceramics made by various AM and their properties are presented, as well as applications, with special attention paid to a performance comparison of printed parts shaped by different AM methods and by conventional manufacturing.

7.3.1 Structural Ceramics and Applications

Table 7.9 presents a list and the properties of structural ceramics. These properties are specific to individual ceramic materials and cannot directly be compared due to different measurement methods. These properties make them excellent candidates for load-bearing and extreme environment applications.

7.3.1.1 Oxides and Applications

Zirconia ***(ZrO_2)***—One of the most studied substances in either conventional or AM routes, being considered to have high flexural and compressive strength, hardness,

Table 7.9 Comparison for properties of advanced ceramic materials obtained by conventional manufacturing technologies [140–149]

Advanced ceramic		Properties				
		Density/g cm^{-3}	Flexural strength/MPa	Elastic modulus/GPa	Hardness/GPa	Fracture toughness/MPa $m^{1/2}$
Oxides	ZrO_2	6.05	~1000[3P]	250	5–15	6–8
	Al_2O_3	3.95	~379[3P]	375	~18	4
	SiO_2	2.65	–	73	–	0.79
Carbides	SiC	3.16	550[3P]	410	25	4.6
	B_4C	2.52	900[3P]	460	37	~2.2
	WC	15.63	1830[3P]	530–700	~17	12
	ZrC	6.6	400[3P]	480	25.5	2
Nitrides	Si_3N_4	3.29	689–830[3P]	310	15	5.7–6.0
	AlN	3.26	320[3P]	330	1100	2.6
	BN	2.1	75.8–113[3P]	46.9–73.8	74	2.6
Borides	TiB_2	4.52	~600[4P]	565	~24	~6.2
	HfB_2	11.2	350–450[4P]	480	21–28	~7.5
	ZrB_2	6.08	300–400[3P]	450	14–23	2–3
PDCs[1]	SiOC	2.55	922[3P]	200	~5–9	–
	SiCN	2.6	–	109–118	7.9–12.8	2.1
	SiBCN	2.49	~202[3P]	~74	~1.6	~2.8

[1]Polymer derived ceramics; [4P]Four-point bending test; [3P]Three-point bending test

elasticity modulus and good chemical resistance, as well as strong wear and good biocompatibility, and the highest fracture toughness among the most used advanced ceramics [150]. These outstanding properties make it of considerable interest in structural and biomedical applications. In general, zirconia can be partially stabilized in tetragonal or cubic phases with a certain concentration of dopant (Y_2O_3, MgO, CaO, CeO_2, etc.) [151–154]. Tetragonal zirconia polycrystal (3Y-TZP), typically doped with 3 mol.% yttria, has become the market standard, which is attributed to its popular applications, such as refractories [155, 156], catalyst carriers [157], electrochemical sensors [158], dental implants [159, 160], bone repair [161], anode for solid oxide fuel cells [162, 163], thermal barrier coatings (TBC) [164] and high-temperature photonics [165]. Nevertheless, in light of biomedical implants, zirconia does suffer from low-temperature degradation (LTD), which is promoted by a moist environment, facilitating the degradation of mechanical strength and increasing the presence of cracks due to spontaneous and progressive transformation of the tetragonal phase to monoclinic [166–170]. Inokoshi et al. studied the influence of sintering conditions on the LTD of dental zirconia, indicating that higher sintering temperatures and longer dwelling times lead to a higher sensitivity to LTD of Y-TZP ceramics as larger grain sizes are more susceptible to transformation [171]. Thus, it is necessary to prevent LTD when zirconia is used, especially for dental implants.

Table 7.10 and Fig. 7.10 present the properties and representative images of zirconia samples via different AM technologies, respectively. Vat-polymerization is a relatively popular technique for making complex ZrO_2 parts using a photosensitive zirconia suspension. A laser pattern or beam is used to cure the suspension and then the final parts are obtained followed by debinding and sintering. Mario et al. demonstrated the ability of DLP to manufacture near-fully dense zirconia components; initially preparing 44 vol.% zirconia acrylate-based slurry with 3 wt.% of dispersants; the sintered parts presented an average density of 99.1% and the flexural strength and Weibull coefficient were 741 MPa and 11.4, respectively [29]. SLA was used by Fu et al. to produce zirconia, with the highlight being the effect of laser energy on dimensional resolution and mechanical properties. They concluded that the use of higher laser energy led to greater reliability and the relative density could reach up to 97%. The manufactured parts had a hardness of 13.1 GPa, a 3 PB flexural strength of 1044 MPa and a fracture toughness of 5.62 MPa $m^{1/2}$ [172]. With respect to the DIW of zirconia, Liao et al. developed a colloidal ink with 47 vol.% powder and use it to form a 3D structure with a feature size of 200 μm [173]. Furthermore, Erwin et al. demonstrated the possibility of the generation of large zirconia parts prepared from an aqueous zirconia suspension; the sintered structures, with 94% density despite high shrinkage, offered hardness up to 13.29 GPa and flexural strength of approx. 242 MPa [174].

Binder jetting is expected to fabricate a wide range of ceramic materials, but few works have been reported in the fabrication of zirconia parts. In contrast, inkjet printing has received a great deal of attention for shaping zirconia objects. For example, in 2009, Ebert et al. prepared aqueous zirconia inks with 27 vol.% TZ-3YS-E powder, the as-sintered zirconia showed a high density of 96.9%, with flexural strength and fractural toughness being 763 MPa and 6.7 ± 1.6 MPa $m^{1/2}$ [71].

Venkatesh et al. used inkjet printing to process bi-layer zirconia thermal barrier coating on porous alumina and nonporous steel substrates, also exploring a different method of assessing porosity of coatings via simpler hardness measurements [187]. An emerging technology, nanoparticle jetting, was also used to manufacture zirconia parts with the absence of a long debinding time. Yongduk et al. explored the printing of zirconia microwave dielectric resonator antennas, with a great focus on the shrinkage after sintering, even though it could not impact the dielectric permittivity [188].

Selective laser melting has also been studied to make ZrO_2 samples. In SLM, the preheating of zirconia powder is a crucial step for obtaining crack-free parts. Liu et al. discussed the effect of preheating temperature on the final density, with results showing that increasing the preheating temperature to 1500, 2000 and 2500 °C was more conducive to the formation of dense zirconia parts and tetragonal crystals [92]. Another approach to process zirconia ceramics is SLS. Ferrage et al. used direct laser sintering to manufacture dense zirconia components; to improve the absorbance properties of the powder from 2 to 60%, a small amount of graphite was blended into the powder. It was found that the SLS enabled the reproducible

Table 7.10 Comparison of properties of additively manufactured zirconia specimens

AM	D_{50}/μm	Density (%)	Flexural strength/MPa	Fracture toughness/MPa m$^{1/2}$	Hardness/GPa	Linear shrinkage (%)	Year/References
VP	0.1–0.8	98.7	1088^{3P}	–	12.6	–	2020 [175]
	0.82	98.8	803.7 ± 59^{3P}	–	–	~22	2020 [176]
	–	92.79	674^{3P}	–	–	19.85	2019 [177]
	0.2	97.14	–	6.038^{In}	13.06	35.26	2018 [178]
	–	99.45	731.11^{3P}	–	–	24.7	2018 [179]
	0.20	99.3	1154 ± 182^{3P}	6.37 ± 0.25^{In}	13.90 ± 0.62	–	2017 [180]
	10	92	539.1^{3P}	–	13.02	32.7	2014 [181]
DIW	0.040	98.1	488.96 ± 79.84^{4P}	2.63 ± 0.2^{SEVNE}	11.52 ± 0.57	–	2020 [48]
	0.372	>94	242.8 ± 11.4^{3P}	–	13.29	33	2018 [174]
	0.052	>99	278 ± 59^{4P}	3.61 ± 0.08^{In}	15.3 ± 0.4	24	2017 [182]
	0.5	97.6	450^{3P}	–	14.4	–	2017 [183]
BJ	–	86.8	438^{3P}	–	–	–	2016 [184]
IJP	0.63	96.1	843^{4P}	–	–	–	2012 [185]
	0.09	96.9	763^{4P}	6.7 ± 1.6^{SEVNE}	–	–	2009 [71]
SLS/SLM	38.8	86.65	279.5 ± 10.5^{3P}	–	–	–	2018 [186]
	22.5–45	90–91	363.5	–	–	–	2015 [92]

^{3P} Three-point bending test; ^{4P}Four-point bending test; ^{In}Indentation fracture toughness; ^{SEVNB}Single edge V-notched beam

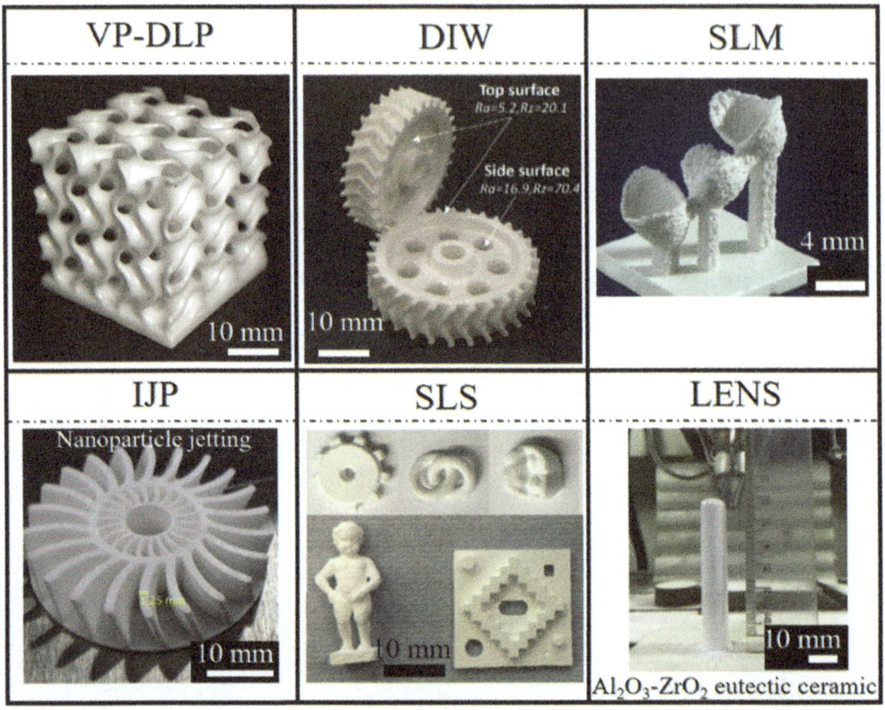

Fig. 7.10 Representative zirconia parts produced by several AM technologies [105, 188–191]

printing of simple parts with a relative density of 96.5% [192]. In 2014, an indirect selective laser sintering method was also used to fabricate zirconia parts by Khuram et al. [189]. In that work, mixed polypropylene (PP)-zirconia composite powder was prepared using thermal induced phase separation, with the post-processing method including warming isostatic pressing and pressure infiltration to raise the sintered density. LENS was favorable for producing thermal barrier coatings; Vasmi et al. used this process to deposit functionally graded YSZ coating on a stainless-steel substrate [193]. In 2019, Fan et al. fabricated a YSZ thin wall structure using a LENS technique; the surface roughness varied from 20 μm to 40 μm, with the achieved hardness and elastic modulus being 19.8 GPa and 236.1 GPa, respectively [194]. Currently, additive manufacturing, like SLA and DLP, produces zirconia that exhibits a comparable performance with conventional forming methods, such as dry pressing and injection molding. For example, Xing et al. fabricated zirconia components using an SLA process. A relative density of up to 99.3% was obtained, hardness of 13.90 ± 0.62 GPa, 3 PB flexural strength of 1154 ± 182 MPa, fracture toughness of 6.37 ± 0.25 MPa m$^{1/2}$, which are very close to those achieved by injection molding [180, 195, 196]. Additionally, the surface roughness was 1.07 μm; this value can meet the assembling requirement. However, anisotropic performance was observed in 3D printed zirconia parts, it is a common issue have to be addressed.

Alumina (***Al_2O_3***)—It is one of the most widely used engineering ceramics because of its high strength and hardness, superior wear and chemical resistance and low density [197]. Alumina components made by conventional manufacturing or additive manufacturing have been extensively applied in a very large number of practical applications, as detailed in dental implants, crowns and bridges [198], joint replacement areas [199], sensors [200], casting core [201, 202], bone scaffolds [203], ceramic tools [204, 205], electronic packaging [206], heating insulators [207], water purification [208] and microwave devices [209].

Table 7.11 and Fig. 7.11 present the properties and representative images of alumina samples via different AM technologies, respectively. Similar to zirconia, SLA and DLP are accepted as the most widely used approaches to manufacture complex alumina parts. The fabrication of alumina is easier than that of zirconia due to its low density and refractive index. Zeng et al. initially prepared 80 wt.% alumina suspension for printing alumina ceramic lattice structures via a DLP method [210]. Moreover, Schwentenwein et al. used a commercial DLP printer to produce dense alumina ceramics, with over 99.3% relative density, 4 PB flexural strength of 427 MPa, and good surface roughness. It has been proved that VP-based 3D printing provides very similar mechanical properties to conventionally manufactured alumina components [211].

Hugh et al., at Sandia National Labs, reported on a robocasting process to manufacture alumina parts and evaluated final properties, which were also compared to alumina parts processed traditionally [212]. They concluded that the mechanical

Table 7.11 Comparison of properties of additively manufactured alumina specimens

AM	$D_{50}/\mu m$	Density (%)	Flexural strength/MPa	Hardness/GPa	Linear shrinkage (%)	Year/References
VP	0.4	98	–	20 ± 1	–	2020 [215]
	0.43	99	475 ± 45^{3P}	–	–	2018 [216]
	0.2	99.3	–	17.5	–	2016 [205]
	–	99.3	427^{4P}	–	–	2014 [217]
	0.15	98	476 ± 54.4^{3P}	–	–	2013 [218]
DIW	0.4–0.7	>95	$200–300^{3P}$	–	19	2020 [219]
	0.48	98	156.6 ± 17.5^{3P}	–	–	2016 [36]
BJ	1.0	–	75.2^{3P}	–	10	2020 [220]
IJP	40	65.7	–	–	–	2018 [56]
SLS /SLM	0.5	98	363.5^{3P}	–	–	2011 [96]
	0.4	~88	255 ± 17^{4P}	–	–	2007 [221]
	0.3	89	148 ± 22^{3P}	–	–	2013 [90]
LENS	44–74	98	–	–	–	2008 [222]
	40–90	99.5	350^{3P}	–	–	2018 [223]

^{3P} Three-point bending test; ^{4P}Four-point bending test

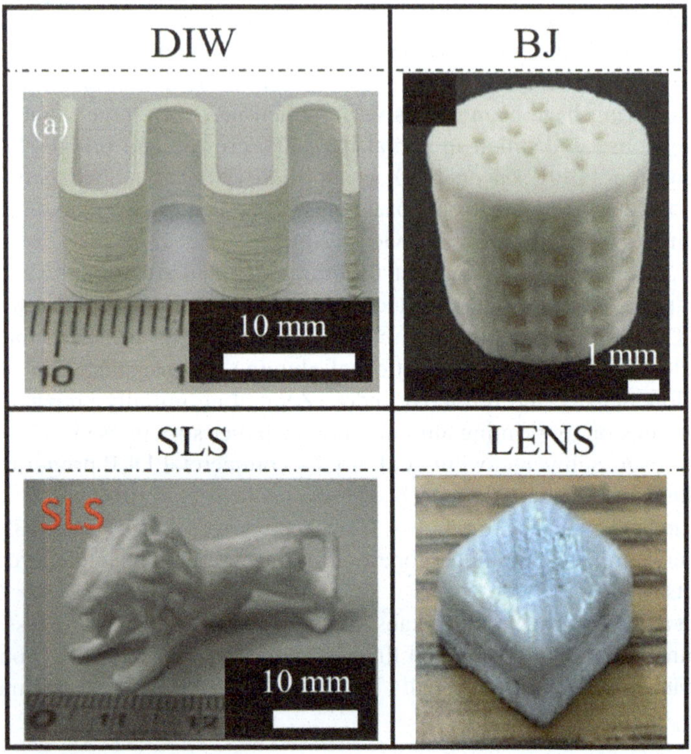

Fig. 7.11 Representative alumina samples made by different AM technologies [96, 203, 225, 226]

performances of alumina made by robocasting were slightly anisotropic and dependent on the build path used. In addition, Rueschhoff et al. printed dense alumina components using 53–56 vol.% aqueous inks, the flexural strength being 134–157 MPa [36]. Huang et al. used 50 vol.% aqueous alumina paste to investigate the influence of process parameters on part quality. The flexural strength, calculated by a four point bend test, were 219 and 198 MPa for the two different deposited paths [213]. Recently, Zhang et al., at the National University of Singapore, proposed a modified, self-curable, epoxide-amine robocasting system to shape ceramics. This method was different from the previous studies and provided a new way of producing alumina parts [214].

In parallel to robocasting, binder jetting is also suitable for preparing porous or lattice structures. Geuntak et al. demonstrated the use of binder jetting in the production of alumina scaffold with a mechanical strength of 30.2 MPa, and controllable porosity of 50% [203]. Derby, in 2011, reported that inkjet printing is a powerful microfabrication tool for the printing of ceramic parts due to its small drop size [77]. Chen et al. used the inkjet printing method to produce defect-free porous Al_2O_3 ceramic multilayer components. They also evaluated the mechanical properties, with the elastic modulus and hardness being 45–250 GPa and 0.61–6.57 GPa, respectively [70].

Alumina samples have already been produced through indirect SLS processes, using ceramic-polymer binder composite powder particles as feedstock. Wang et al. used this method to fabricate alumina components; initially, Al_2O_3 powder was coated by PVA binder as the raw material, and alumina parts with a relative density of 94.6% were made by a homemade SLS printer following a cold isostatic pressing process [224]. In addition, Khuram et al. mixed alumina and PP powder to produce Al_2O_3-PP composite particles, post-processing such as pressure infiltration and warm isostatic pressing was employed to offer a final relative density of 88% [90]. In conclusion, most indirect SLS processes require an additional process to densify the final parts, which is the same as for binder jetting processes. However, Sing et al. used a slurry-based SLS to fabricate high-strength alumina parts without a densification process; after debinding and sintering, a flexural strength of 363.5 MPa and relative density of 98% were achieved [96]. Direct AM technology, SLM and LENS have been used to fabricate alumina parts without any post-processing stage. Qiu et al. studied the balling phenomenon and cracks during the SLM process, concluding that a slightly decreased scanning speed and prolonging the laser dwell time can alleviate the balling and a high scanning speed may cause the generation of transverse cracks [106]. Therefore, the laser parameters still require more research and optimization to consistently yield defect-free parts. Li et al. studied the effect of laser parameters on single track-single layer quality [225].

Silica (SiO_2)—It is the major ingredient in the production of most glass and has attracted considerable interest owing to its outstanding optical transparency, good heat resistance and chemical stability, biocompatibility, as well as low thermal expansion coefficient [227–229]. These unmatched properties make silica suitable for diverse applications such as ceramic cores and sand molds [230–232], optical components [233, 234], biomedical applications [235–237], drug delivery [238, 239], energy system [240, 241].

VP is still a popular approach to fabricate SiO_2 parts. Indeed, the fabrication of silica parts is not challenging work thanks to its comparatively low refractive index and high light transmittance [242–244]. Table 7.12 and Fig. 7.12 present the properties and representative images of silica samples processed via different AM

Table 7.12 Comparison of properties of additively manufactured silica specimens

AM	D_{50}/μm	Density (%)	Flexural strength/MPa	Hardness/HV0.05	Linear shrinkage (%)	Year/References
VP	0.1	>99	–	793.18	24	2020 [248]
	5	73	58.9 ± 2.9^{3P}	–	–	2006 [249]
DIW	2.7	–	280 ± 20^{3P}	–	–	2015 [250]
SLS/SLM	10	99.6	34^{3P}	–	–	2011 [251]

^{3P} Three-point bending test; ^{4P}Four-point bending test

Fig. 7.12 Representative SiO_2 samples produced by several AM technologies [248, 255, 263]

technologies, respectively. An excellent VP technology for glass was reported by Kotz et al.; transparent fused silica glass with excellent resolution was fabricated via a bottom-up printer. However, in order to improve the efficiency and eliminate the separation force resulting from the bottom-up setup [245] Liu et al. used a top-down printer developing a fast heat treatment of less than 16 h [246]. Wang et al. reported a sintered silica part made using an SLA machine with a density and bending strength of 1.57 g/cm^3 and 13.31 MPa, respectively, which are comparable values to those of conventional manufactured silica [247].

Generally, robocasting is applied to produce bioactive glass scaffolds for biomedical applications. Siamak et al. initially prepared 42 vol.% paste mixing bioactive glass with a chemical composition of 53% SiO_2, 20% CaO, 12% K_2O, 6% Na_2O, 5% MgO and 4% P_2O_5 (wt.%) and additives, to fabricate the controlled porous scaffold with compressive strength of approx. 114 MPa and flexural strength of approx. 280 MPa [250]. Amy et al. also used this method of AM to create patient-specific bioactive glass scaffolds for bone repair [252]. An alternative method was used by Klein et al., in 2015, who developed a high temperature robocasting system to direct melt the silica for the fabrication of optically transparent glass with complex architecture [253].

It is well-known that binder jetting is suitable for making sand molds due to its advantages of large build volume, the absence of support structure and short printing time. Snelling et al., at Georgia Southern University, have focused on the fabrication of sand molds via binder jetting in recent years, investigating potential differences of four different commercially available binders jetting systems to fabricate sand molds; they confirmed that binder content can affect the part quality [254]. Recently, the researchers compared two binder jetting materials, ExOne and ZCast sands and evaluated their properties. They found that final part quality, such as microstructure, porosity, density and surface roughness depended on the build materials and printing process [255]. Another AM technology, inkjet printing, has had a strong interest in the printing of ceramic membranes, coating or biosensors on substrates because of its fine features. Dzik et al. deposited titania-silica composite coatings with thicknesses ranging from 40 to 400 nm onto glass and PET substrates by inkjet printing to improve mechanical, optical and photocatalytic properties [256, 257]. Martin et al.

applied inkjet printing to fabricate silica-based gas separation membranes on alumina substrates [258].

Again, powder-based SLS is specifically for making porous structures. For example, Chang et al. demonstrated the possibility of the fabrication of porous silica through mixing 0.1–0.3 wt.% carbon into the silica powder [259]. Silica glass is a promising candidate for SLM due to its extremely low thermal expansion coefficient and low melting point as compared to zirconia and alumina. Khmyrov et al. carried out SLM work for the production of crack-free parts with optimized laser powder and scanning speed [260]. Gan et al. studied the effect of process conditions on the part properties, despite a slight improvement of flexural strength, from 2.10 to 4.33 MPa, there was still a potential performance gap between SLM and conventional routes [261]. In the area of LENS of silica, Heer et al. demonstrated the ability of this technology to deposit silica barrier on metal substrate, aiming to improve surface hardness and wear resistance of metal [262].

Other Oxides—Titanium dioxide (TiO_2) and cerium oxide (CeO_2) are relatively new developments based on AM technology, having developed over the past few years. TiO_2 is very suitable for industrial use at present and also probably in the future among many candidates for photocatalysts as a result of its efficient photoactivity and high stability [264]. In addition, it is used for biomedical applications due to its good compatibility [265]. Other applications such as photonic crystal, water splitting and solar fuel have also been investigated [266–268]. Akria et al. demonstrated the fabrication of periodic structures of TiO_2 via micro-stereolithography and studied photonic band gap properties in the millimeter waveguides; 40 vol.% ceramic slurry was formulated and the printed component showed that the relative density and band gap were 96% and 110 GHz [269]. TiO_2 hierarchical structures were printed by Fabiola et al., who focused on a solar heterogeneous photocatalytic system [270]. Aleni et al. reported the DIW to manufacture porous TiO_2 structure with a porosity of up to 65% and pore size of 180 μm, and the measured elastic modulus and compressive strength were 0.5 GPa and 12–18 MPa, respectively [271]. Elkoro et al. successfully printed porous Au/TiO_2 parts for photo generating hydrogen application. With the impregnation of gold nanoparticles, the hydrogen photoproduction rate was improved by 2–3 orders of magnitude. Moreover, the diameter of the filament could also influence the catalytic performance in an inverse way [272]. Few works have been reported to create TiO_2 parts via other ceramic AM technologies; Maryam et al. used an inkjet printer to fabricate a thin TiO_2 micro-emulsion film onto a glass slide [273].

CeO_2 has gained a major position in the list of environment applications and energy conversion systems due to its particular redox properties [274]. Few papers can be found on the fabrication of ceria via additive manufacturing. Ceria pastes were initially prepared, and a catalytic support was fabricated with Ni coating by the robocasting process, as shown in Fig. 7.13, by Ilaria et al. [275]. The result indicated that the 3D printed ceria structures exhibited a much higher catalytic activity than that of the cordierite honeycomb on a reactor volume basis. Xu et al. fabricated thin Gd-doped CeO_2 (GDC) layers on NiO-YSZ substrates via inkjet printing [276].

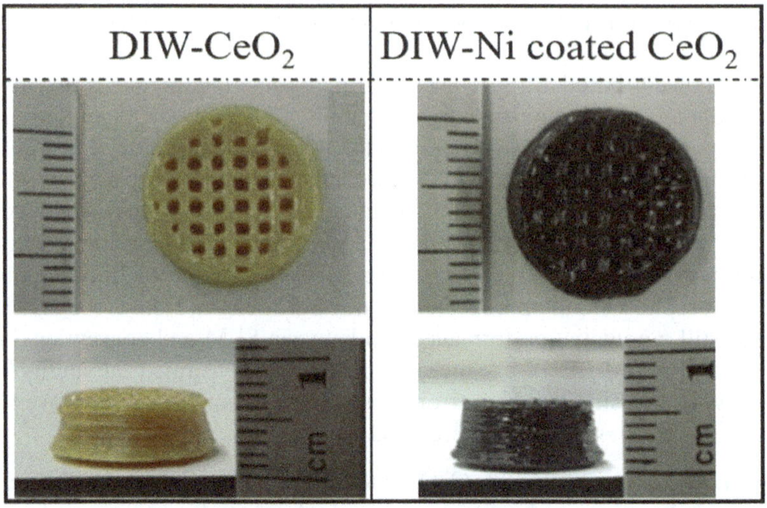

Fig. 7.13 Representative ceria samples produced by the DIW [276]

7.3.1.2 Carbides and Applications

Silicon carbide—SiC is recognized as an excellent structural ceramic in various aggressive environments, such as aerospace [277], mechanical components [278], biomedical [279, 280], automotive [281], energy systems [282], semiconductor industries [283, 284] and nuclear applications [285]. This is due to its excellent performance in terms of high resistance to temperature and thermal shock, high strength, extreme hardness, high thermal conductivity, low bulk density and wide bandgap [286, 287]. In addition, SiC is also a very beneficial second phase reinforcement in the form of fibers, whiskers and particulates, basically blending into ceramic, polymer and metal matrix composites for enhancement of their physical, chemical and mechanical properties [288, 289].

Silicon carbide ceramics produced by AM have been extensively studied, their properties and representative samples are summarized in Table 7.13 and Fig. 7.14, respectively. SLA and DLP are still popular method due to their good accuracy; even

Table 7.13 Comparison of properties of additively manufactured silicon carbide specimens

AM	D_{50}/μm	Density (%)	Flexural strength/MPa	Linear shrinkage (%)	Year/References
VP	–	93.5	165.2^{3P}	–	2020 [298]
	1.1	82.6	184.2^{3P}	4.537	2019 [292]
DIW	0.5	–	300^{3P}	2	2019 [299]
SLS	59.03	–	239.4^{3P}	–	2020 [300]

^{3P}Three-point bending test

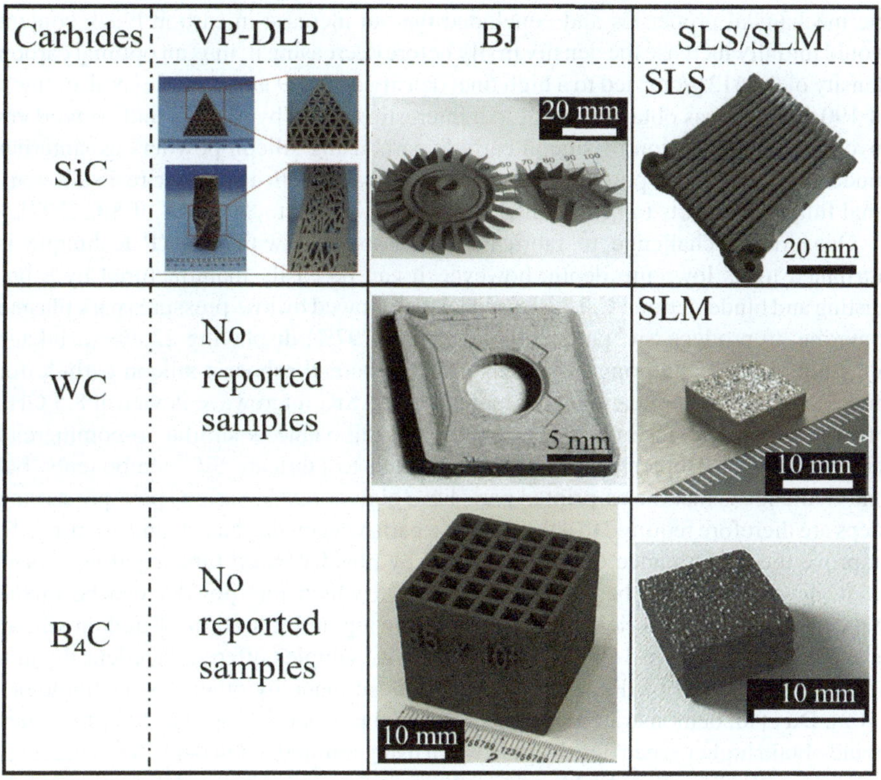

Fig. 7.14 Representative carbide samples produced by several AM technologies [292, 297, 302, 305–308]

though SiC is very difficult and challenging to manufacture due to its large refractive index and low transmittance [290, 291]. For example, He et al. demonstrated the feasibility of the production of SiC components using the DLP technique, with precursor infiltration and pyrolysis being used to improve the density and strength of printed parts [292]. To eliminate the scattering effects, Park et al., from Chungnam National University, fabricated 3D SiC nano/micro ceramic using a photosensitive precursor system, which was subsequentially pyrolyzed to SiC; this work provided a new route for the generation of SiC components [293]. Another AM technique, selective laser sintering has been extensively applied to shape SiC ceramic parts. In 1995, Nelson et al., from the University of Texas in Austin, originally used an SLS technique to fabricate SiC parts, with PMMA as the binder, concluding that the green part density and strength were related to a combination of laser parameters [294]. Liu et al. initially prepared SiC-polymer-carbon black composite powder and then manufactured SiC components with complex shapes by SLS incorporating cold isostatic pressing (CIP) and reaction sintering, aiming to improve the final density [295]. After a few years, they continued to study the effects of carbon content on the mechanical properties and concluded that an increase in carbon black content would initially increase the density firstly before decreasing it; thus an optimal carbon density of 0.6612 g/cm^3 led to a high final density of 2.899 g/cm^3. A flexural strength of 190.19 MPa was obtained [296]. An interesting study by Meyers and co-workers involved the fabrication of silicon carbide parts using silicon powders as sintering binder; afterwards, the porous parts were impregnated with a phenolic resin allowing final fully dense parts to be obtained containing with up to 84 vol.% of SiC [297].

One critical challenge to fabricate SiC associated with the VP technique is attributed to its low cure depth, however, it can be easily manufactured by robocasting and binder jetting. Cai et al. used DIW, followed by low-pressure spark plasma sintering, to produce SiC parts with a density of 97%; displaying 22.8% shrinkage after liquid-phase sintering [43]. Larson et al. studied robocast silicon carbide for microwave optics applications; the manufactured SiC microwave device at 8.5 GHz offered a very low insertion loss of 0.05 dB, this value is similar to commercial standards [301]. Binder jetting is also accepted to fabricate SiC components, but one challenge is that the as-printed parts have high porosity, various post-processing steps are therefore required to achieve dense parts via powder-based binder jetting. To improve the performance of parts obtained by powder-based binder jetting, Zocca et al. developed slurry-based binder jetting, in which fine powder can be mixed with solvents. They fabricated SiC parts following an infiltration of molten silicon at a temperature above 1400 °C; the final SiSiC samples offered excellent flexural strength of up to 479 MPa, even exceeding those made by other AM technologies [302]. Du et al. demonstrated that when bimodal powder was used as feedstock they could obtain higher green density than when using unimodal powder [303]. Direct 3D printing methods, such as SLM and LENS, are less reported for the fabrication of SiC components. However, silicon carbide particles are typically used as reinforcement in metal-based composites made by SLM or LENS [107, 304].

Tungsten carbide—Tungsten carbide (WC) is one of the cemented carbide materials and it is widely used in the mining industries, as wear resistant components, for aerospace engineering, in the oil and gas exploration industries and as reinforcement due to its high melting point (>2900 °C), extremely high hardness, strength and wear resistance [309–311]. Only a few AM technologies have been successfully used with tungsten carbide parts compared to oxides up until now. The representative images of WC samples produced by BJ and SLM are shown in Fig. 7.14. It is not a preferred material for VP processes since WC has a large refractive index, thus the fabrication of is better by other AM technologies that are not limited by the materials' physical properties. Uhlmann et al. reported an SLS method to produce a WC driller, studying the effect of laser parameters on the resultant microstructures; the results showed that a high energy density resulted in embrittlement of the processed WC, whilst a low energy density may have caused unwanted residual porosity [312]. The ability of AM of cemented WC was examined by Ku and colleagues in 2019, who also investigated the effect of different processing conditions on the final density and microstructure, with the final density reaching as high as 95% of the theoretical value [313]. Benichou and Lauder disclosed the WC/Co ink composition for the 3D inkjet printing process, with the ink comprising of a liquid dispersion of tungsten carbide and cobalt particles [314]. Robocasting provides a new route to manufacture tungsten carbide. For example, Carrillo et al. used this approach to fabricate porous WC parts using precursor paste [315]. Binder jetting is another versatile method to manufacture WC ceramics. For example, Kernan et al. printed tungsten carbide parts with the addition of 10 wt.% cobalt using a slurry-based bed; the results showed that the sintered density reached the theoretical values and that the microstructure was similar to that of conventionally manufactured parts [305]. Using the same method, Enneti et al. investigated the sintering densification, shrinkage and mechanical properties of printed WC–12%Co, subsequently assessing its wear properties. The results confirmed that printed parts displayed comparative wear resistance to standard cemented carbides [316, 317]. Scheithauer et al. manufactured hard metal components using WC–10% Co thermoplastic suspensions, achieving dense samples after debinding and sintering [318]. In summary, although the geometry of printed WC is simple, the mechanical properties, dimensional accuracy, and sintered density are comparable to those of conventionally shaped WC.

Boron carbide—Boron carbide (B_4C), another refractory ceramic, possesses high hardness, a high melting point and low density (2.52 g/cm^3), making it an ideal candidate for extreme environment applications [319]. Common applications include: grinding and cutting tools, lightweight ceramic armor [320], high temperature thermocouples, nuclear reactors, rocket propellant, neutron absorber [321] and neutron collimator [322], as well as wear resistant components (blasting nozzles, die tips and grinding wheels) [323–326]. The machining of B_4C is time-consuming and costly due to its extremely high hardness. Ultimately, there is a need to fabricate net-shape B_4C components by AM. The representative sample images of B_4C produced by BJ and SLM are shown in Fig. 7.14. Costakis Jr et al. reported on boron carbide green bodies being printed via DIW, using 54 vol.% aqueous suspensions and sintered at 2000 °C, to produce final parts with 82% theoretical density [46]. Recently, Chandrasekaran et al. used the same approach to shape complex porous B_4C objects using a high solid loading of ~60 vol.% paste; afterwards, the porous parts were infiltrated with molten aluminum to obtain a dense B_4C cermet [327]. Davydova et al. proved the feasibility of the generation of simple B_4C objects with cobalt binder by SLM, studying the microstructure, porosity, compressive strength and microhardness of the specimens, with the obtained samples showing a porosity of ~37% and a compressive strength of 110 MPa, respectively [308]. Cramer et al. used the binder jetting method to manufacture a boron carbide collimator with a complex geometry; printed samples displayed high density and hardness as well as favorable neutron scattering results [307]. It is clear that most WC parts were fabricated using materials-independent AM methods, similar to WC. DIW and BJ are expected to produce a lattice structure, thus the performance of additive manufactured bulk B_4C is not comparable to that produced by conventional manufacturing.

7.3.1.3 Nitrides and Applications

Silicon nitride—Silicon nitride (Si_3N_4), one of the major advanced high temperature ceramics, possesses high strength and hardness, abrasion resistance and relatively high toughness as well as good creep resistance [328, 329]. These promising thermal and mechanical properties at elevated temperatures make it the main candidate for high temperature applications [330], such as turbocharger rotors, gas turbine engine components [331] and cutting tools, semiconductors [332], heat exchangers [333], engine valves and other wear-resistant components [334, 335]. In addition, Si_3N_4 possesses good biocompatibility, hydrophilic behavior, and resistance to bacterial adhesion. All of these make it an attractive choice for bio-applications [336, 337].

Considering its distinctive combination of material properties, a wide range of AM have been used to fabricate silicon nitride components. Table 7.14 and Fig. 7.15 present the properties and representative images of Si_3N_4 samples via different AM technologies, respectively. VP based and robocasting are the typically used ceramic AM technologies and have been widely employed to fabricate silicon nitride with complex and precise structures. For example, Ventura et al. used SLA to fabricate

Table 7.14 Comparison of properties of additively manufactured silicon nitride specimens

AM	D_{50}/μm	Density	Flexural strength/MPa	Fracture toughness/MPa m$^{1/2}$	Hardness	Linear shrinkage	Year/References
VP	0.8	3.38 g/cm^3	–	~5.82^{In}	~14.63 GPa	–	2020 [341]
	–	99.8%	764^{3P}	–	1500 HV10	–	2020 [340]
DIW	0.5	95%	–	–	–	18–24%	2020 [342]
	0.7	3.23 g/cm^3	552 ± 68^{4P}	–	–	–	2016 [51]
IJP	0.5	3.18 g/cm^3	600^{3P}	4.4^{In}	17 GPa	–	2008 [74]

^{3P} Three-point bending test; ^{4P}Four-point bending test; ^{In}Indentation fracture toughness

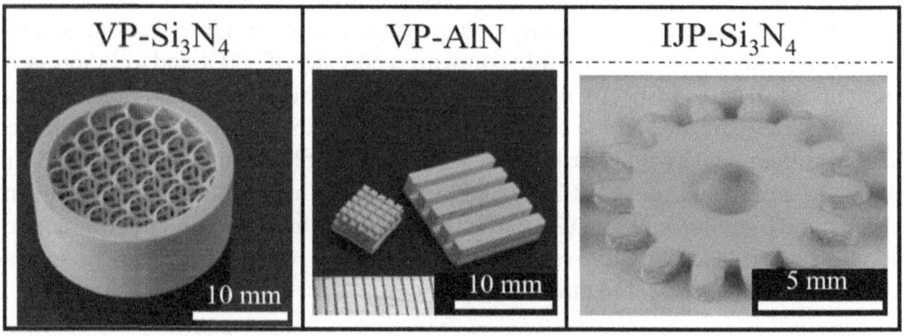

Fig. 7.15 Representative nitride samples produced by several AM technologies [74, 340, 349]

functional silicon nitride components, reporting that the sintered samples were found to be more than 99% dense, with high flexural strength above 800 MPa, which is very close to that of hot pressed specimens [338]. As mentioned, the fabrication of Si_3N_4 is a challenging goal through VP because the large refractive index difference leads to insufficient cure depth. To address this issue, Huang et al. oxidized powder at 1150–1200 °C for 3 h in air; the results showed that the amorphous silica layer attached onto the surface of silicon nitride. Afterwards, the suspensions were prepared by mixing resin and treated powder and then they successfully printed complex-shaped parts [339]. Altun et al. demonstrated the ability of the DLP to shape silicon nitride-based ceramics having high relative density, flexural strength, and Vickers's hardness [340]. Liu et al. created dense Si_3N_4 ceramics; the mechanical properties of parts were comparable to those of the samples made using conventional technology [341].

DIW has been used to make Si_3N_4 ceramics. For example, He et al., in the Sandia National Lab, used robotics to control the deposition of ceramics paste and obtained dense parts after post-processing [343]. Rahaman et al. prepared an aqueous paste containing ceramic powder and sintering additives to build dense parts with the help of hot isostatic pressing [344]. Binder jetting provides a great advantage for the fabrication of porous ceramics. Rabinskiy et al. reported the production of porous silicon nitride ceramics by binder jetting [345]. This study produced ceramics exhibiting a fibrous α-Si_3N_4 structure and porosity up to 70%. Moreover, the fabrication of Si_3N_4 through direct SLM can prove challenging due to its poor sinterability. Thus, Minasyan et al. proposed a new approach to fabricate Si_3N_4 parts; they printed SiO_2 of a required shape using SLS, and then nitrided the as-printed silica parts with the aim of forming final Si_3N_4 components [346]. Wei et al. used SLS directly to fabricate pre-sintered ceramics with high porosity of 80% [347]. In terms of inkjet printing, Cappi et al. demonstrated its ability to manufacture high performance Si_3N_4 ceramics; the suspension was initially prepared by dispersing 30 vol% Si_3N_4 powder into an aqueous medium, with the samples being deposited using a drop on demand printer [74]. After a few years, they used inkjet printing to produce complex-shaped functional ceramic parts without delamination or other defects [348]. As yet, LENS and SLS or SLM technologies have not been developed to fabricate highly dense silicon nitride ceramics.

Aluminum Nitride—Aluminum nitride (AlN), as a family of advanced ceramics, has excellent properties, including very high thermal conductivity, a wide band gap, high electrical insulation capacity, low thermal expansion, piezoelectric properties and good biocompatibility [350]. These attractive properties allow it to be used for a large range of applications, such as MEMS [351, 352], electronic packaging [353], piezoelectric transducers [354–356] and heat exchangers [357]. Up to now, only SLA, DLP, and BJ have been adopted in the fabrication of AlN ceramics. Figure 7.15 presents the representative images of AlN samples via VP and BJ. Paulina et al. used a CeraFab 7500 printer to shape aluminum nitride microchannel heat exchangers [358]. The same approach was applied by Lin et al., the sintered body displayed >99% density. A flexural strength of 398 ± 10.4 MPa and hardness of 10.47 ± 0.08 GPa were measured [349]. The reported mechanical properties of AlN produced by SLA or DLP are very similar to those obtained by conventional methods. Other AM methods have also been used to fabricate AlN objects. For example, Díaz-Moreno et al. studied a novel approach to print AlN components using binder jetting combined with HIP for 8 h at a temperature of 1900 °C [359]. The thermal conductivity of printed samples, finding it to be 4.82 W/mK at room temperature, is not a desirable value compared to its standard value (321 W/mK) due to its low relative density of 60.1%.

7.3.1.4 Borides and Applications

Zirconium boride (ZrB_2) and Hafnium diboride (HfB_2)—They are typically ultra-high-temperature ceramics (UHTCs), which have high melting points (>3000 °C), chemical inertness, good oxidation resistance and high thermal and electrical conductivities [360, 361]. Their excellent high temperature properties have drawn considerable interest for engineering applications in hypersonic vehicles, atmospheric re-entry and rocket propulsion [362, 363]. Research on the AM of borides has emerged in recent years, yet it is still far behind the AM of oxides ceramics. Representative sample image of AM-produced borides as shown in Fig. 7.16. For example, Feilden et al. fabricated HfB_2 using DIW. After pressureless sintering, the bulk parts reached

Fig. 7.16 Representative borides samples made by different AM technologies [364]

densities of 94–97% and strengths of 196 ± 5 MPa up to 1950 °C, which was comparable to UHTC parts fabricated by conventional means [364]. McMillen et al. used robocasting to build complex zirconium diboride components. The part density was up to 99% with incorporation of 4 wt.% B_4C particles; the relative density exceeded 99% [365]. Moreover, Leu et al. used stearic acid as the binder to print ZrB_2 ceramics via SLS; after the thermal process, the specimens achieved 80% theoretical density and the average strength of 195 MPa [366].

7.3.1.5 Polymer Derived Ceramics and Applications

Polymer derived ceramics (PDCs), preceramic precursors, have been proposed for over four decades for the fabrication of silicon-based advanced ceramics. These ceramics are obtained from the chemical pyrolysis of polymer precursors under inert atmospheres. Most preceramic precursors studies represent inorganic/organometallic systems, such as polyorganosiloxanes, polyorganosilazanes, and polyorganosilylcarbodiimides, typically transforming to binary and ternary silicon-based ceramics, with SiC, Si_3N_4, silicon oxynitride (SiON), silicon oxycarbide (SiOC) and silicon carbonitride (SiCN) [144]. In general, PDCs exhibit enhanced thermo-mechanical properties in creep and oxidation at high temperature of 1500 °C and thus can be applied for high temperature environment applications, such as gas separation membranes [367], heat sensors [368], high temperature acoustics and heat insulation [369], as well as barrier coating [370]. PDCs routes offer a significant advantage over the more conventional powder-based AM processing route. However, the main drawback of PDC routes is that they can suffer from significant shrinkage during the polymer pyrolysis stage. Indeed, the escape of substantial amounts of gases during chemical transformation process is expected to cause high mass loss and significant volume shrinkage. Therefore, it is better to make cellular ceramic components rather than dense materials.

SLA, DLP, and DIW have demonstrated success in the shape of polymer derived ceramics, as summarized in Table 7.15 and Fig. 7.17. Research into the AM of polymer-derived ceramics was first reported by Eckel et al. [371], who fabricated complex shaped SiOC ceramics after pyrolysis at 1000 °C in argon using stereolithography 3D printing of siloxane-based polymers. As-sintered SiOC was accompanied by 42% mass loss and 30% linear shrinkage, and the compressive strength of the resulting honeycomb structure with a density of 0.8 g/cm^3 was 163 MPa, which was ~ 10 times higher than for commercially available ceramic foams, such as those made from SiC, aluminosilicate and silicon oxycarbide foams. Moreover, Hundley et al. reported the fabrication of fully dense SiOC impellers via SLA, in which a UV-curable mixture of (mercaptopropyl) methylsiloxane and vinylmethoxysiloxane was converted to amorphous silicon oxycarbide after thermal pyrolysis [372]; the final parts exhibited uniform linear and volumetric shrinkage of 29.7% ± 1.2% and 64.9% ± 1.6%, respectively. Li et al. prepared three types of polysiloxane as photocurable precursor resins and obtained lattice structured SiOC ceramics using a DLP printer; the measured results showed that the linear shrinkage and mass loss

Table 7.15 The reported work of PDCs made by VP and DIW

AM	Polymer precursors	PDC	Linear shrinkage (%)	Year/References
VP	Polysilazane	SiCN	~30	2020 [376]
	Polysiloxane	SiOC	30	2020 [377]
	Polysiloxane	SiOC	>50	2020 [378]
	Zirconium *n*-propoxide	ZrOC	42	2019 [375]
	Polyvinylsilazane and SiO_2	SiCN	20	2019 [379]
	Polyborosilazane	SiBCN	–	2018 [23]
	HBPCS and SiHx	SiOC	–	2018 [380]
	Polysiloxane	SiOC	29.7	2017 [372]
	Modified polycarbosilane	SiC/SiOC	–	2017 [381]
	Polysiloxane	SiOC	30	2016 [371]
DIW	Polymethylsiloxane, SILRES® MK	SiOC	~20	2020 [382]
	Polysiloxane	SiOC	25	2020 [383]
	Polycarbosilane	SiCO	8.3	2019 [384]
	Polysiloxane	SiOC	42 vol.% shrinkage	2016 [35]

were 42.01% and 70.37%, respectively. In addition, the authors indicated that the lattice produced in the study showed an exceptional specific compressive strength to density ratio of up to 5.74×10^4 N m/kg, which was distinctly higher than that of other cellular materials [373]. The compressive strength was ~2 times higher than the SiC honeycomb reported by Agrafiotis et al. [374]. With the same AM technology, Fu et al. prepared porous ZrOC ceramics with an octet truss structure-based DLP 3D printing, offering the possibility to expand the range of available ceramics precursors. As prepared samples with ~55% porosity exhibited linear shrinkage of 42% and mass loss of 79% [375]. Another PDC, SiCN, was successfully fabricated by Xiao et al., they used carbon nanotubes as light absorbers to effectively control curing depth and improve printing resolution by mixing with commercially available polysilazane preceramic polymer. The subsequently pyrolyzed the mixture to obtain the ultra-light and high-strength SiCN microwave-absorbing components for aerospace applications, obtaining a flexural strength of ~55 MPa and elastic modulus of ~33 GPa, respectively [376]. Additionally, Li et al. investigated the preceramic polymer formulation and yielded of complicated SiBCN ceramic parts with high precision and a linear shrinkage of 28.5% [23]. The printed ceramics also exhibited excellent thermal stability and resistance to high temperature oxidation up to 1500 °C. Particularly, TPP, one of the VP-based AM technologies, is particularly relevant for the fabrication of high-resolution polymer-derived ceramics. For example, Konstantinous et al. demonstrated the feasibility of making high resolution micro-PDC parts without cracks and shape distortion via two-photo polymerization; as final structure exhibited a low linear shrinkage of 30% [377].

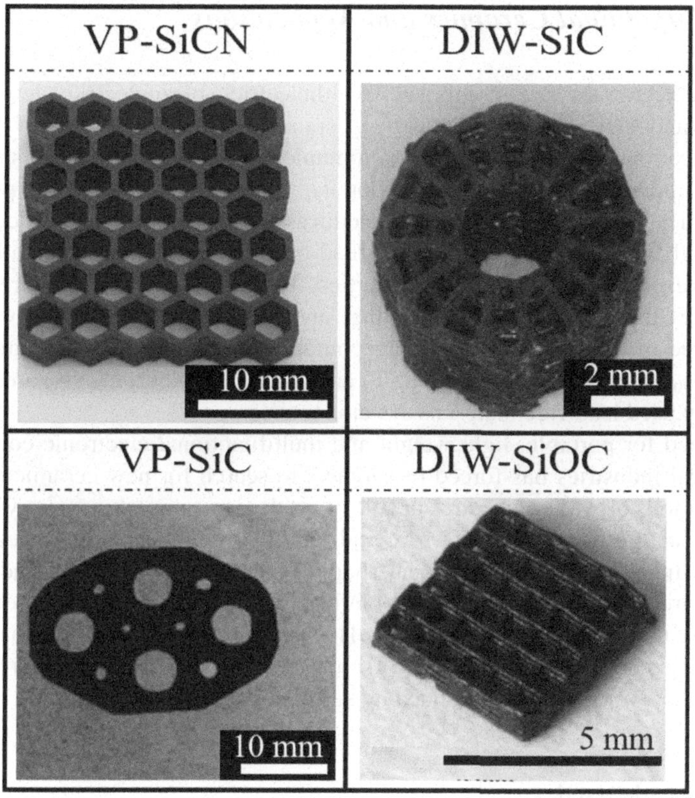

Fig. 7.17 Representative PDC samples made by different AM technologies [35, 385–387]

Robocasting has also been used to fabricate polymer derived ceramics, but the printing precision is not better than that achieved through VP. Pierin et al. fabricated micro-sized SiOC ceramic components by robocasting; they tried to improve compressive strength and reduce the shrinkage of final parts through adding a low amount of graphene oxide [35]. Using the same strategy, Xiong et al. prepared 3D SiCO-based ceramics after pyrolysis of polycarbosilane-based lattices; the linear shrinkage and weight loss decreased from 18.2% to 8.3% and 17.5% to 10.6% thanks to the addition of SiC particles [384]. Huang et al. manufactured SiOC ceramics with hierarchical and tunable porosity using an ink containing PMMA particles; the prepared porous parts exhibited a high strength of 2.92 MPa and porosity of 86.5% [382].

7.3.2 *Functional Ceramics and Applications*

Functional ceramics are comprised of dielectric, piezoelectric, ferroelectric, magnetic and electronically conductive ceramics and have been used for practical applications, including wireless communication and energy, sensors and electronic components [388]. The most popular functional ceramics are piezoelectric ceramics, where an electric field produces in response to mechanical forces. Traditionally, piezoelectric materials are lead-based titanate-zirconate family (PZT), featuring excellent piezoelectric properties. However, the toxic nature of lead complicates their applications, recycling, and disposal. Hence, the development of lead free piezoceramics has been encouraged in recent years. Barium titanate (BTO), $Ba_{0.85}Ca_{0.15}Zr_{0.1}Ti_{0.9}O_3$ (BCZT) and $(K_{0.5}Na_{0.5})NbO_3$ (KNN) are the most commonly used lead-free materials systems to date.

The need for portable, light weight and multifunctional electronic components in electrical industries has forced researchers to search for new ceramic manufacturing methods of AM, which are capable of producing complex geometry ceramics with fine features. In addition, AM could eliminate the mechanical stress arising from post-machining of conventional shaping processes, which can lead to grain pull-out, strength degradation, and depolarization of the near surface region, further influencing the electric properties [389]. The most commonly used AM methods for piezoelectric ceramics are DIW, SLS, BJ and VP. Figure 7.18 shows representative functional ceramic samples produced by various AM technologies.

7.3.2.1 Lead-Based

Lead Zirconate Titanate ($PbZr_xTi_{1-x}O_3$, PZT)—PZT is of considerable commercial importance for a host of piezoelectric applications, such as ultrasonic transducer, piezoelectric transformer, filter and energy storage thanks to its excellent piezoelectric constant ($d_{33} \approx 600$ pCN^{-1}) and electromechanical coupling coefficients ($k_t \approx 0.55$) [395]. However, the shortcoming is that PZT is toxic with the addition of lead oxide.

AM has been conducted to fabricate PZT for three decades. In 2002, Dufaud et al. started to explore the feasibility of fabrication of PZT ceramics via VP, mainly investigating the preparation of PZT suspensions, with the PZT transducer and rods being finally printed. However, dielectric and piezoelectric properties have not yet been evaluated [396]. Chen et al. used SLA to fabricate PZT ultrasound transducer arrays with the piezoelectric constant of 212–345 pCN^{-1} and dielectric constant 760–1390 [395]. Other 3D printing methods, IJP, SLS and LENS have also been used. For example, Wang et al. prepared a suitable PZT suspension for inkjet printing, and linear shrinkage of the inkjet printed object in the Z direction was 25 ± 1% [81]. Lejeune et al. used this process to build a PZT piezoelectric micro-pillar array structure. The results revealed that PZT micro-pillar structures exhibit good definition, which is illustrated in Fig. 7.18 [390]. Guo et al. used SLS approach to indirectly

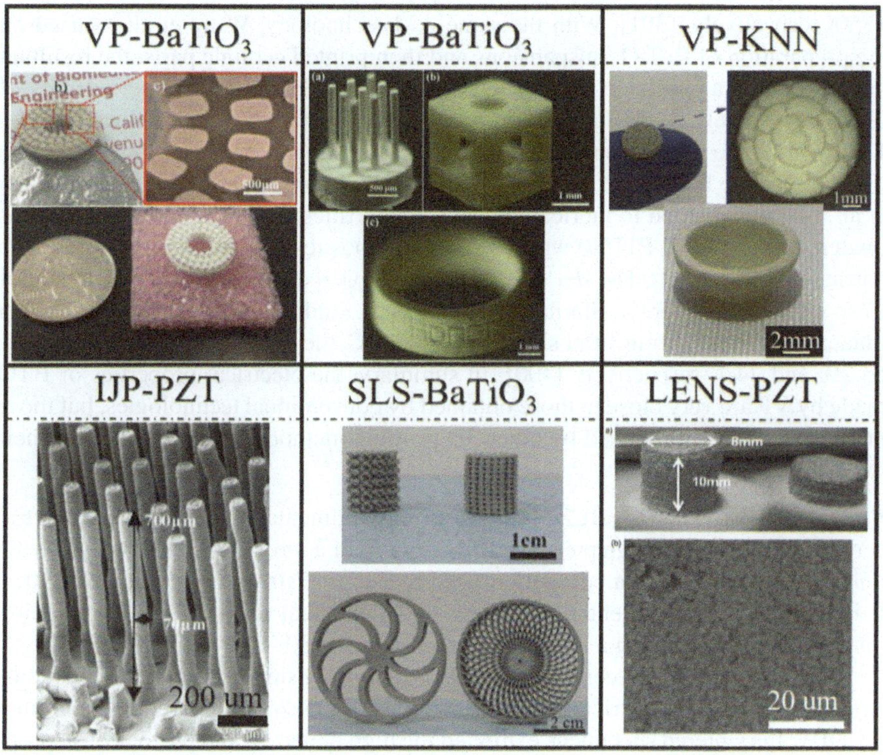

Fig. 7.18 Representative functional ceramic samples made by different AM technologies [111, 390–394]

fabricate PZT ceramics; a sacrificial mold was firstly fabricated by SLS, and then aqueous suspension was cast in the mold. The indirect formed PZT offered comparable electrical properties compared to die pressed PZT samples [397]. LENS has also been used to fabricate dense lead zirconate titanate (PZT) on a metallic substrate; the LENS-fabricated PZT structures at the process parameters of 150 W, 5 mm/s and 1.3 g/min exhibited reasonable dielectric properties for sensors and transducers application [111].

7.3.2.2 Lead-Free Based

Barium titanate (BTO)—It has become one of the most popular electronic function ceramics and is broadly used for functional applications, such as temperature coefficient resistors, and ceramic capacitors due to its excellent dielectric, ferroelectric, and piezoelectric properties [398, 399]. Chen et al. reported DLP method to fabricate $BaTiO_3$ ultrasonic transducer realizing the energy focusing and ultrasonic sensing; with piezoelectric constant and relative permittivity being 160 pC/N and

1350, respectively. [391]. With the same AM technology, Wang et al. focused on the preparation of $BaTiO_3$ suspensions and then printed ceramic parts, the resultant parts exhibited a piezoelectric constant d_{33} of 163 pC/N and relative permittivity of 2762, respectively. The relative permittivity is larger than the empirical value (ε_r = 1700), while the piezoelectric constant is very close to that value (d_{33} = 190 pC/N) [392]. Selective laser sintering is another popular ceramic 3D printing approach. Qi et al. used this method to fabricate $BaTiO_3$ composites, in which PA11 was used as binder. As fabricated P11/80 wt.% $BaTiO_3$ composites had an improved dielectric constant of 19.3 at 110 Hz, d_{33} = 4.7 pC/N and piezoelectric voltage coefficient of 27.6 × 10^{-3} Vm/N [393]. Similarly, Gaytan et al. studied the fabrication of barium titanate via binder jetting; after sintering at 1400 °C, the relative density and d_{33} were 65.2% and 74.1, respectively [400]. In summary, the electrical properties of BTO made by VP are very close to those obtained by conventional technologies, but those properties of parts produced by other 3D printing methods are inferior due to their low relative density.

$Ba_{0.85}Ca_{0.15}Zr_{0.1}Ti_{0.9}O_3$ (BCZT)—It is an environmentally friendly piezoelectric ceramics, offering the high piezoelectric response of approximately 630 pC/N as the polarization direction can be easily rotated by external stress or electric field [401]. It is considered as a potential candidate for biomedical applications due to being non-cytotoxic and having a low Curie temperature of 80–100 °C [402, 403].

Fabrication of BCZT ceramics has been reported using different AM methods. Bo et al. reported the fabrication of lead-free BCZT piezoelectric ceramics through the DIW; the sintered specimens exhibited the high dielectric and piezoelectric properties, with tan δ of 0.021, remnant polarization of 4.56 μC/cm^2 and d_{33} of 100 ± 4 pC/N [49]. Following that, the authors developed a novel DIW method, directly depositing BCZT pillars in a converse Z direction [404].

$(K_{0.5}Na_{0.5})NbO_3$ (KNN)—The piezoelectric coefficient of KNN systems lies in the range of 80–160 pC/N, and their Curie temperature exceeds 400 °C. However, the densification (more than 90–95% of theoretical density) of KNN remains a challenge due to the phase instability of $KNbO_3$ and $NaNbO_3$ above 1040 and 1140, respectively [405]. Research on the AM of KNN has emerged recently, such as that by Chen et al., who demonstrated the success of a DLP printer in fabricating ultrasonic transducers. As sintered KNN ceramics exhibit superior piezoelectric and ferroelectric properties (Curie temperature, T_C = 230 °C; d_{33} = 170 pC/N; permittivity at 1 kHz, ε_r = 2150; the dielectric loss at 1 MHz, tanδ = 0.08), which are comparable to those of KNN produced by traditional methods [394]. Besides, DIW is also utilized for the fabrication of KNN ceramics; Li et al. used this versatile method to print Li, Ta and Sb co-doped KNN with woodpile structures, with a high piezoelectric coefficient. The samples had a relative density of 98%, piezoelectric coefficient d_{33} of 128 pC/N, dielectric loss of 1775, remnant polarization P_r of 18.8 μC/cm^2 and coercive field of 8.5 kV/cm after sintering at 1100 °C for 2 h [406].

7.3.3 *Bioceramics and Applications*

Bioceramics, is supposed to be the subject of extremely importance as it has specific biological or physiological functions for biomedical applications. They exhibit many good properties that are capable of replacing diseased tissues, fixing bone grafts, fractures or prostheses, to bone directly, such as biocompatibility, anti-bactericidal effects, high wear resistance and good physical and chemical stability [407]. The success of bioceramics is mostly dependent on their compatibility with the physiological environment, achieving a stable attachment to the native tissues within a reasonable period and showing limited toxicity. Nowadays, it is extensively used in hard tissue repair and regeneration, mainly for teeth and bone. Bioceramics are further divided into three groups: bioinert, bioactive, and bioresorbable ceramics [408].

7.3.3.1 Bioinert Ceramics

Bioinert ceramics have outstanding chemical stability when they are long-term exposed to the physiological environment; a very thin fibrous membrane is generated in the response area. Common bioinert ceramics include carbon, alumina, zirconia and titanium oxide (TiO_2), etc. Carbon has been found to be very useful for blood interfacing implants, such as heart valves [409]. It can also be used as a reinforcement for composite implants owing to its excellent strength as fibers [410]. Alumina and zirconia are widely employed in load-bearing hip and knee prostheses along with dental implants due to their high wear resistance and strength [411]. Clinical studies indicate that alumina-alumina of articulating surfaces can greatly reduce wear debris for present total hip systems [412]. In addition, TiO_2 can also act as an additive on alumina or zirconia nanocomposites for potential dental applications [413].

7.3.3.2 Bioactive Ceramics

Bioactive ceramics can chemically bind with surrounding tissue when they are implanted in living tissue. Actually, bioactive ceramics are subjected to induce specific biological activity for repairing damaged tissues; a hydroxyl-carbonate apatite layer on the surface forms [414]. Some bioactive ceramics have already been used for bone repair, realizing tight fixation with living bone. Commonly used bioactive ceramics include some calcium phosphate (Ca–P) based bioceramics such as hydroxyapatite, and glassy materials such as bioglass and A-W glass [415]. The first bioactive ceramic known as bioglass in the Na_2O–CaO–SiO_2–P_2O_5 system was developed by Hench in the early 1970s [416]. The compositional features of bioglass make the surface highly reactive when exposed to an aqueous medium. According to different ratios of components, bioglass can be divided into a series of bioactive glasses and glass–ceramics. The clinical use of bioglass in periodontal repair, maxillofacial reconstruction, and bone tissue regeneration have all been reported [417]. Hydroxyapatite (HA) is still popular in the biomedical field due to its compositional similarity to native bone. Significant studies of HA reported in literature have been found over more than 40 years regarding their materials, fabrication, and final properties. Currently, HA is employed for a different application as implant materials. For example, repair of body defects in the dental and orthopedic fields, maxilla-facial reconstruction and middle ear reconstruction [418, 419].

7.3.3.3 Bioresorbable Ceramics

Bioresorbable ceramics are a class of specific bioactive ceramics that can be gradually absorbed in vivo and replaced by new tissues. The number of articles devoted to bioresorbable ceramics has increased dramatically in recent decades since the goal of tissue repair is to allow the regeneration of the body's own tissue and organs. The degradation rate varies from material to material. For example, calcium sulfate dehydrate (CSD) disappears within weeks, whereas β-tricalcium phosphate (β-TCP) may take a longer time [420]. Additionally, the degradation rate is affected by several factors, such as biomaterials properties, scaffold structures and environmental conditions [421]. Ideally, the degradation rate should be equal to the rate of tissue generation to avoid nonmatched damage. Tricalcium phosphate and calcium sulfate are commonly used bioresorbable ceramics for biomedical applications. Indeed, almost all bioresorbable ceramics are variations of calcium phosphate. Nowadays, biodegradable ceramics have been employed for implant applications, such as drug-delivery devices and bone damage [422, 423].

Bioceramics are a relatively new development in AM. It offers great advantages of small-scale parts manufacturing with complex structures, high resolution and low cost, which are beneficial to the specific needs of patients for bone replacements. Several AM technologies, such as VP, binder jetting, SLS and direct ink writing have been used for fabricating bioceramics. DLP and SLA exhibit the advantage in that they can shape bioceramics, particularly in dental implants because dental prostheses require high dimensional tolerance and surface quality. Wenjun et al. reported an SLA method to fabricate zirconia dental ceramic under optimized printing parameters, the flexural strength and density reached 978 and 98.42%, respectively, which is adequate for clinical application [424]. Wei et al. proved the feasibility of using DLP to make calcium phosphate ceramics; they provided an optimized formula to fabricate with 60 wt.% solid loading using a DLP system. The in vitro cell culture showed the ceramics have a good cell viability, attachment, and proliferation. Also, when implanted in vivo, abundant blood vessels and obvious ectopic bone formation were clearly observed [425]. Zhou et al. printed high quality spine shaped β-TCP bioceramics [426]. Robocasting, binder jetting and SLS are competitive methods to produce bioceramics owing to their low cost. Moreover, they are favorable for the fabrication of porous structures with controlled pore size and porosity. Zhou et al. developed a binder jetting formulation via mixing HA powders with high molecular weight polyvinyl alcohol (PVOH), suggesting a high level of printability,

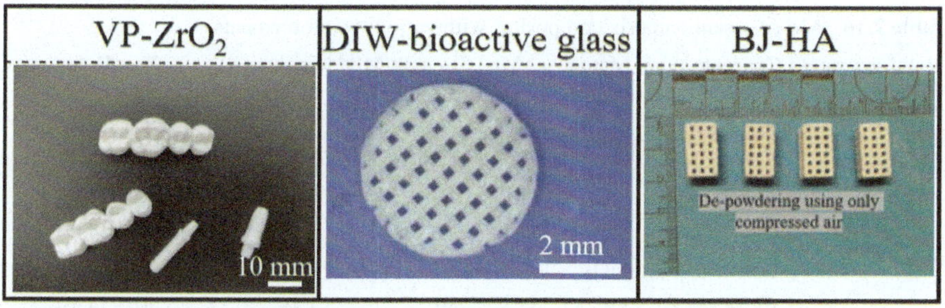

Fig. 7.19 Representative bioceramic samples produced by several AM technologies [427, 430, 431]

high geometrical accuracy and improved compressive properties [427]. Marques et al. used robocasting to fabricate Sr and Ag coped biphasic calcium phosphate scaffolds with different pore sizes and rod diameters; the results showed that the additives of Sr and Ag could improve the mechanical properties, as well as good antimicrobial activities [428]. Liu et al. designed and fabricated a porous 45S5 glass–ceramic scaffold via direct SLS; the printed samples were dense and had superior fracture toughness with a higher degree of crystallinity [429]. Figure 7.19 presents the representative images of 3D printed bioceramics, clearly exhibiting the difference in dimensional accuracy, surface quality and resolution.

7.4 Additive Manufacturing of Ceramic Matrix Composites

Ceramic matrix composites, CMCs, are composite materials that were developed to enhance the ductility and multifunctional properties of monolithic ceramics (e.g., combinations of thermal and electrical conductivity). They are most commonly made to remedy the inherent low toughness of ceramics and are reinforced by fibers, whisker, carbon nanotubes, graphene and particulates. AM of ceramic matrix composites allows us to fabricate composites with unique combinations of desired properties and geometries. However, AM of ceramic matrix composites remains at a very early stage of research and development due to its significant challenges. In the section, a comprehensive overview of the state of the art of ceramic matrix composites via AM is provided. The reported additively manufactured ceramic matrix composites reinforced with various fillers are summarized in Table 7.16.

7.4.1 Chopped Fiber

Chopped fibers, also known as short fibers, including ceramic fibers and carbon fiber [457], are typically used to enhance the mechanical properties of monolithic

Table 7.16 AM of ceramic matrix composites with various reinforcements

Reinforcement	Concentration	Matrix	AM	Strength/MPa	Fracture toughness/MPa $m^{1/2}$	Year/References
Carbon fiber	20.86 vol.%	SiC	VP	262.6	1.8	2020 [432]
	17.5 vol.%	SiC	DIW	274	5.82	2019 [433]
	–	SiC	SLS	237	3.56	2019 [434]
	1.0 wt.%	SiO_2	VP	54.5	–	2017 [435]
	10–15 vol.%	SiC	SLS	249	3.48	2018 [436]
	10–15 vol.%	SiC	VP	350	4.49	2014 [437]
Chopped fiber	>30 vol.%	SiOC	DIW	~4	–	2017 [38]
Alumina fiber	10.0 wt.%	SiO_2	VP	33.5	–	2017 [435]
SiC whisker	–	SiOC	VP	225–325	>3	2020 [438]
	5–15 vol.%	Al_2O_3	VP	405	7.1	2020 [439]
	–	SiC	DIW	390	–	2020 [440]
	25 vol.%	SiC	BJ	200	3.4	2019 [441]
SiO_2 particle SiC whisker	20 wt.% 1 wt.%	SiCN	VP	65.8	–	2020 [385]
Si_3N_4 whisker	60 wt.%	SiBCN	VP	~180	–	2020 [442]
CNT	3–10 wt.%	Al_2O_3	DIW	–	Б = 10^{-1} S/m	2020 [443]
Graphene	50 wt.%	SiC	DIW	193	Б = 3769 S/m	2019 [444]
	20 vol.%	SiC	DIW	10–50	Б = 611 S/m	2016 [445]
	0.5 wt.%	$CaSiO_3$	SLS	160.7	1.39	2014 [446]
Graphene oxide	–	Kaolin	DIW	8.21	–	2019 [447]
	0.2, 0.4 wt.%	HAP	BJ	3.5	–	2014 [448]
Zirconia	25 wt.%	Al_2O_3	VP	516.7	7.76	2020 [449]
	15 vol.%	Al_2O_3	VP	575	7.4	2020 [450]
	20 wt.%	Al_2O_3	VP	–	5.2	2018 [451]
	20 wt.%	Al_2O_3	VP	530.25	5.72	2017 [452]
	6 wt.%	HAP	VP	~80	–	2019 [453]
Alumina	20 wt.%	ZrO_2	VP	~800	–	2020 [19]
	20 wt.%	ZrO_2	SLM	31	–	2020 [454]
	–	ZrO_2	VP	>500	–	2019 [455]
	–	ZrO_2	VP	500	–	2019 [455]
	3.9 wt.%	ZrO_2	VP	–	6.42	2019 [456]

Б Electrical conductivity

ceramic materials, such as strength and fracture toughness. Although fiber-reinforced composite (fiber/CMCs) sounds promising, there are three major challenges that need to be resolved. Namely, it is difficult to achieve uniform dispersion in the matrix; the degradation of the reinforcing elements at high temperature and the development of an interface that is either too weak or too strong with matrix [458, 459]. Some have reported that fiber-reinforced ceramic matrix composites have proven that utilizing fiber as a reinforcement enables positive reinforcing effects if these challenges could be addressed using conventional methods [460–462]. The positive results are attributed to the toughening mechanism of fiber debonding and pull-out, as well as crack deflection, further leading to an improvement in the mechanical properties [463–465].

AM of chopped fiber-reinforced ceramic matrix composites allows the production of highly customized parts with significantly enhanced mechanical properties, compared to monolithic ceramic components. Great achievements in the AM of fiber-reinforced ceramic matrix composites have been witnessed and printed samples are shown in Fig. 7.20. Lu et al. used SLA for printing high performance carbon fiber enhanced SiC composites (C_f/SiC) parts [437]. The results showed that the addition of carbon fiber significantly improved flexural strength and fracture toughness to 350 MPa and 4.49 MPa $m^{1/2}$, respectively. Zhang et al. implemented the same approach to manufacture lightweight C_f/SiC composites; the samples exhibited dense and maximum flexural strength of 262.6 MPa [432]. Polozov et al. demonstrated the

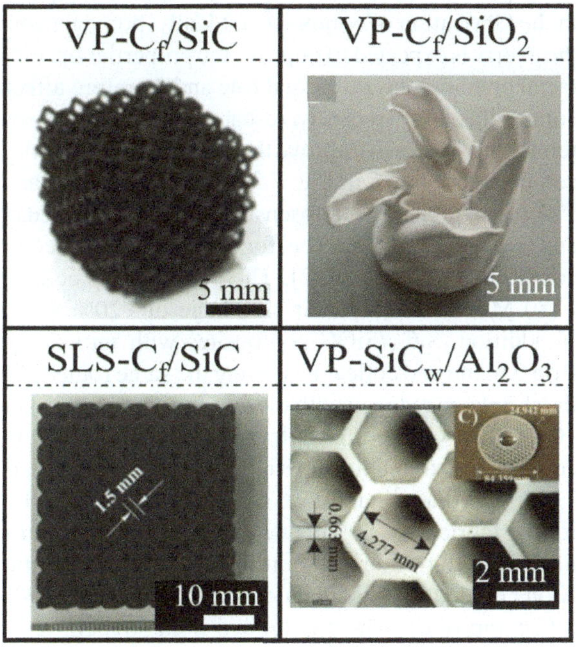

Fig. 7.20 Representative chopped fiber and whisker reinforced ceramic composites made by different AM technologies [432, 435, 436, 439]

feasibility of fabricating dense SiC_f/SiC composites using binder jetting; indicating that polymer infiltration and pyrolysis were helpful to improve the final density [466].

Robocasting has received considerable attention in the generation of ceramic matrix composites, primarily as it can extrude multimaterial using different nozzles. Franchin et al. formulated a composite ink containing 30 vol.% chopped carbon fiber and commercially preceramic polymer; a robocasting printer was employed to fabricate complex $C_f/SiOC$ structures with a porosity of ~75% and compressive strength of ~4 MPa, respectively [38]. Moreover, C_f/SiC composite was successfully shaped by Lu et al. using the same method; they studied the effect of the CVI time on interfacial thicknesses, with the results showing the 3 PB flexural strength and fracture toughness as 274 MPa and 5.82 MPa $m^{1/2}$, respectively [433].

7.4.2 *Whisker*

Whisker is a nano-sized filament of material that is structured as a single, defect-free crystal fiber compared to conventional fiber. Some typical whisker materials are graphite, alumina, silicon carbide, silicon nitride and silicon, which offer extraordinary strength and modulus as high as 2.5 GPa and 1000 GPa, far exceeding those in their bulk form, making them widely used in composites [465, 467–470]. As discussed in fiber-reinforced CMCs, although significant advances have been made in the processing of whisker/ceramic matrix composites, several problems still exist in the fabrication of composites. Ideally, homogeneous dispersion of the whiskers in the matrix is expected to improve the properties of composites. Additionally, the interfacial bonding plays a crucial role and therefore affects the fracture toughness of the composites. More recently, research has been focused on the AM of ceramic whisker reinforced CMCs to improve the mechanical properties. AM printed whisker/ceramic composite is shown in Fig. 7.20, and more details are discussed below.

Li et al. prepared photosensitive slurry composed of SiBCN preceramic polymers and 60 wt.% Si_3N_4 whiskers for the production of complex structured $SiBCN/Si_3N_{4w}$ composites via DLP [442]. The printed parts achieved a high bending strength of ~180 MPa and low linear shrinkage of ~20% with the introduction of whiskers. In addition, $SiC_w/SiOC$ composites with intricate geometries were fabricated by Brinckmann and colleagues, using the same routes [24]. They also demonstrated that the shrinkage reduced with the addition of SiC whiskers and hardness improved from 10.8 to 12.1 GPa. Xing et al. used SLA to manufacture SiC_w/Al_2O_3 composites; the composites exhibited a superior fracture toughness and a higher hardness, which were 7.1 ± 1.2 MPa $m^{1/2}$ and 17.6 ± 0.78 GPa, respectively [439]. DIW shows advantages for the fabrication of whisker or fiber-reinforced ceramic matrix composites with a filler of preferred orientation in the extruded direction and large-scale production. Xiong et al. used this method to manufacture highly aligned SiC whisker-enhanced SiC composites, which exhibited high strength resulting from the pull-out of the whiskers [471]. Similarly, well-aligned SiC_w/SiC porous composites were developed by Zhu et al.; the 3D printed porous composites displayed comparable compressive strength of 390 MPa both out-plane and in-plane, suggesting reduced anisotropy of macro-mechanical properties, which is different to other porous ceramics [440].

7.4.3 Carbon Nanotubes

Carbon nanotubes (CNTs) are seamless cylinders of one or more layers of graphene (denoted single-wall, SWCNTs, or multi-wall, MWCNTs) [472, 473]. A great deal of effort has been made to develop CMCs with CNTs as reinforcement via conventional processing [474, 475]. However, such composites made by AM are still at an early stage. More recently, Liu et al. reported robocasting to make CNTs/Al_2O_3 composites; following debinding and sintering, the composites exhibited higher electrical conductivity than that of pure alumina [443]. Qi et al. printed CNT-coated PA11/$BaTiO_3$ composites with an SLS process; the addition of CNTs into the powder improved the laser absorption, endowing the composite powders with better sinterability. SLS-printed ternary composites exhibited a high dielectric permittivity of 16.2 at a frequency of 1 kHz, while, the piezoelectric strain coefficient d_{33} increased from 1.7 to 2.1 pC/N [476].

7.4.4 Graphene

Graphene, a 2D nano-scale material, has remarkable mechanical properties, excellent optical properties and superior electrical and thermal properties, making it a potentially good reinforcement for producing multifunctional and structural ceramic matrix composites for a wide range of technical applications, including biomedical electronic, automotive, energy storage, defense and aerospace applications [477, 478]. The properties of composites reinforced by graphene are directly dependent upon the dispersion of the graphene in the ceramic matrix, and the retention of the graphene structure at high sintering temperatures [479, 480]. The main two types of AM technologies, selective laser sintering and robocasting, have been successfully used with graphene-based CMCs (Fig. 7.21). These routes are briefly introduced below.

Azhari et al. reported using an SLS process to fabricate graphene oxide (GO) reinforced hydroxyapatite; the printed GO/HP composites exhibited improved compressive mechanical strength of 0.4 wt.% GO [448]. In addition, Jin et al. used SLS to fabricate functional polyamide11/$BaTiO_3$/graphene piezoelectric nanocomposite with enhanced dielectric and piezoelectric properties thanks to its special architecture and morphology. Robocasting has been proven effective for the fabrication of ceramic matrix composites [481]. Román-Manso et al. initially formulated homogeneous graphene-SiC pseudoplastic inks, using DIW to shape cellular structures of graphene/ceramic composites with electrical conductivity of 611 S m^{-1}

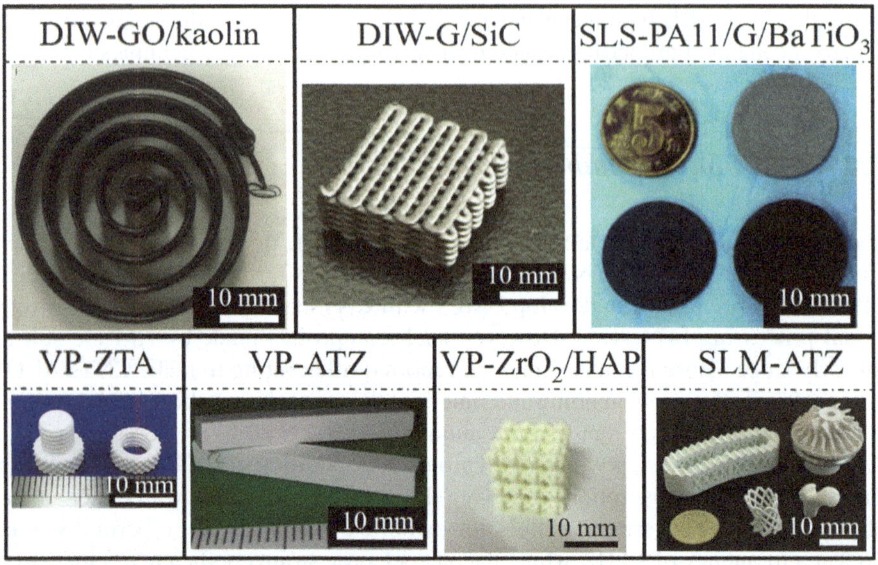

Fig. 7.21 Representative graphene and particulates reinforced ceramic composites made by different AM technologies [444, 447, 449, 453–455, 481]

[445]. Moreover, Xiao et al. developed a novel approach to produce multifunctional graphene/silicon carbide composites via DIW. The combination of printed 3D graphene scaffold and SiC matrix offers excellent electrical conductivity due to its great number of conductive channels [444]. Applying the same routes, Román-Manso et al. fabricated 3D graphene oxide (GO) lattice skeleton with infiltration of liquid organic-polysilazane, subsequently pyrolyzed, to form conductive PDCs composites [482].

7.4.5 Particulates

Second-phase particulates, such as SiC, TiC and WC particles, are popular candidates for producing metal matrix composites based on additive manufacturing technologies, which lead to defect-free metal parts [483]. However, zirconia particles (ZrO_{2p}) or alumina particles (Al_2O_{3p}) are more common as reinforcement to enhance the mechanical properties of technical ceramics. Indeed, particulates play a number of roles in composites, which can deflect cracks out of their paths, obstruct crack propagation, and promote the densification process [484, 485]. However, the particulate-reinforced composite still fails in a catastrophic failure manner, which contrasts with the failure of fiber-reinforced composites.

In the current study, great attempts have been to produce ZrO_{2p}/CMC via AM, as shown in Fig. 7.21. Zheng et al. used zirconia particles to enhance alumina with a weight ratio of alumina/zirconia 3:1. The composites exhibited an enhanced relative density (99.4%), 3 PB flexural strength (516.7 MPa) and indentation fracture toughness (7.76 MPa $m^{1/2}$) compared with pure alumina [449]. In addition, Wu et al. mixed 20 wt.% zirconia powder by total ceramic weight to alumina matrix powder and then prepared a homogeneously mixed slurry for the production of ZTA. A Vickers hardness of 17.76 GPa, flexural strength of 530.25 MPa, and fracture toughness of 5.72 MPa $m^{1/2}$ were obtained; these properties are comparable to those obtained by conventional ceramic processing [452]. With the same 3D printing method, ZTA composites with 15 vol.% zirconia content were successfully prepared by Xing et al. [450]. The highest flexural strength of 575 ± 87 MPa, fracture toughness of 7.4 ± 1.02 MPa $m^{1/2}$, hardness of 19.20 ± 0.89 GPa, and relative density of up to 99.4% were achieved. Stanciuc et al. used robocasting 3D printing technology to fabricate ZTA scaffolds, along with inks with a weight ratio of 84:16. They demonstrated the potential advance to produce multi-scale topography components for biomedical applications [486].

Alumina-reinforced ceramic matrix composites were well-printed via additive manufacturing, as shown in Fig. 7.21. Wu et al. proposed an integrated approach combining VP and liquid precursor infiltration to fabricate alumina toughened zirconia (ATZ) composites. As-fabricated ATZ composites with 3.9 wt.% alumina exhibited the highest relative density of 98.11%, hardness of 12.65 ± 0.24 GPa, and fracture toughness of 6.42 ± 0.33 MPa $m^{1/2}$; the enhanced mechanical properties could be attributed to the crack deflection and energy absorption [456]. Furthermore, Schwarzer et al. printed functionally graded ATZ components for implant applications [455]. The relative density and strength of sintering ATZ composites exceeded 99% and 500 MPa, respectively. Similar ATZ work was developed by Borlaf et al., who fabricated pure zirconia and ATZ composites via DLP process and evaluated their mechanical properties; the highest relative density and 4 PB flexural strength could reach ~99% and ~800 MPa, respectively [19]. In addition to SLA, Verga et al. attempted SLS to print ATZ parts with a flexural strength of 31 MPa, proposing a novel way to enhance the laser-materials interaction based on the pyrolysis of the granulates binder, reaching the conclusion that the improved absorption significantly promotes the processing of ATZ using a CW Nd-YAG laser [454]. Cappi et al. studied the potential of direct inkjet printing to produce complex-shaped Si_3N_4/$MoSi_2$ multi-layered components using Si_3N_4 and $MoSi_2$ suspensions, respectively [348].

So far, a large range of advanced ceramics have been successfully fabricated via various AM technologies, as detailed in Sects. 7.3 and 7.4, which provides extensive ceramic knowledge for AM specialists to select suitable ceramic materials for a specific application. In addition, they may consider applying their AM insight to explore more technical ceramic materials, although they first need to understand the advantages and limitations associated with processing and shaping high performance ceramics better. In terms of the ceramic matrix composites, the fabrication of continuous fiber-reinforced CMCs with complex architectures is not easily realized through current AM technologies. Thus, the development of a novel AM method or combining other tools, is urgent because the mechanical properties of continuous fiber-reinforced CMCs are better than those of chopped fiber-reinforced composites.

7.5 Summary and Outlook

Additive manufacturing has significant potential to become a game-changer in the ceramics forming industry by offering a significant opportunity, such as unparalleled design intricacy, integration of structure and function, materials/structural grading, mold-free fabrication, cost-effective low volume production, time-saving prototype manufacturing and production of intractable materials. Numerous cases exist which demonstrate the benefits of ceramic AM. However, the latter has proven to be significantly more challenging than that of polymers or metals, which creates barriers to its industrial adoption.

7.5.1 Benefits of Ceramic AM

7.5.1.1 Embracing Complexity

AM technologies, such as VP, binder jetting, selective laser sintering, and disrupting the conventional design rules, provide enhanced design freedom for ceramic engineers. They increase the flexibility with regard to the design of components and enable innovative and maybe more effective structural designs, which cannot be produced using subtractive or formative manufacturing methods. Beyond machine and mold driven considerations, ceramic engineers are able to design highly complex-shaped ceramic products. Generally speaking, AM enables a shift from manufacturing-derived design towards design-derived manufacturing.

Case study—ceramic heat exchanger: conventional heat exchangers are mostly made by injection molding or slip casting, thus, suggesting a low heat transfer surface and simple shape. AM has provided a method by which the exchanger can be redesigned and engineered to improve its heat transfer capability, creating a more complicated structure with an increased transfer surface.

7.5.1.2 Rapid Prototype Manufacturing

Traditionally, the time between product design and full-scale production has been long and slow. In particular, ceramic shaping via injection molding is very expensive. Now, AM is a new tool; companies can shorten the time between idea and customer delivery and cost due to the mold-free/tool-free manufacturing process. Consequently, designers can provide multiple design iterations that can be fabricated with rapid response. Speeding up time to market is a key benefit of 3D printing.

Case study—design iteration of ceramic blood pump: the researchers at the Vienna University of Technology were currently developing a ceramic blood pump, with the design having been subjected to more than 15 alterations. Ceramic 3D printing offers huge advantages in terms of saving cost and lead times in new product iterations.

7.5.1.3 Multifunctionality

Ceramic AM, such as robocasting or hybrid AM technology, can be used to effectively create components made of several materials in a single run, such as ceramic-ceramic, ceramic–metal and ceramic-polymer. Multi-materials printing enables the speedy manufacture of substrate materials with an integrated functional layer (conductive traces, resistors, sensors), which are capable of realizing the integration of structure and function with reduced process time, part numbers and lower cost fabrication. For example, by means of 3D printing, a complex-shaped ceramic substrate combined with an even more complex metal housing, with customized mechanical, electrical, optical, and thermal functionalities can be obtained. Particularly notable are the special requirements of ceramic substrates with respect to sintering temperature and shrinkage match.

Case study—LTCC circuits: electronic circuits are often exposed to special loads. LTCC multilayer ceramic technology offers unique solutions for demanding applications in harsh environments. Conventionally, the manufacturing of LTCC multilayer circuits, as shown in Fig. 7.22a, is costly and time consuming as several processes are required such as micromachining, patterning, stacking, and laminating. AM provides an efficient solution to fabricate intricate multilayer circuits embedded in low temperature co-fired ceramic (LTCC) with the integration of structure and function, which saves on lead times and costs, as well as assembling errors [487].

7.5.1.4 Productivity of Intractable Materials

AM provides the new possibility of making objects out of hard ceramic materials that would require more machining steps using conventional machining, especially in the case of ultra-hard ceramics such as WC, B_4C and SiC. Conventionally, the fabrication of ceramics with extremely high hardness leads to post-process machining that is time-and cost-intensive. AM can help reduce the need for machining, which can account for more than 70% of manufacturing costs.

Case study—Silicon Carbide seal ring: SiC is an inorganic material with high hardness, strength, wear resistance and extreme brittleness, making SiC machining difficult and costly through a conventional powder metallurgy route. As shown in Fig. 7.22b, conventionally, silicon carbide seal ring is fabricated through dry pressing, sintering, grinding, and polishing. To ensure a high yield, it is necessary to strictly control the grinding rate and tooling conditions during the grinding and polishing process owing to ultra-high hardness. However, 3D printing removes the post machining steps in the fabrication route of SiC seal ring, thus leading to a time/cost saving process and ensuring a high yield. In addition, the complex shaped SiC component can be formed via AM.

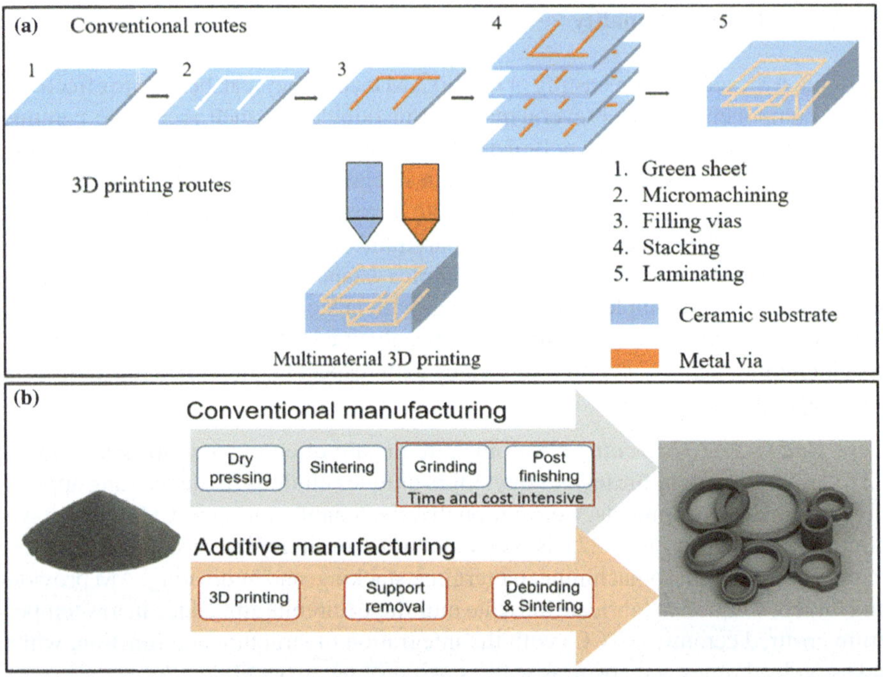

Fig. 7.22 The fabrication of **a** LTCC multilayer circuits and **b** SiC seal rings

7.5.1.5 High Materials Efficiency

Being a near-net-shape manufacturing process, unlike subtractive manufacturing in which waste materials are removed to reveal a ceramic product, AM only creates what is needed for the ceramic components, with minimal support structure. The layer-by-layer shaping manner of AM helps reduce material waste and thus saves costs.

Case study—zirconia dental prosthesis: restorative dentistry has significantly metamorphosed over recent years. Currently, computer-aided design (CAD) and computer-aided manufacturing (CAM) technologies are the main tools to achieve a consistent and high-quality dental prosthesis. However, the milling process can produce a waste of approximately 70% of the commercial zirconia disk, generating a significant economic loss. So, CAD/CAM are not environmentally friendly during the milling process.

7.5.2 Challenges/Limitations and Potential Solutions of Ceramic AM

7.5.2.1 Limited Material Selection

There is no doubt that VP, including SLA and DLP, is the most popular ceramic additive manufacturing technology, but its variants are dependent on the properties of the ceramic materials, such as materials color, refractive index, size, and shape of the powder. For instance, non-oxide ceramics (e.g., SiC and B_4C) may prove significantly more difficult to fabricate by VP processes. In addition, the shape of refractory ceramics (e.g., WC) may be hindered by SLM or LENS. Thus, ceramists and AM specialists should pay more attention to understanding the science underpinning these AM technologies and materials properties, as well as making more efforts to extend the range of ceramics that can be produced using AM.

7.5.2.2 Size Limitation

So far, small-sized ceramic components with acceptable properties have already been shaped and a number of commercial ceramic printers exist on the market. However, large ceramic parts (e.g., few hundred mm to a few meters) that are crack-free are yet to be produced. In terms of direct ceramic AM technology, such as SLM and LENS, large parts tend to produce cracks due to thermal stresses as a result of the poor thermal shock resistance and the low toughness of ceramics. Moreover, the binder decomposition process of VP, DIW, SLS and BJ limits the maximum cross-sectional area of the parts as the binder must be able to escape from the bulk. This creates inherent limitations in terms of part size. Additionally, the machine build volume also limits the printing size. Fortunately, more ceramic printer manufacturers are focusing on the building of large-size setups. For example, the French company 3DCeram is undoubtedly one of the historical players in the market of ceramic AM. Its latest machine, the C3600 Ultimate, was presented at Formnext 2019 and is aiming for mass production; it has the largest build volume of $600 \times 600 \times 300$ mm. The German manufactured Voxeljet VX1000 is best known for its 3D printing capabilities; with a volume of $1000 \times 600 \times 500$ mm, it can create complex ceramic parts via a binder jetting process.

7.5.2.3 Cost-Intensive and Time-Consuming

Currently, the most fundamental barrier to AM adoption and exploitation is the associated high costs, which comprise the equipment, materials and process costs. Further to this, in comparison to the AM of metals and polymers, due to the immature state of the ceramic AM market, the material and machine suppliers are also in their infancy, with very little competition, resulting in expensive commercial offerings.

Indeed, the tool-free nature of AM indicates that substantial lead time and cost savings can be made in the generation of advanced ceramic components, especially in the case of small-scale series production because the time and costs for the production of the tool can be saved. However, for large-scale series, conventional manufacturing methods, like injection molding, are more cost and time saving than AM. At present, the production of ceramic components via AM is still in a small-scale stage; mass production through AM is impossible or very challenging.

Furthermore, most AM processes require post-treatment for the removal of organic components and a densification step to prepare the dense parts, and they basically require more time than that of conventional processes. Typically, debinding time depends on the binder content and geometry of the part. Therefore, researchers should make more efforts to decrease binder content or develop a fast binder removal process to save debinding time. For example, XJet developed nanoparticle jetting technologies with solvent-based suspensions to print green body, following a simple and relatively short overnight sintering process to obtain the final component, significantly shortening the lead time. Additionally, direct AM processes are able to produce ceramic parts without post-thermal processing, thus leading to a short time, although, these technologies currently tend to form cracks during the printing process.

7.5.2.4 Low Reliability

Although the physicochemical and mechanical properties of ceramic components produced by AM are in many cases comparable with their conventionally processed counterparts, properties can vary with process parameters and locally within a part. There remain several limitations that have so far prevented the widespread adoption of AM by industrial organizations. As the technology develops, they turn to asking for not only exceptional properties, but also high production reliability. Yet, the AM manufactured ceramic components remain low reliability. Due to easily induced defects such as porosity, cracks and inhomogeneity during part fabrication, AM printed ceramics exhibit inferior mechanical properties and low reliability. It is therefore a major task to reduce flaws in AM processed ceramics through making more efforts on feedstock preparation, process optimization, post-processing and real-time monitoring. In addition, the control of shrinkage plays a vital role in production reliability in dimension accuracy; the presence of shrinkage makes dimensional and topological control significantly more difficult. Therefore, it is essential that research efforts focus not only on improving the performance reliability, but also on enhancing the dimensional accuracy reliability under a more systematic discussion of the reasons behind possible inconsistent manufacturing quality.

7.5.2.5 Limitation of Ceramic Matrix Composites

Ceramic matrix composites are a growing research area and are being utilized for an increasingly wide range of key industry sectors (e.g., aerospace, defense, energy, medical, automotive and electronic) due to their exceptional mechanical and physical properties. AM offers a great potential to fabricate complex shaped CMC without the use of tool or mold. However, it is very challenging to obtain final CMCs, as mentioned in Sect. 7.4. Currently, the printed green body of composites obtained via an indirect AM route requires binder removal in air flow at a temperature of ~600 °C, in which some reinforcements, such as graphene and CNTs, can be subject to low degradation temperature. More recently, research has been focused on the use of high temperature fillers, including SiC fibers or whiskers, to avoid high temperature degradation issues. Conventionally, continuous fiber-reinforced CMCs have demonstrated excellent toughening results, although CMC manufacturing is impossible or more difficult with current AM technologies. Therefore, it is highly recommended to explore the fabrication of CMC reinforced by continuous fiber via AM.

7.5.3 Future Perspective

Today, ceramic AM offers the ceramics industry a new forming tool for making ceramic parts. The emergence of AM represents a huge potential to not only transform the way the ceramic industry conducts prototyping, but also how it can extend to new applications. However, AM is not a threat to the conventional ceramics industry; it is not recommended that AM should be compared with conventional forming technologies in terms of cost, lead time and performance, because AM is not intended to replace traditional manufacturing technology. Additive fabrication will push the current limits of conventional manufacturing, thus allowing the ceramic industry to move towards opening up new markets. With the steady improvement of the performance and reliability of additively manufactured ceramic components, as well as reduced lead times, and costs, interest in ceramic AM is growing rapidly and industrial adoption is slowly increasing.

Acknowledgements The authors would like to thank to Hongqiao Qu, Xiaoteng Chen and Yue Wang of the Southern University of Science and Technology for their kind assistance.

References

1. Shigeyuhi S (2013) Advanced ceramics materials, applications, processing, and properties. Handb Adv Ceram
2. Guo JK, Li J, Kou HM (2017) Advanced ceramic materials. Mod Inorg Synth Chem Second Ed
3. Halloran JW (2016) Ceramic stereolithography: additive manufacturing for ceramics by photopolymerization. Annu Rev Mater Res 46:19–40
4. Schwentenwein M, Homa J (2014) How can CIM benefit from additive manufacturing? CFI Ceram. Forum Int

5. Moritz T, Maleksaeedi S (2018) Ceramic components. Addit Manuf Mater Process Quantif Appl 105–161 (Elsevier Inc)
6. Chartrain NA, Williams CB, Whittington AR (2018) A review on fabricating tissue scaffolds using vat photopolymerization. Acta Biomater 74:90–111
7. Chen Z, Li Z, Li J, Liu C, Lao C, Fu Y, Liu C, Li Y, Wang P, He Y (2019) 3D printing of ceramics: a review. J Eur Ceram Soc 39:661–687
8. Melchels FPW, Feijen J, Grijpma DW (2010) A review on stereolithography and its applications in biomedical engineering. Biomater 31:6121–6130
9. Lee MP, Cooper GJT, Hinkley T, Gibson GM, Padgett MJ, Cronin L (2015) Development of a 3D printer using scanning projection stereolithography. Sci Rep 5:9875
10. Truxova V, Safka J, Seidl M, Kovalenko I, Volesky L, Ackermann M (2020) Ceramic 3d printing: comparison of SLA and DLP technologies. MM Sci J 3905–3911
11. Zhou X, Hou Y, Lin J (2015) A review on the processing accuracy of two-photon polymerization. AIP Adv 5:030701
12. Cui H, Hensleigh R, Yao D, Maurya D, Kumar P, Kang MG, Priya S, Zheng X (Rayne) (2019) Three-dimensional printing of piezoelectric materials with designed anisotropy and directional response. Nat Mater 18:234–241
13. Bomze D, Schwentenwein M, Schweiger J, Russmüller G, Ioannidis A (2019) 3D-Printing of high-strength and bioresorbable ceramics for dental and maxillofacial surgery applications–the LCM process. Technol Insights 7:38–43
14. Wang Z, Huang C, Wang J, Zou B (2019) Development of a novel aqueous hydroxyapatite suspension for stereolithography applied to bone tissue engineering. Ceram Int 45:3902–3909
15. Sun J, Binner J, Bai J (2018) Effect of surface treatment on the dispersion of nano zirconia particles in non-aqueous suspensions for stereolithography. J Eur Ceram Soc 39:1660–1667
16. Bae CJ, Ramachandran A, Chung K, Park S (2017) Ceramic stereolithography: additive manufacturing for 3D complex ceramic structures. J Korean Ceram Soc 54:470–477
17. Ligon SC, Liska R, Stampfl J, Gurr M, Mülhaupt R (2017) Polymers for 3D printing and customized additive manufacturing. Chem Rev 117:10212–10292
18. Zhou W, Li D, Wang H (2010) A novel aqueous ceramic suspension for ceramic stereolithography. Rapid Prototyp J 16:29–35
19. Borlaf M, Szubra N, Serra-Capdevila A, Kubiak WW, Graule T (2020) Fabrication of ZrO_2 and ATZ materials via UV-LCM-DLP additive manufacturing technology. J Eur Ceram Soc 40:1574–1581
20. Schmidt J, Altun AA, Schwentenwein M, Colombo P (2019) Complex mullite structures fabricated via digital light processing of a preceramic polysiloxane with active alumina fillers. J Eur Ceram Soc 39:1336–1343
21. Zanchetta E, Cattaldo M, Franchin G, Schwentenwein M, Homa J, Brusatin G, Colombo P (2016) Stereolithography of SiOC ceramic microcomponents. Adv Mater 28:370–376
22. Jana P, Santoliquido O, Ortona A, Colombo P, Sorarù GD (2018) Polymer-derived SiCN cellular structures from replica of 3D printed lattices. J Am Ceram Soc 101:2732–2738
23. Li S, Duan W, Zhao T, Han W, Wang L, Dou R, Wang G (2018) The fabrication of SiBCN ceramic components from preceramic polymers by digital light processing (DLP) 3D printing technology. J Eur Ceram Soc 38:4597–4603
24. Brinckmann SA, Patra N, Yao J, Ware TH, Frick CP, Fertig RS (2018) Stereolithography of SiOC polymer-derived ceramics filled with SiC micronwhiskers. Adv Eng Mater 20:1800593
25. Lee JH, Prud'homme RK, Aksay IA (2001) Cure depth in photopolymerization: experiments and theory. J Mater Res 16:3536–3544
26. Griffith ML, Halloran JW (1997) Scattering of ultraviolet radiation in turbid suspensions. J Appl Phys 81:2538
27. Mitteramskogler G, Gmeiner R, Felzmann R, Gruber S, Hofstetter C, Stampfl J, Ebert J, Wachter W, Laubersheimer J (2014) Light curing strategies for lithography-based additive manufacturing of customized ceramics. Addit Manuf 1:110–118
28. Gentry SP, Halloran JW (2015) Light scattering in absorbing ceramic suspensions: effect on the width and depth of photopolymerized features. J Eur Ceram Soc 35:1895–1904

29. Borlaf M, Serra-Capdevila A, Colominas C, Graule T (2019) Development of UV-curable ZrO_2 slurries for additive manufacturing (LCM-DLP) technology. J Eur Ceram Soc 39:3797–3803
30. Elsayed H, Schmidt J, Bernardo E, Colombo P (2019) Comparative analysis of wollastonite-diopside glass-ceramic structures fabricated via stereo-lithography. Adv Eng Mater 21:1801160
31. Li H, Liu Y, Liu Y, Wang J, Zeng Q, Hu K, Lu Z (2020) Influence of vacuum debinding temperature on microstructure and mechanical properties of three-dimensional-printed alumina via stereolithography. 3D Print Addit Manuf 7:8–18
32. Sun J, Binner J, Bai J (2020) 3D printing of zirconia via digital light processing: optimization of slurry and debinding process. J Eur Ceram Soc 40:5837–5844
33. Cesarano III J, Calvert DP (2000) Freeforming objects with low binder slurry. United States Pat
34. Cesarano J, King BH, Denham HB, Cesarano III J, Denham HB (1998) Recent developments in robocasting of ceramics and multimaterial deposition. Proc Solid Free Fabr Symp
35. Pierin G, Grotta C, Colombo P, Mattevi C (2016) Direct Ink Writing of micrometric SiOC ceramic structures using a preceramic polymer. J Eur Ceram Soc 36:1589–1594
36. Rueschhoff L, Costakis W, Michie M, Youngblood J, Trice R (2016) Additive manufacturing of dense ceramic parts via direct ink writing of aqueous alumina suspensions. Int J Appl Ceram Technol 13:821–830
37. Martínez-Vázquez FJ, Perera FH, Miranda P, Pajares A, Guiberteau F (2010) Improving the compressive strength of bioceramic robocast scaffolds by polymer infiltration. Acta Biomater 6:4361–4368
38. Franchin G, Wahl L, Colombo P (2017) Direct ink writing of ceramic matrix composite structures. J Am Ceram Soc 100:4397–4401
39. Fiocco L, Elsayed H, Badocco D, Pastore P, Bellucci D, Cannillo V, Detsch R, Boccaccini AR, Bernardo E (2017) Direct ink writing of silica-bonded calcite scaffolds from preceramic polymers and fillers. Biofabrication 9:025012
40. Lewis JA, Smay JE, Stuecker J, Cesarano J (2006) Direct ink writing of three-dimensional ceramic structures. J Am Ceram Soc 89:3599–3609
41. Yang L, Tang S, Li G, Qian L, Mei J, Jiang W, Fan Z (2019) Layered extrusion forming of complex ceramic structures using starch as removable support. Ceram Int 45:21843–21850
42. Martínez-Vázquez FJ, Pajares A, Miranda P (2018) A simple graphite-based support material for robocasting of ceramic parts. J Eur Ceram Soc 38:2247–2250
43. Cai K, Román-Manso B, Smay JE, Zhou J, Osendi MI, Belmonte M, Miranzo P (2012) Geometrically complex silicon carbide structures fabricated by robocasting. J Am Ceram Soc 95:2660–2666
44. Bocanegra-Bernal MH, Matovic B (2009) Dense and near-net-shape fabrication of Si_3N_4 ceramics. Mater Sci Eng A 500:130–149
45. Travitzky N, Bonet A, Dermeik B, Fey T, Filbert-Demut I, Schlier L, Schlordt T, Greil P (2014) Additive manufacturing of ceramic-based materials. Adv Eng Mater 16:729–754
46. Costakis WJ, Rueschhoff LM, Diaz-Cano AI, Youngblood JP, Trice RW (2016) Additive manufacturing of boron carbide via continuous filament direct ink writing of aqueous ceramic suspensions. J Eur Ceram Soc 36:3249–3256
47. M'Barki A, Bocquet L, Stevenson A (2017) Linking rheology and printability for dense and strong ceramics by direct ink writing. Sci Rep 7:6017
48. Yu T, Zhang Z, Liu Q, Kuliiev R, Orlovskaya N, Wu D (2020) Extrusion-based additive manufacturing of yttria-partially-stabilized zirconia ceramics. Ceram Int 46:5020–5027
49. Nan B, Olhero S, Pinho R, Vilarinho PM, Button TW, Ferreira JMF (2019) Direct ink writing of macroporous lead-free piezoelectric $Ba_{0.85}Ca_{0.15}Zr_{0.1}Ti_{0.9}O_3$. J Am Ceram Soc 102:3191–3203
50. Elsayed H, Chmielarz A, Potoczek M, Fey T, Colombo P (2019) Direct ink writing of three dimensional Ti_2AlC porous structures. Addit Manuf 28:365–372

51. Zhao S, Xiao W, Rahaman MN, O'Brien D, Seitz-Sampson JW, Sonny Bal B (2017) Robocasting of silicon nitride with controllable shape and architecture for biomedical applications. Int J Appl Ceram Technol 14:117–127
52. Gibson I, Rosen D, Stucker B, Khorasani M (2021) Binder jetting. Addit Manuf Technol 237–252
53. Sachs E, Cima M, Williams P, Brancazio D, Cornie J (1992) Three dimensional printing: rapid tooling and prototypes directly from a CAD model. J Manuf Sci Eng Trans ASME 114:481–488
54. Diaz-Moreno CA, Rodarte C, Ambriz S, Bermudez D, Roberson D, Terrazas C, Espalin D, Ferguson R, Shafirovich E, Lin Y, Wicker RB (2018) Binder jetting of high temperature and thermally conductive (Aluminum Nitride) ceramic. Solid Free Fabr Symp 143–159
55. Ziaee M, Crane NB (2019) Binder jetting: a review of process, materials, and methods. Addit Manuf 28:781–801
56. Kunchala P, Kappagantula K (2018) 3D printing high density ceramics using binder jetting with nanoparticle densifiers. Mater Des 155:443–450
57. Gonzalez JA, Mireles J, Lin Y, Wicker RB (2016) Characterization of ceramic components fabricated using binder jetting additive manufacturing technology. Ceram Int 42:10559–10564
58. Asadi-Eydivand M, Solati-Hashjin M, Farzad A, Abu Osman NA (2016) Effect of technical parameters on porous structure and strength of 3D printed calcium sulfate prototypes. Robot Comput Integr Manuf 37:57–67
59. Bai Y, Wagner G, Williams CB (2015) Effect of bimodal powder mixture on powder packing density and sintered density in binder jetting of metals. Solid Free Fabr Proc
60. Zheng J, Carlson WB, Reed JS (1995) The packing density of binary powder mixtures. J Eur Ceram Soc 15:479–483
61. Cima MJ, Oliveira M, Wang H (2001) Slurry-based 3DP and fine ceramic components. Proc Solid Free Fabr
62. Sachs EM, Cima MJ, Caradonna MA, Grau J, Serdy JG, Saxton PC, Uhland SA, Moon J (2003) Jetting layers of powder and the formation of fine powder beds thereby. US Pat 6 596 224
63. Du W, Ren X, Ma C, Pei Z (2017) Binder jetting additive manufacturing of ceramics: a literature review. ASME Int Mech Eng Congr Expo Proc
64. Tancred DC, McCormack BAO, Carr AJ (1998) A synthetic bone implant macroscopically identical to cancellous bone. Biomater 19:2303–2311
65. Fielding GA, Bandyopadhyay A, Bose S (2012) Effects of silica and zinc oxide doping on mechanical and biological properties of 3D printed tricalcium phosphate tissue engineering scaffolds. Dent Mater 28:113–122
66. Lv X, Ye F, Cheng L, Fan S, Liu Y (2019) Binder jetting of ceramics: powders, binders, printing parameters, equipment, and post-treatment. Ceram Int 45:12609–12624
67. Carrijo MMM, Lorenz H, Filbert-Demut I, De Oliveira Barra GM, Hotza D, Yin X, Greil P, Travitzky N (2016) Fabrication of Ti_3SiC_2-based composites via three-dimensional printing: Influence of processing on the final properties. Ceram Int 42:9557–9564
68. Blazdell PF, Evans JRG, Edirisinghe MJ, Shaw P, Binstead MJ (1995) The computer aided manufacture of ceramics using multilayer jet printing. J Mater Sci Lett 14:1562–1565
69. Derby B (2015) Additive mmanufacture of ceramics components by inkjet printing. Engineering 1:113–123
70. Chen Z, Brandon N (2016) Inkjet printing and nanoindentation of porous alumina multilayers. Ceram Int 42:8316–8324
71. Ebert J, Özkol E, Zeichner A, Uibel K, Weiss Ö, Koops U, Telle R, Fischer H (2009) Direct inkjet printing of dental prostheses made of zirconia. J Dent Res 88
72. Noshchenko O, Kuscer D, Mocioiu OC, Zaharescu M, Bele M, Malič B (2014) Effect of milling time and pH on the dispersibility of lead zirconate titanate in aqueous media for inkjet printing. J Eur Ceram Soc 34:297–305
73. Mott M, Evans JRG (2004) Solid freeforming of silicon carbide by inkjet printing using a polymeric precursor. J Am Ceram Soc 84:307–313

74. Cappi B, Özkol E, Ebert J, Telle R (2008) Direct inkjet printing of Si_3N_4: characterization of ink, green bodies and microstructure. J Eur Ceram Soc 28:2625–2628
75. Gingter P, Wätjen AM, Kramer M, Telle R (2015) Functionally graded ceramic structures by direct thermal inkjet printing. J Ceram Sci Technol 6:119–124
76. Levi H (2018) Additive manufacturing in technical ceramics. Interceram-Int Ceram Rev 67:12–13
77. Derby B (2011) Inkjet printing ceramics: From drops to solid. J Eur Ceram Soc 31:2543–2550
78. Fromm JE (1984) Numerical calculation of the fluid dynamics of drop-on-demand jets. IBM J Res Dev 28:322–333
79. Bergeron V, Bonn D, Martin JY, Vovelle L (2000) Controlling droplet deposition with polymer additives. Nature 405:772–775
80. Noguera R, Lejeune M, Chartier T (2005) 3D fine scale ceramic components formed by ink-jet prototyping process. J Eur Ceram Soc 25:2055–2059
81. Wang T, Derby B (2005) Ink-jet printing and sintering of PZT. J Am Ceram Soc 88:2053–2058
82. Ramakrishnan N, Rajesh PK, Ponnambalam P, Prakasan K (2005) Studies on preparation of ceramic inks and simulation of drop formation and spread in direct ceramic inkjet printing. J Mater Process Technol 169:372–381
83. Wätjen AM, Gingter P, Kramer M, Telle R (2014) Novel prospects and possibilities in additive manufacturing of ceramics by means of direct inkjet printing. Adv Mech Eng 6
84. Deckard C (1997) Method for producing parts by selective sintering. US Patent 5,597,589
85. Deckard CR., Beaman JJ., Darrah JF (1992) Method and apparatus for producing parts by selective sintering. US Patent 4,863,538
86. Bremen S, Meiners W, Diatlov A (2012) Selective laser melting: a manufacturing technology for the future? Laser Tech J 9:33–38
87. Singh R, Gupta A, Tripathi O, Srivastava S, Singh B, Awasthi A, Rajput SK, Sonia P, Singhal P, Saxena KK (2020) Powder bed fusion process in additive manufacturing: an overview. Mater Today Proc 26:3058–3070
88. Chen AN, Wu JM, Liu K, Chen JY, Xiao H, Chen P, Li CH, Shi YS (2018) High-performance ceramic parts with complex shape prepared by selective laser sintering: a review. Adv Appl Ceram 117:100–117
89. Ferrage L, Bertrand G, Lenormand P, Grossin D, Ben-Nissan B (2017) A review of the additive manufacturing (3DP) of bioceramics: alumina, zirconia (PSZ) and hydroxyapatite. J Aust Ceram Soc 53:11–20
90. Shahzad K, Deckers J, Kruth JP, Vleugels J (2013) Additive manufacturing of alumina parts by indirect selective laser sintering and post processing. J Mater Process Technol 213:1484–1494
91. Bose S, Ke D, Sahasrabudhe H, Bandyopadhyay A (2018) Additive manufacturing of biomaterials. Prog Mater Sci 93:45–111
92. Liu Q, Danlos Y, Song B, Zhang B, Yin S, Liao H (2015) Effect of high-temperature preheating on the selective laser melting of yttria-stabilized zirconia ceramic. J Mater Process Technol 222:61–74
93. Goodridge RD, Wood DJ, Ohtsuki C, Dalgarno KW (2007) Biological evaluation of an apatite-mullite glass-ceramic produced via selective laser sintering. Acta Biomater 3:221–231
94. Waetjen AM, Polsakiewicz DA, Kuhl I, Telle R, Fischer H (2009) Slurry deposition by airbrush for selective laser sintering of ceramic components. J Eur Ceram Soc 29:1–6
95. Sing SL, Yeong WY, Wiria FE, Tay BY, Zhao Z, Zhao L, Tian Z, Yang S (2017) Direct selective laser sintering and melting of ceramics: a review. Rapid Prototyp J 23:611–623
96. Tang HH, Chiu ML, Yen HC (2011) Slurry-based selective laser sintering of polymer-coated ceramic powders to fabricate high strength alumina parts. J Eur Ceram Soc 31:1383–1388
97. Doyle H, Lohfeld S, McHugh P (2015) Evaluating the effect of increasing ceramic content on the mechanical properties, material microstructure and degradation of selective laser sintered polycaprolactone/β-tricalcium phosphate materials. Med Eng Phys 37:767–776
98. Shishkovsky I, Scherbakov V (2012) Selective laser sintering of biopolymers with micro and nano ceramic additives for medicine. Phys Procedia 39:491–499

99. Marin E, Zanocco M, Boschetto F, Santini M, Zhu W, Adachi T, Ohgitani E, McEntire BJ, Bal BS, Pezzotti G (2020) Silicon nitride laser cladding: a feasible technique to improve the biological response of zirconia. Mater Des 191:108649
100. Mudge RP, Wald NR (2007) Laser engineered net shaping advances additive manufacturing and repair. Weld J 44–48
101. Liu K, Shi Y, Li C, Hao L, Liu J, Wei Q (2014) Indirect selective laser sintering of epoxy resin-Al_2O_3 ceramic powders combined with cold isostatic pressing. Ceram Int 40:7099–7106
102. Liu J, Zhang B, Yan C, Shi Y (2010) The effect of processing parameters on characteristics of selective laser sintering dental glass-ceramic powder. Rapid Prototyp J 16:138–145
103. Chiu K, Chen K, Wang Y, Lin F, Huang J (2020) Formability of Fe-doped bioglass scaffold via selective laser sintering. Ceram Int 88:1–8
104. Song S, Gao Z, Lu B, Bao C, Zheng B, Wang L (2020) Performance optimization of complicated structural SiC/Si composite ceramics prepared by selective laser sintering. Ceram Int 46:568–575
105. Yves-Christian H, Jan W, Wilhelm M, Konrad W, Reinhart P (2010) Net shaped high performance oxide ceramic parts by selective laser melting. Phys Procedia 5:587–594
106. Di QY, Wu JM, Chen AN, Chen P, Yang Y, Liu RZ, Chen G, Chen S, Shi YS, Li CH (2020) Balling phenomenon and cracks in alumina ceramics prepared by direct selective laser melting assisted with pressure treatment. Ceram Int 46:13854–13861
107. Zhao X, Gu D, Ma C, Xi L, Zhang H (2019) Microstructure characteristics and its formation mechanism of selective laser melting SiC reinforced Al-based composites. Vacuum 160:189–196
108. Li W, Soshi M (2019) Modeling analysis of grain morphologies in directed energy deposition (DED) coating with different laser scanning patterns. Mater Lett 251:8–12
109. Weidong Z, Qibin L, Min Z, Xudong W (2008) Biocompatibility of a functionally graded bioceramic coating made by wide-band laser cladding. J Biomed Mater Res Part A 87A:429–433
110. Gill D, Smugeresky J, Atwood C (2006) Laser engineered net Shaping^{TM}(LENS®) for the repair and modification of NWC metal components. Prod.Sandia.Gov
111. Bernard SA, Balla VK, Bose S, Bandyopadhyay A (2010) Direct laser processing of bulk lead zirconate titanate ceramics. Mater Sci Eng B Solid-State Mater Adv Technol 172:85–88
112. Hu Y, Ning F, Cong W, Li Y, Wang X, Wang H (2018) Ultrasonic vibration-assisted laser engineering net shaping of ZrO_2-Al_2O_3 bulk parts: Effects on crack suppression, microstructure, and mechanical properties. Ceram Int 44:2752–2760
113. Li Q, Lei Y, Fu H (2014) Laser cladding in-situ NbC particle reinforced Fe-based composite coatings with rare earth oxide addition. Surf Coatings Technol 239:102–107
114. Ma G, Yan S, Wu D, Miao Q, Liu M, Niu F (2017) Microstructure evolution and mechanical properties of ultrasonic assisted laser clad yttria stabilized zirconia coating. Ceram Int 43:9622–9629
115. Niu FY, Wu DJ, Yan S, Ma GY, Zhang B (2017) Process optimization for suppressing cracks in laser engineered net shaping of Al_2O_3 ceramics. JOM 69:557–562
116. Niu F, Wu D, Zhou S, Ma G (2014) Power prediction for laser engineered net shaping of Al_2O_3 ceramic parts. J Eur Ceram Soc 34:3811–3817
117. Ma G, Yan S, Niu F, Zhang Y, Wu D (2017) Microstructure and mechanical properties of solid Al_2O_3-ZrO_2 (Y_2O_3) eutectics prepared by laser engineered net shaping. J Laser Appl 29:022305
118. Roy M, Vamsi Krishna B, Bandyopadhyay A, Bose S (2008) Laser processing of bioactive tricalcium phosphate coating on titanium for load-bearing implants. Acta Biomater 4:324–333
119. Fan Z, Zhao Y, Tan Q, Mo N, Zhang MX, Lu M, Huang H (2019) Nanostructured Al_2O_3-YAG-ZrO_2 ternary eutectic components prepared by laser engineered net shaping. Acta Mater 170:24–37
120. Jian X, Meek TT, Han Q (2006) Refinement of eutectic silicon phase of aluminum A356 alloy using high-intensity ultrasonic vibration. Scr Mater 54:893–896

121. Jian X, Xu H, Meek TT, Han Q (2005) Effect of power ultrasound on solidification of aluminum A356 alloy. Mater Lett 59:190–193
122. Laugier P, Haïat G (2011) Introduction to the physics of ultrasound. Bone Quant Ultrasound 29–45
123. Cong W, Ning F (2017) A fundamental investigation on ultrasonic vibration-assisted laser engineered net shaping of stainless steel. Int J Mach Tools Manuf 121:61–69
124. Yan S, Wu D, Niu F, Huang Y, Liu N, Ma G (2018) Effect of ultrasonic power on forming quality of nano-sized Al_2O_3-ZrO_2 eutectic ceramic via laser engineered net shaping (LENS). Ceram Int 44:1120–1126
125. Zhu Z, Dhokia VG, Nassehi A, Newman ST (2013) A review of hybrid manufacturing processes—state of the art and future perspectives. Int J Comput Integr Manuf 26:596–615
126. Bandyopadhyay A, Heer B (2018) Additive manufacturing of multi-material structures. Mater Sci Eng R Rep 129:1–16
127. Chong L, Ramakrishna S, Singh S (2018) A review of digital manufacturing-based hybrid additive manufacturing processes. Int J Adv Manuf Technol 95:2281–2300
128. Merklein M, Junker D, Schaub A, Neubauer F (2016) Hybrid additive manufacturing technologies—an analysis regarding potentials and applications. Phys Procedia 83:549–559
129. Li J, Wasley T, Nguyen TT, Ta VD, Shephard JD, Stringer J, Smith P, Esenturk E, Connaughton C, Kay R (2016) Hybrid additive manufacturing of 3D electronic systems. J Micromechanics Microengineering 26:105005
130. Karunakaran KP, Suryakumar S, Pushpa V, Akula S (2010) Low cost integration of additive and subtractive processes for hybrid layered manufacturing. Robot Comput Integr Manuf 26:490–499
131. Gong M, Meng Y, Zhang S, Zhang Y, Zeng X, Gao M (2020) Laser-arc hybrid additive manufacturing of stainless steel with beam oscillation. Addit Manuf 33:101180
132. Li L, Haghighi A, Yang Y (2018) A novel 6-axis hybrid additive-subtractive manufacturing process: design and case studies. J Manuf Process 33:150–160
133. Chen N, Frank M (2019) Process planning for hybrid additive and subtractive manufacturing to integrate machining and directed energy deposition. Procedia Manuf 34:205–213
134. Silva M, Felismina R, Mateus A, Parreira P, Malça C (2017) Application of a hybrid additive manufacturing methodology to produce a metal/polymer customized dental implant. Procedia Manuf 12:150–155
135. Hertle S, Kleffel T, Wörz A, Drummer D (2020) Production of polymer-metal hybrids using extrusion-based additive manufacturing and electrochemically treated aluminum. Addit Manuf 33:101135
136. Sheydaeian E, Sarikhani K, Chen P, Toyserkani E (2017) Material process development for the fabrication of heterogeneous titanium structures with selective pore morphology by a hybrid additive manufacturing process. Mater Des 135:142–150
137. Liravi F, Toyserkani E (2018) A hybrid additive manufacturing method for the fabrication of silicone bio-structures: 3D printing optimization and surface characterization. Mater Des 138:46–61
138. Taboas JM, Maddox RD, Krebsbach PH, Hollister SJ (2003) Indirect solid free form fabrication of local and global porous, biomimetic and composite 3D polymer-ceramic scaffolds. Biomater 24:181–194
139. Raynaud J, Pateloup V, Bernard M, Gourdonnaud D, Passerieux D, Cros D, Madrangeas V, Chartier T (2020) Hybridization of additive manufacturing processes to build ceramic/metal parts: Example of LTCC. J Eur Ceram Soc 40:759–767
140. Ayode Otitoju T, Ugochukwu Okoye P, Chen G, Li Y, Onyeka Okoye M, Li S (2020) Advanced ceramic components: materials, fabrication, and applications. J Ind Eng Chem 85:34–65
141. Golla BR, Mukhopadhyay A, Basu B, Thimmappa SK (2020) Review on ultra-high temperature boride ceramics. Prog Mater Sci 111:100651
142. Tetard D, Tixier C, Faure C, Chabas E, Aneziris G (2008) Mechanical properties and oxidation behaviour of electroconductive ceramic composites. In: 10th international conference of the European ceramic society, pp 1315–1320

143. Munro RG (2000) Material properties of titanium diboride. J Res Natl Inst Stand Technol 105:709–720
144. Colombo P, Mera G, Riedel R, Sorarù GD (2010) Polymer-derived ceramics: 40 years of research and innovation in advanced ceramics. J Am Ceram Soc 93:1805–1837
145. Hasegawa Y, Iimura M, Yajima S (1980) Synthesis of continuous silicon carbide fibre-Part 2 conversion of polycarbosilane fibre into silicon carbide fibres. J Mater Sci 15:720–728
146. Sasikumar PVW, Blugan G, Casati N, Kakkava E, Panusa G, Psaltis D, Kuebler J (2018) Polymer derived silicon oxycarbide ceramic monoliths: microstructure development and associated materials properties. Ceram Int 44:20961–20967
147. Moraes KV, Interrante LV (2003) Processing, fracture toughness, and vickers hardness of allylhydridopolycarbosilane-derived silicon carbide. J Am Ceram Soc 86:342–346
148. Nishimura T, Haug R, Bill J, Thurn G, Aldinger F (1998) Mechanical and thermal properties of Si-C-N material from polyvinylsilazane. J Mater Sci 33:5237–5241
149. Yang ZH, Jia DC, Duan XM, Sun KN, Zhou Y (2011) Effect of Si/C ratio and their content on the microstructure and properties of SiBCN ceramics prepared by spark plasma sintering techniques. Mater Sci Eng A 528:1944–1948
150. Bravo-Leon A, Morikawa Y, Kawahara M, Mayo MJ (2002) Fracture toughness of nanocrystalline tetragonal zirconia with low yttria content. Acta Mater 50:4555–4562
151. Heuer C, Storti E, Graule T, Aneziris CG (2020) Electrospinning of Y_2O_3- and MgO-stabilized zirconia nanofibers and characterization of the evolving phase composition and morphology during thermal treatment. Ceram Int 46:12001–12008
152. Wäsche R, Sato K, Brandt G, Schmid T, Sasaki S, Woydt M (2019) Wear behaviour of MgO stabilized zirconia in hot steam environment up to 400 °C. Wear 426–427:428–432
153. Miller SP, Dunlap BI, Fleischer AS (2013) Effects of dopant clustering in cubic zirconia stabilized by yttria and scandia from molecular dynamics. Solid State Ionics 253:130–136
154. Zeng Z, Liu Y, Qian F, Guo J, Wu M (2019) Role of a low level of La_2O_3 dopant on the tetragonal-to-monoclinic phase transformation of ceria-yttria co-stabilized zirconia. J Eur Ceram Soc 39:4338–4346
155. Suk MO, Park JH (2009) Corrosion behaviors of zirconia refractory by CaO-SiO_2-MgO-CaF_2 slag. J Am Ceram Soc 92:717–723
156. Sibil A, Douillard T, Cayron C, Godin N, R'mili M, Fantozzi G (2011) Microcracking of high zirconia refractories after t→m phase transition during cooling: an EBSD study. J Eur Ceram Soc 31:1525–1531
157. Yung MM, Holmgreen EM, Ozkan US (2007) Cobalt-based catalysts supported on titania and zirconia for the oxidation of nitric oxide to nitrogen dioxide. J Catal 247:356–367
158. Miura N, Nakatou M, Zhuiykov S (2003) Impedancemetric gas sensor based on zirconia solid electrolyte and oxide sensing electrode for detecting total NOx at high temperature. Sensors Actuators, B Chem 93:221–228
159. Özkurt Z, Kazazoğlu E (2011) Zirconia dental implants: a literature review. J Oral Implantol 37:367–376
160. Denry I, Kelly JR (2008) State of the art of zirconia for dental applications. Dent Mater 24:299–307
161. Afzal A (2014) Implantable zirconia bioceramics for bone repair and replacement: a chronological review. Mater Express 4:1–12
162. Goodenough JB, Huang YH (2007) Alternative anode materials for solid oxide fuel cells. J Power Sources 173:1–10
163. Schefold J, Brisse A, Zahid M (2009) Electronic conduction of yttria-stabilized zirconia electrolyte in solid oxide cells operated in high temperature water electrolysis. J Electrochem Soc 156
164. Baiamonte L, Marra F, Pulci G, Tirillò J, Sarasini F, Bartuli C, Valente T (2015) High temperature mechanical characterization of plasma-sprayed zirconia-yttria from conventional and nanostructured powders. Surf Coatings Technol 277:289–298
165. Leib EW, Pasquarelli RM, Do Rosário JJ et al (2015) Yttria-stabilized zirconia microspheres: novel building blocks for high-temperature photonics. J Mater Chem C 4:62–74

166. Kontonasaki E, Giasimakopoulos P, Rigos AE (2020) Strength and aging resistance of monolithic zirconia: an update to current knowledge. Jpn Dent Sci Rev 56:1–23
167. Chevalier J, Gremillard L, Deville S (2007) Low-temperature degradation of zirconia and implications for biomedical implants. Annu Rev Mater Res 37:1–32
168. Lughi V, Sergo V (2010) Low temperature degradation aging of zirconia: a critical review of the relevant aspects in dentistry. Dent Mater 26:807–820
169. Tredici IG, Sebastiani M, Massimi F, Bemporad E, Resmini A, Merlati G, Anselmi-Tamburini U (2016) Low temperature degradation resistant nanostructured yttria-stabilized zirconia for dental applications. Ceram Int 42:8190–8197
170. Zhang F, Batuk M, Hadermann J, Manfredi G, Mariën A, Vanmeensel K, Inokoshi M, Van Meerbeek B, Naert I, Vleugels J (2016) Effect of cation dopant radius on the hydrothermal stability of tetragonal zirconia: grain boundary segregation and oxygen vacancy annihilation. Acta Mater 106:48–58
171. Inokoshi M, Zhang F, De Munck J, Minakuchi S, Naert I, Vleugels J, Van Meerbeek B, Vanmeensel K (2014) Influence of sintering conditions on low-temperature degradation of dental zirconia. Dent Mater 30:669–678
172. Fu X, Zou B, Xing H, Li L, Li Y, Wang X (2019) Effect of printing strategies on forming accuracy and mechanical properties of ZrO_2 parts fabricated by SLA technology. Ceram Int 45:17630–17637
173. Liao J, Chen H, Luo H, Wang X, Zhou K, Zhang D (2017) Direct ink writing of zirconia three-dimensional structures. J Mater Chem C 24:5867–5871
174. Peng E, Wei X, Garbe U, Yu D, Edouard B, Liu A, Ding J (2018) Robocasting of dense yttria-stabilized zirconia structures. J Mater Sci 53:247–273
175. Revilla-León M, Methani MM, Morton D, Zandinejad A (2020) Internal and marginal discrepancies associated with stereolithography (SLA) additively manufactured zirconia crowns. J Prosthet Dent 124:730–737
176. Wang L, Liu X, Wang G, Tang W, Li S, Duan W, Dou R (2020) Partially stabilized zirconia moulds fabricated by stereolithographic additive manufacturing via digital light processing. Mater Sci Eng A 770:138537
177. Jang KJ, Kang JH, Fisher JG, Park SW (2019) Effect of the volume fraction of zirconia suspensions on the microstructure and physical properties of products produced by additive manufacturing. Dent Mater 35:97–106
178. He R, Liu W, Wu Z, An D, Huang M, Wu H, Jiang Q, Ji X, Wu S, Xie Z (2018) Fabrication of complex-shaped zirconia ceramic parts via a DLP- stereolithography-based 3D printing method. Ceram Int 44:3412–3416
179. Wang JC, Dommati H (2018) Fabrication of zirconia ceramic parts by using solvent-based slurry stereolithography and sintering. Int J Adv Manuf Technol 44:3412–3416
180. Xing H, Zou B, Li S, Fu X (2017) Study on surface quality, precision and mechanical properties of 3D printed ZrO_2 ceramic components by laser scanning stereolithography. Ceram Int 43:16340–16347
181. Jiang CP, Hsu HJ, Lee SY (2014) Development of mask-less projection slurry stereolithography for the fabrication of zirconia dental coping. Int J Precis Eng Manuf 15:2413–2419
182. Ghazanfari A, Li W, Leu MC, Watts JL, Hilmas GE (2017) Additive manufacturing and mechanical characterization of high density fully stabilized zirconia. Ceram Int 43:6082–6088
183. Shao H, Zhao D, Lin T, He J, Wu J (2017) 3D gel-printing of zirconia ceramic parts. Ceram Int 43:13938–13942
184. Zhao H, Ye C, Fan Z, Shi Y (2016) 3D printing of ZrO_2 ceramic using nano-zirconia suspension as a binder. In: 4th international conference on sensors, measurement and intelligent materials, pp 654–657
185. Özkol E, Zhang W, Ebert J, Telle R (2012) Potentials of the "direct inkjet printing" method for manufacturing 3Y-TZP based dental restorations. J Eur Ceram Soc 32:2193–2201
186. Chen F, Wu J-M, Wu H-Q, Chen Y, Li C-H, Shi Y-S (2018) Microstructure and mechanical properties of 3Y-TZP dental ceramics fabricated by selective laser sintering combined with cold isostatic pressing. Int J Light Mater Manuf 1:239–245

187. Venkatesh S, Rahul SH, Balasubramanian K (2020) Inkjet printing yttria stabilized zirconia coatings on porous and nonporous substrates. Ceram Int 46:3994–3999
188. Oh Y, Bharambe V, Mummareddy B, Martin J, McKnight J, Abraham MA, Walker JM, Rogers K, Conner B, Cortes P, Macdonald E, Adams J (2019) Microwave dielectric properties of zirconia fabricated using NanoParticle Jetting^{TM}. Addit Manuf 27:586–594
189. Shahzad K, Deckers J, Zhang Z, Kruth JP, Vleugels J (2014) Additive manufacturing of zirconia parts by indirect selective laser sintering. J Eur Ceram Soc 34:81–89
190. Li W, Ghazanfari A, McMillen D, Leu MC, Hilmas GE, Watts J (2018) Characterization of zirconia specimens fabricated by ceramic on-demand extrusion. Ceram Int 44:12245–12252
191. Yan S, Huang Y, Zhao D, Niu F, Ma G, Wu D (2019) 3D printing of nano-scale Al_2O_3-ZrO_2 eutectic ceramic: principle analysis and process optimization of pores. Addit Manuf 28:120–126
192. Ferrage L, Bertrand G, Lenormand P (2018) Dense yttria-stabilized zirconia obtained by direct selective laser sintering. Addit Manuf 21:472–478
193. Balla VK, Bandyopadhyay PP, Bose S, Bandyopadhyay A (2007) Compositionally graded yttria-stabilized zirconia coating on stainless steel using laser engineered net shaping (LENS^{TM}). Scr Mater 57:861–864
194. Fan Z, Zhao Y, Lu M, Huang H (2019) Yttria stabilized zirconia (YSZ) thin wall structures fabricated using laser engineered net shaping (LENS). Int J Adv Manuf Technol 105:4491–4498
195. Lee SY (2004) Sintering behavior and mechanical properties of injection-molded zirconia powder. Ceram Int 30:579–584
196. Guazzato M, Albakry M, Ringer SP, Swain MV (2004) Strength, fracture toughness and microstructure of a selection of all-ceramic materials. Part II Zirconia-based dental ceramics. Dent Mater 20:449–456
197. Fan Z, Lu M, Huang H (2018) Selective laser melting of alumina: A single track study. Ceram Int 44:9484–9493
198. Galante R, Figueiredo-Pina CG, Serro AP (2019) Additive manufacturing of ceramics for dental applications: a review. Dent Mater 35:825–846
199. Kuntz M (2006) Validation of a new high performance alumina matrix composite for use in total joint replacement. Semin Arthroplasty 17:141–145
200. Gorokh G, Mozalev A, Solovei D, Khatko V, Llobet E, Correig X (2006) Anodic formation of low-aspect-ratio porous alumina films for metal-oxide sensor application. Electrochim Acta 52:1771–1780
201. Pradyumna R, Baig MAH (2012) Ceramic cores for turbine blades : a tooling perspective. Int J Mech Ind Eng 2:1–7
202. Huang S, Ye C, Zhao H, Fan Z (2019) Additive manufacturing of thin alumina ceramic cores using binder-jetting. Addit Manuf 29:100802
203. Lee G, Carrillo M, McKittrick J, Martin DG, Olevsky EA (2020) Fabrication of ceramic bone scaffolds by solvent jetting 3D printing and sintering: towards load-bearing applications. Addit Manuf 33:101107
204. Xu C, Ai X, Huang C (2001) Fabrication and performance of an advanced ceramic tool material. Wear 249:503–508
205. Zhou M, Liu W, Wu H, Song X, Chen Y, Cheng L, He F, Chen S, Wu S (2016) Preparation of a defect-free alumina cutting tool via additive manufacturing based on stereolithography–optimization of the drying and debinding processes. Ceram Int 42:11598–11602
206. Yang Z, Yin Z, Wang D, Wang H, Song H, Zhao Z, Zhang G, Qing G, Wu H, Jin H (2020) Effects of ternary sintering aids and sintering parameters on properties of alumina ceramics based on orthogonal test method. Mater Chem Phys 241:122453
207. Kobayashi M, Goto T, Aoba T, Miura H (2019) Three-dimensional structure of high-performance heat insulator produced with micro and nano particle alumina. Mater Charact 154:424–436
208. Sarkar S, Bandyopadhyay S, Larbot A, Cerneaux S (2012) New clay-alumina porous capillary supports for filtration application. J Memb Sci 392–393:130–136

209. Hu Y, Li D, Dai W, Wang M, Wang H, Sun K (2012) Fabrication of three-dimensional electromagnetic band-gap structure with alumina based on stereolithography and gelcasting. J Manuf Syst 31:22–25
210. Zeng Q, Yang C, Tang D, Li J, Feng Z, Liu J, Guan K (2019) Additive manufacturing alumina components with lattice structures by digital light processing technique. J Mater Sci Technol 35:2751–2755
211. Schwentenwein M, Schneider P, Homa J (2014) Lithography-based ceramic manufacturing: a novel technique for additive manufacturing of high-performance ceramics. Adv Sci Technol 88:60–64
212. Denham HB, Cesarano J, King BH, Calvert P (1998) Mechanical behavior of robocast alumina. Solid Free Fabr Proc 589–596
213. Huang T, Mason MS, Zhao X, Hilmas GE, Leu MC (2009) Aqueous based freeze form extrusion fabrication of alumina components. Rapid Prototyp J 15:88–95
214. Zhang D, Jonhson W, Herng TS, Ang YQ, Yang L, Tan SC, Peng E, He H, Ding J (2020) A 3D-printing method of fabrication for metals, ceramics, and multi-materials using a universal self-curable technique for robocasting. Mater Horizons 7:1083–1090
215. Curto H, Thuault A, Jean F, Violier M, Dupont V, Hornez JC, Leriche A (2020) Coupling additive manufacturing and microwave sintering: a fast processing route of alumina ceramics. J Eur Ceram Soc 40:2548–2554
216. Xing Z, Liu W, Chen Y, Li W (2018) Effect of plasticizer on the fabrication and properties of alumina ceramic by stereolithography-based additive manufacturing. Ceram Int 44:19939–19944
217. Schwentenwein M, Homa J (2015) Additive manufacturing of dense alumina ceramics. Int J Appl Ceram Technol 12:1–7
218. Wang JC (2013) A novel fabrication method of high strength alumina ceramic parts based on solvent-based slurry stereolithography and sintering. Int J Precis Eng Manuf 14:485–491
219. Orlovská M, Chlup Z, Bača JM, Kitzmantel M (2020) Fracture and mechanical properties of lightweight alumina ceramics prepared by fused filament fabrication. J Eur Ceram Soc 40:4837–4843
220. Huang SJ, Ye CS (2020) Preparation and performance of binder jetting porous alumina ceramic. IOP Conf Ser Mater Sci Eng 770:012057
221. Liu ZH, Nolte JJ, Packard JI, Hilmas G, Dogan F, Leu MC (2007) Selective laser sintering of high-density alumina ceramic parts. Proc 35th Int MATADOR 2007 Conf 351–354
222. Balla VK, Bose S, Bandyopadhyay A (2008) Processing of bulk alumina ceramics using laser engineered net shaping. Int J Appl Ceram Technol 5:234–242
223. Niu F, Wu D, Lu F, Liu G, Ma G, Jia Z (2018) Microstructure and macro properties of Al_2O_3 ceramics prepared by laser engineered net shaping. Ceram Int 44:14303–14310
224. Wang Z, Shi Y, He W, Liu K, Zhang Y (2015) Compound process of selective laser processed alumina parts densified by cold isostatic pressing and solid state sintering: experiments, full process simulation and parameter optimization. Ceram Int 41:3245–3253
225. Li Y, Hu Y, Cong W, Zhi L, Guo Z (2017) Additive manufacturing of alumina using laser engineered net shaping: effects of deposition variables. Ceram Int 43:7768–7775
226. Yang L, Zeng X, Zhang Y (2019) 3D printing of alumina ceramic parts by heat-induced solidification with carrageenan. Mater Lett 255:126564
227. Arthur CL, Pawliszyn J (1990) Solid phase microextraction with thermal desorption using fused silica optical fibers. Anal Chem 62:2145–2148
228. Chen QZ, Thompson ID, Boccaccini AR (2006) 45S5 Bioglass®-derived glass-ceramic scaffolds for bone tissue engineering. Biomater 27:2414–2425
229. Hill R (1996) An alternative view of the degradation of bioglass. J Mater Sci Lett 15:1122–1125
230. Bae CJ, Kim D, Halloran JW (2019) Mechanical and kinetic studies on the refractory fused silica of integrally cored ceramic mold fabricated by additive manufacturing. J Eur Ceram Soc 39:618–623

231. Chen X, Han J, Zhang W, Zhang L, Liu C (2019) Silica-based ceramic core for aviation applications: Facile pore filling and flexural strength improvement. Int J Appl Ceram Technol 16:2128–2189
232. Wereszczak AA, Breder K, Ferber MK, Kirkland TP, Payzant EA, Rawn CJ, Krug E, Larocco CL, Pietras RA, Karakus M (2002) Dimensional changes and creep of silica core ceramics used in investment casting of superalloys. J Mater Sci 37:4235–4245
233. Kawachi M (1990) Silica waveguides on silicon and their application to integrated-optic components. Opt Quantum Electron 22:391–416
234. Fanderlik I (1991) Silica glass and its application. Glas Sci Technol
235. Tang L, Cheng J (2013) Nonporous silica nanoparticles for nanomedicine application. Nano Today 8:290–312
236. Rahaman MN, Day DE, Sonny Bal B, Fu Q, Jung SB, Bonewald LF, Tomsia AP (2011) Bioactive glass in tissue engineering. Acta Biomater 7:2355–2373
237. Jones JR, Ehrenfried LM, Hench LL (2006) Optimising bioactive glass scaffolds for bone tissue engineering. Biomater 27:964–973
238. Chen JF, Ding HM, Wang JX, Shao L (2004) Preparation and characterization of porous hollow silica nanoparticles for drug delivery application. Biomater 25:723–727
239. Vivero-Escoto JL, Slowing II, Lin VSY, Trewyn BG (2010) Mesoporous silica nanoparticles for intracellular controlled drug delivery. Small 6:1952–1967
240. Nomura M, Ono K, Gopalakrishnan S, Sugawara T, Nakao SI (2005) Preparation of a stable silica membrane by a counter diffusion chemical vapor deposition method. J Memb Sci 251:151–158
241. Park HB, Lee YM (2003) Pyrolytic carbon-silica membrane: a promising membrane material for improved gas separation. J Memb Sci 213:263–272
242. Sun C, Zhang X (2002) The influences of the material properties on ceramic micro-stereolithography. Sens Actuators, A Phys 101:364–370
243. Chartier T, Badev A, Abouliatim Y, Lebaudy P, Lecamp L (2012) Stereolithography process: Influence of the rheology of silica suspensions and of the medium on polymerization kinetics-cured depth and width. J Eur Ceram Soc 32:1625–1634
244. Griffith ML, Halloran JW (2005) Freeform fabrication of ceramics via stereolithography. J Am Ceram Soc 79:2601–2608
245. Kotz F, Arnold K, Bauer W, Schild D, Keller N, Sachsenheimer K, Nargang TM, Richter C, Helmer D, Rapp BE (2017) Three-dimensional printing of transparent fused silica glass. Nature 544:337–339
246. Liu C, Qian B, Liu X, Tong L, Qiu J (2018) Additive manufacturing of silica glass using laser stereolithography with a top-down approach and fast debinding. RSC Adv 8:16344–16348
247. Wang YY, Li L, Wang ZY, Liu FT, Zhao JH, Zhang PP, Lu C (2018) Fabrication of dense silica ceramics through a stereo lithography-based additive manufacturing. Solid State Phenom 456–462
248. Cai P, Guo L, Wang H, Li J, Li J, Qiu Y, Zhang Q, Lue Q (2020) Effects of slurry mixing methods and solid loading on 3D printed silica glass parts based on DLP stereolithography. Ceram Int 46:16833–16841
249. Esposito Corcione C, Montagna F, Greco A, Licciulli A, Maffezzoli A (2006) Free form fabrication of silica moulds for aluminium casting by stereolithography. Rapid Prototyp J 12:184–188
250. Eqtesadi S, Motealleh A, Pajares A, Miranda P (2015) Effect of milling media on processing and performance of 13–93 bioactive glass scaffolds fabricated by robocasting. Ceram Int 41:1379–1389
251. Wang J, Bai P, Zhang Z, Li Y (2011) Processing and characterization of core-shell PA12/silica composites produced by selective laser sintering. Adv Mater Res 160–162:756–761
252. Nommeots-Nomm A, Lee PD, Jones JR (2018) Direct ink writing of highly bioactive glasses. J Eur Ceram Soc 38:837–844
253. Klein J, Stern M, Franchin G et al (2015) Additive manufacturing of optically transparent glass. 3D Print Addit Manuf 2:92–105

254. Snelling D, Williams CB, Druschitz AP (2014) A comparison of binder burnout and mechanical characteristics of printed and chemically bonded sand molds. Solid Free Fabr Symp
255. Snelling DA, Williams CB, Druschitz AP (2019) Mechanical and material properties of castings produced via 3D printed molds. Addit Manuf 27:199–207
256. Dzik P, Veselý M, Kete M, Pavlica E, Štangar UL, Neumann-Spallart M (2015) Properties and application perspective of hybrid titania-silica patterns fabricated by inkjet printing. ACS Appl Mater Interfaces 7:16177–16190
257. Li Y, Dahhan O, Filipe CDM, Brennan JD, Pelton RH (2018) Optimizing piezoelectric inkjet printing of silica sols for biosensor production. J Sol-Gel Sci Technol 87:657–664
258. Bram M, Dornseiffer J, Hoffmann J, Van Gestel T, Meulenberg WA, Stöver D (2015) Inkjet printing of microporous silica gas separation membranes. J Am Ceram Soc 98:2388–2394
259. Chang S, Li L, Lu L, Fuh JYH (2017) Selective laser sintering of porous silica enabled by carbon additive. Materials 10:1313
260. Hostetler JM, Goldstein JT, Urbas AM, Gutierrez RE, Bender TE, Wojnar CS, Kinzel EC (2016) Selective laser sintering of low density, low coefficient of thermal expansion silica parts. Solid Free Fabr 2016 Proc 27th Annu Int Solid Free Fabr 2016 Proc 26th Annu Int Solid Free Fabr Symp—An Addit Manuf Conf Rev Pap 978–988
261. Gan MX, Wong CH (2017) Properties of selective laser melted spodumene glass-ceramic. J Eur Ceram Soc 37:4147–4154
262. Heer B, Bandyopadhyay A (2018) Silica coated titanium using laser engineered net shaping for enhanced wear resistance. Addit Manuf 23:303–311
263. Elsayed H, Picicco M, Dasan A, Kraxner J, Galusek D, Bernardo E (2019) Glass powders and reactive silicone binder: interactions and application to additive manufacturing of bioactive glass-ceramic scaffolds. Ceram Int 45:13740–13746
264. Hashimoto K, Irie H, Fujishima A (2007) TiO_2 photocatalysis: a historical overview and future prospects. Assoc Asia Pacific Phys Soc Bull 44:8269
265. Khan S, Ul-Islam M, Khattak WA, Ullah MW, Park JK (2015) Bacterial cellulose-titanium dioxide nanocomposites: nanostructural characteristics, antibacterial mechanism, and biocompatibility. Cellulose 22:565–579
266. Wang A, Chen SL, Dong P, Zhou Z (2011) Preparation of photonic crystal heterostructures composed of two TiO_2 inverse opal films with different filling factors. Synth Met 161:504–507
267. Weiwu C, Kirihara S, Miyamoto Y (2007) Fabrication of three-dimensional micro photonic crystals of resin-incorporating TiO_2 particles and their terahertz wave properties. J Am Ceram Soc 90:92–96
268. Lee CY, Taylor AC, Beirne S, Wallace GG (2017) 3D-printed conical arrays of TiO_2 electrodes for enhanced photoelectrochemical water splitting. Adv Energy Mater 7:1701060
269. Kaneko M, Kirihara S (2010) Millimeter wave control using TiO_2 photonic crystal with diamond structure fabricated by micro-stereolithography. Mater Sci Forum 632:293–298
270. Mendez-Arriaga F, de la Calleja E, Ruiz-Huerta L, Caballero-Ruiz A, Almanza R (2019) TiO_2 3D structures for environmental purposes by additive manufacturing: photoactivity test and reuse. Mater Sci Semicond Process 100:35–41
271. Aleni AH, Kretzschmar N, Jansson A, Ituarte IF, St-Pierre L (2020) 3D printing of dense and porous TiO_2 structures. Ceram Int 46:16725–16732
272. Elkoro A, Soler L, Llorca J, Casanova I (2019) 3D printed microstructured Au/TiO_2 catalyst for hydrogen photoproduction. Appl Mater Today 16:265–272
273. Hosseini Zori M, Soleimani-Gorgani A (2012) Ink-jet printing of micro-emulsion TiO_2 nano-particles ink on the surface of glass. J Eur Ceram Soc 32:4271–4277
274. Trovarelli A (1996) Catalytic properties of ceria and CeO_2-containing materials. Catal Rev-Sci Eng 38:439–520
275. Lucentini I, Serrano I, Soler L, Divins NJ, Llorca J (2020) Ammonia decomposition over 3D-printed CeO_2 structures loaded with Ni. Appl Catal A Gen 591:117382
276. Xu Y, Farandos N, Rosa M, Zielke P, Esposito V, Vang Hendriksen P, Jensen SH, Li T, Kelsall G, Kiebach R (2018) Continuous hydrothermal flow synthesis of Gd-doped CeO_2 (GDC) nanoparticles for inkjet printing of SOFC electrolytes. Int J Appl Ceram Technol 15:315–327

277. Senesky DG, Cheng KB, Pisano AP, Jamshidi B (2009) Harsh environment silicon carbide sensors for health and performance monitoring of aerospace systems: a review. IEEE Sens J 9:1472–1478
278. Rajan N, Mehregany M, Zorman CA, Stefanescu S, Kicher TP (1999) Fabrication and testing of micromachined silicon carbide and nickel fuel atomizers for gas turbine engines. J Microelectromechanical Syst 8:251–257
279. Wu XL, Xiong SJ, Zhu J, Wang J, Shen JC, Chu PK (2009) Identification of surface structures on 3C-SiC nanocrystals with hydrogen and hydroxyl bonding by photoluminescence. Nano Lett 9:4053–4060
280. Saddow SE (2016) Silicon carbide materials for biomedical applications. Silicon Carbide Biotechnol A Biocompatible Semicond Adv Biomed Devices Appl Second Ed 1–5
281. Okada A (2008) Automotive and industrial applications of structural ceramics in Japan. J Eur Ceram Soc 28:1097–1104
282. Syväjärvi M, Ma Q, Jokubavicius V et al (2016) Cubic silicon carbide as a potential photovoltaic material. Sol Energy Mater Sol Cells 145:104–108
283. Yang J (2013) A harsh environment wireless pressure sensing solution utilizing high temperature electronics. Sensors (Switzerland) 13:2719–2734
284. Casady JB, Johnson RW (1996) Status of silicon carbide (SiC) as a wide-bandgap semiconductor for high-temperature applications: a review. Solid State Electron 39:1409–1422
285. Koyanagi T, Terrani K, Harrison S, Liu J, Katoh Y (2021) Additive manufacturing of silicon carbide for nuclear applications. J Nucl Mater 543:152577
286. Li S, Luo X, Zhao L, Wei C, Gao P, Wang P (2020) Crack tolerant silicon carbide ceramics prepared by liquid-phase assisted oscillatory pressure sintering. Ceram Int 46:18965–18969
287. Stephen E, Saddow, Anant A (2014) Advances in silicon carbide processing and applications. Igarss
288. Mo R, Yin X, Ye F, Liu X, Ma X, Li Q, Zhang L, Cheng L (2019) Electromagnetic wave absorption and mechanical properties of silicon carbide fibers reinforced silicon nitride matrix composites. J Eur Ceram Soc 39:743–754
289. Potluri R (2019) Mechanical properties of pineapple leaf fiber reinforced epoxy infused with silicon carbide micro particles. J Nat Fibers 16:137–151
290. Ding G, He R, Zhang K, Xie C, Wang M, Yang Y, Fang D (2019) Stereolithography-based additive manufacturing of gray-colored SiC ceramic green body. J Am Ceram Soc 102:7198–7209
291. Park HK, Shin M, Kim B, Park JW, Lee H (2018) A visible light-curable yet visible wavelength-transparent resin for stereolithography 3D printing. NPG Asia Mater 10:82–89
292. He R, Ding G, Zhang K, Li Y, Fang D (2019) Fabrication of SiC ceramic architectures using stereolithography combined with precursor infiltration and pyrolysis. Ceram Int 45:14006–14014
293. Park S, Lee DH, Ryoo HI, Lim TW, Yang DY, Kim DP (2009) Fabrication of three-dimensional SiC ceramic microstructures with near-zero shrinkage via dual crosslinking induced stereolithography. Chem Commun 28:4880–4882
294. Christian Nelson J, Vail NK, Barlow JW, Beaman JJ, Bourell DL, Marcus HL (1995) Selective laser sintering of polymer-coated silicon carbide powders. Ind Eng Chem Res 34:1641–1651
295. Liu K, Wu T, Bourell DL, Tan Y, Wang J, He M, Sun H, Shi Y, Chen J (2018) Laser additive manufacturing and homogeneous densification of complicated shape SiC ceramic parts. Ceram Int 44:21067–21075
296. Liu K, Wang J, Wu T, Sun H (2020) Effects of carbon content on microstructure and mechanical properties of SiC ceramics fabricated by SLS/RMI composite process. Ceram Int 46:22015–22023
297. Meyers S, De Leersnijder L, Vleugels J, Kruth JP (2018) Direct laser sintering of reaction bonded silicon carbide with low residual silicon content. J Eur Ceram Soc 38:3709–3717
298. Ding G, He R, Zhang K, Zhou N, Xu H (2020) Stereolithography 3D printing of SiC ceramic with potential for lightweight optical mirror. Ceram Int 46:18785–18790

299. Zhang H, Yang Y, Liu B, Huang Z (2019) The preparation of SiC-based ceramics by one novel strategy combined 3D printing technology and liquid silicon infiltration process. Ceram Int 45:10800–10804
300. Xu TT, Cheng S, Jin LZ, Zhang K, Zeng T (2020) High-temperature flexural strength of SiC ceramics prepared by additive manufacturing. Int J Appl Ceram Technol 17:438–448
301. Larson CM, Choi JJ, Gallardo PA, Henderson SW, Niemack MD, Rajagopalan G, Shepherd RF (2016) Direct ink writing of silicon carbide for microwave optics. Adv Eng Mater 18:39–45
302. Zocca A, Lima P, Diener S, Katsikis N, Günster J (2019) Additive manufacturing of SiSiC by layerwise slurry deposition and binder jetting (LSD-print). J Eur Ceram Soc 39:3527–3533
303. Du W, Singh M, Singh D (2020) Binder jetting additive manufacturing of silicon carbide ceramics: development of bimodal powder feedstocks by modeling and experimental methods. Ceram Int 46:19701–19707
304. Das M, Balla VK, Kumar TSS, Manna I (2013) Fabrication of biomedical implants using laser engineered Nnet shaping (LENS^{TM}). Trans Indian Ceram Soc 72:169–174
305. Kernan BD, Sachs EM, Oliveira MA, Cima MJ (2007) Three-dimensional printing of tungsten carbide-10 wt% cobalt using a cobalt oxide precursor. Int J Refract Met Hard Mater 25:82–94
306. Li CW, Chang KC, Yeh AC, Yeh JW, Lin SJ (2018) Microstructure characterization of cemented carbide fabricated by selective laser melting process. Int J Refract Met Hard Mater 75:225–233
307. Cramer CL, Elliott AM, Kiggans JO, Haberl B, Anderson DC (2019) Processing of complex-shaped collimators made via binder jet additive manufacturing of B_4C and pressureless melt infiltration of Al. Mater Des 180:107956
308. Davydova A, Domashenkov A, Sova A et al (2016) Selective laser melting of boron carbide particles coated by a cobalt-based metal layer. J Mater Process Technol 229:361–366
309. Sun SK, Kan YM, Zhang GJ (2011) Fabrication of nanosized tungsten carbide ceramics by reactive spark plasma sintering. J Am Ceram Soc 94:3230–3233
310. Put S, Vleugels J, Van Der Biest O (2001) Functionally graded WC-Co materials produced by electrophoretic deposition. Scr Mater 45:1139–1145
311. Padmakumar M, Guruprasath J, Achuthan P, Dinakaran D (2018) Investigation of phase structure of cobalt and its effect in WC–Co cemented carbides before and after deep cryogenic treatment. Int J Refract Met Hard Mater 74:87–92
312. Uhlmann E, Bergmann A, Gridin W (2015) Investigation on additive manufacturing of tungsten carbide-cobalt by selective laser melting. Procedia CIRP 35:8–15
313. Ku N, Pittari JJ, Kilczewski S, Kudzal A (2019) Additive manufacturing of cemented tungsten carbide with a cobalt-free alloy binder by selective laser melting for high-hardness applications. JOM 71:1535–1542
314. Benichou A, Laufer L (2016) Tungsten-carbide/Cobalt ink composition for 3D inkjet printing. United States Patent
315. Carrillo G, Keck D, Martinez-Duarte R (2019) Mechanical properties and process improvement of tungsten carbide additively manufactured with renewable biopolymers. Procedia Manuf 34:704–711
316. Enneti RK, Prough KC, Wolfe TA, Klein A, Studley N, Trasorras JL (2018) Sintering of WC-12%Co processed by binder jet 3D printing (BJ3DP) technology. Int J Refract Met Hard Mater 71:28–35
317. Enneti RK, Prough KC (2019) Wear properties of sintered WC-12%Co processed via Binder Jet 3D Printing (BJ3DP). Int J Refract Met Hard Mater 78:228–232
318. Scheithauer U, Pötschke J, Weingarten S, Schwarzer E, Vornberger A, Moritz T, Michaelis A (2017) Droplet-based additive manufacturing of hard metal components by thermoplastic 3D Printing (T3DP). J Ceram Sci Technol 8:155–160
319. Dünner P, Heuvel HJ, Hörle M (1984) Absorber materials for control rod systems of fast breeder reactors. J Nucl Mater 124:185–194
320. Liaptsis D, Cooper I, Ludford N, Gunner A, Williams M, Willis D (2010) NDT characterisation of boron carbide for ballistic applications. Non-Destructive Test Conf 2010, NDT 2010 1335:981

321. Kiani MA, Ahmadi SJ, Outokesh M, Adeli R, Mohammadi A (2017) Preparation and characteristics of epoxy/clay/B_4C nanocomposite at high concentration of boron carbide for neutron shielding application. Radiat Phys Chem 141:223–228
322. Stone MB, Siddel DH, Elliott AM, Anderson D, Abernathy DL (2017) Characterization of plastic and boron carbide additive manufactured neutron collimators. Rev Sci Instrum 88:123102
323. Domnich V, Reynaud S, Haber RA, Chhowalla M (2011) Boron carbide: structure, properties, and stability under stress. J Am Ceram Soc 94:3605–3628
324. Suri AK, Subramanian C, Sonber JK, Ch Murthy TSR (2010) Synthesis and consolidation of boron carbide: a review. Int Mater Rev 55:4–40
325. Zhang W, Yamashita S, Kita H (2019) Progress in pressureless sintering of boron carbide ceramics—a review. Adv Appl Ceram 118:222–239
326. Aramian A, Razavi SMJ, Sadeghian Z, Berto F (2020) A review of additive manufacturing of cermets. Addit Manuf 33:101130
327. Chandrasekaran S, Lu R, Landingham R, Cahill JT, Thornley L, Du Frane W, Worsley MA, Kuntz JD (2020) Additive manufacturing of graded B_4C-Al cermets with complex shapes. Mater Des 188:108516
328. Hampshire S (2007) Silicon nitride ceramics-review of structure, processing and properties. J Achiev Mater Manuf Eng 24:43–50
329. Bocanegra-Bernal MH, Matovic B (2010) Mechanical properties of silicon nitride-based ceramics and its use in structural applications at high temperatures. Mater Sci Eng A 527:1314–1338
330. Klemm H (2010) Silicon nitride for high-temperature applications. J Am Ceram Soc 93:1501–1522
331. Soboyejo WO, Obayemi JD, Annan E, Ampaw EK, Daniels L, Rahbar N (2015) Review of high temperature ceramics for aerospace applications. Adv Mater Res 1132:385–407
332. Piprek J (2007) Nitride semiconductor devices: principles and simulation
333. Yin L, Zhou X, Yu J, Wang H (2016) Preparation of high porous silicon nitride foams with ultra-thin walls and excellent mechanical performance for heat exchanger application by using a protein foaming method. Ceram Int 42:1713–1719
334. Doğan CP, Hawk JA (2001) Microstructure and abrasive wear in silicon nitride ceramics. Wear 250:256–263
335. Han W, Li Y, Chen G, Yang Q (2017) Effect of sintering additive composition on microstructure and mechanical properties of silicon nitride. Mater Sci Eng A 700:19–24
336. Bock RM, McEntire BJ, Bal BS, Rahaman MN, Boffelli M, Pezzotti G (2015) Surface modulation of silicon nitride ceramics for orthopaedic applications. Acta Biomater 26:318–330
337. Bodišová K, Kašiarová M, Domanická M, Hnatko M, Lenčéš Z, Nováková ZV, Vojtaššák J, Gromošová S, Šajgalík P (2013) Porous silicon nitride ceramics designed for bone substitute applications. Ceram Int 39:8355–8362
338. Ventura S, Narang S, Guerit P, Liu S, Twait D, Khandelwal P, Cohen E, Fish R (2000) Freeform fabrication of functional silicon nitride components by direct photo shaping. MRS Proceedings 625
339. Huang RJ, Jiang QG, Wu HD, Li YH, Liu WY, Lu XX, Wu SH (2019) Fabrication of complex shaped ceramic parts with surface-oxidized Si_3N_4 powder via digital light processing based stereolithography method. Ceram Int 45:5158–5162
340. Altun AA, Prochaska T, Konegger T, Schwentenwein M (2020) Dense, strong, and precise silicon nitride-based ceramic parts by lithography-based ceramic manufacturing. Appl Sci 10:996
341. Liu Y, Zhan L, He Y, Zhang J, Hu J, Cheng L, Wu Q, Liu S (2020) Stereolithographical fabrication of dense Si_3N_4 ceramics by slurry optimization and pressure sintering. Ceram Int 46:2063–2071
342. Sainz MA, Serena S, Belmonte M, Miranzo P, Osendi MI (2020) Protein adsorption and in vitro behavior of additively manufactured 3D-silicon nitride scaffolds intended for bone tissue engineering. Mater Sci Eng C 115:110734

343. He G, Hirschfeld D, Cesarano III J, Stuecker J (2000) Processing of silicon nitride ceramics from concentrated aqueous suspensions by robocasting. US
344. Rahaman MN, Xiao W (2018) Three-dimensional printing of Si_3N_4 bioceramics by robocasting. Ceram Eng Sci Proc 38
345. Rabinskiy L, Ripetsky A, Sitnikov S, Solyaev Y, Kahramanov R (2016) Fabrication of porous silicon nitride ceramics using binder jetting technology. IOP Conf Ser Mater Sci Eng 140:012023
346. Minasyan T, Liu L, Aghayan M, Kollo L, Kamboj N, Aydinyan S, Hussainova I (2018) A novel approach to fabricate Si_3N_4 by selective laser melting. Ceram Int 44:13689–13694
347. Wei ZH, Cheng LJ, Ma YX, Chen AN, Guo XF, Wu JM, Shi YS (2019) Direct fabrication mechanism of pre-sintered Si_3N_4 ceramic with ultra-high porosity by laser additive manufacturing. Scr Mater 173:91–95
348. Cappi B, Ebert J, Telle R (2011) Rheological properties of aqueous Si_3N_4 and $MoSi_2$ suspensions tailor-made for direct inkjet printing. J Am Ceram Soc 94:111–116
349. Lin L, Wu H, Xu Y, Lin K, Zou W, Wu S (2020) Fabrication of dense aluminum nitride ceramics via digital light processing-based stereolithography. Mater Chem Phys 249:122969
350. Jackson TB, Virkar AV, More KL, Dinwiddie RB, Cutler RA (1997) High-thermal-conductivity aluminum nitride ceramics: the effect of thermodynamic, kinetic, and microstructural factors. J Am Ceram Soc 80:1421–1435
351. Xiong C, Pernice WHP, Sun X, Schuck C, Fong KY, Tang HX (2012) Aluminum nitride as a new material for chip-scale optomechanics and nonlinear optics. New J Phys 14:095014
352. Hou Y, Zhang M, Han G, Si C, Zhao Y, Ning J (2016) A review: aluminum nitride MEMS contour-mode resonator. J Semicond 37:101001
353. Werdecker W, Aldinger F (1984) Aluminum nitride—an alternative ceramic substrate for high power applications in microcircuits. IEEE Trans Components Hybrids Manuf Technol 7:399–404
354. Dubois MA, Muralt P (1999) Properties of aluminum nitride thin films for piezoelectric transducers and microwave filter applications. Appl Phys Lett 74:3032
355. Iqbal A, Mohd-Yasin F (2018) Reactive sputtering of aluminum nitride (002) thin films for piezoelectric applications: a review. Sensors (Switzerland) 18:1797
356. Yarar E, Hrkac V, Zamponi C, Piorra A, Kienle L, Quandt E (2016) Low temperature aluminum nitride thin films for sensory applications. AIP Adv 6:075115
357. Fattahi M, Vaferi K, Vajdi M, Sadegh Moghanlou F, Sabahi Namini A, Shahedi Asl M (2020) Aluminum nitride as an alternative ceramic for fabrication of microchannel heat exchangers: a numerical study. Ceram Int 46:11647–11657
358. Ożóg P, Rutkowski P, Kata D, Graule T (2020) Ultraviolet lithography-based ceramic manufacturing (UV-LCM) of the aluminum nitride (AlN)-based photocurable dispersions. Materials 13:4219
359. Díaz-Moreno CA, Lin Y, Hurtado-Macías A, Espalin D, Terrazas CA, Murr LE, Wicker RB (2019) Binder jetting additive manufacturing of aluminum nitride components. Ceram Int 45:13620–13627
360. Parthasarathy TA, Rapp RA, Opeka M, Kerans RJ (2007) A model for the oxidation of ZrB_2, HfB_2 and TiB_2. Acta Mater 55:5999–6010
361. Vajdi M, Sadegh Moghanlou F, Ahmadi Z, Motallebzadeh A, Shahedi Asl M (2019) Thermal diffusivity and microstructure of spark plasma sintered TiB_2/SiC/Ti composite. Ceram Int 45:8333–8344
362. Gasch M, Ellerby D, Irby E, Beckman S, Gusman M, Johnson S (2004) Processing, properties and arc jet oxidation of hafnium diboride/silicon carbide ultra high temperature ceramics. J Mater Sci 39:5925–5937
363. Daw MS, Lawson JW, Bauschlicher CW (2011) Interatomic potentials for zirconium diboride and hafnium diboride. Comput Mater Sci 50:2828–2835
364. Feilden E, Glymond D, Saiz E, Vandeperre L (2019) High temperature strength of an ultra high temperature ceramic produced by additive manufacturing. Ceram Int 45:18210–18214

365. McMillen D, Li W, Leu MC, Hilmas GE, Watts J (2016) Designed extrudate for additive manufacturing of zirconium diboride by ceramic on-demand extrusion. Solid Free Fabr 2016 Proc 27th Annu Int Solid Free Fabr Symp—An Addit Manuf Conf SFF 2016
366. Leu MC, Adamek EB, Huang T, Hilmas GE, Dogan F (2008) Freeform fabrication of zirconium diboride parts using selective laser sintering. 19th Annu Int Solid Free Fabr Symp SFF 2008
367. Jüttke Y, Richter H, Voigt I, Prasad RM, Bazarjani MS, Gurlo A, Riedel R (2013) Polymer derived ceramic membranes for gas separation. Chem Eng Trans 32:1891–1896
368. Nagaiah NR, Kapat JS, An L, Chow L (2006) Novel polymer derived ceramic-high temperature heat flux sensor for gas turbine environment. J Phys Conf Ser 34:458–463
369. Colombo P (2008) Engineering porosity in polymer-derived ceramics. J Eur Ceram Soc 28:1389–1395
370. Wang K, Unger J, Torrey JD, Flinn BD, Bordia RK (2014) Corrosion resistant polymer derived ceramic composite environmental barrier coatings. J Eur Ceram Soc 34:3597–3606
371. Eckel ZC, Zhou C, Martin JH, Jacobsen AJ, Carter WB, Schaedler TA (2016) Additive manufacturing of polymer-derived ceramics. Science 351:58–62
372. Hundley JM, Eckel ZC, Schueller E, Cante K, Biesboer SM, Yahata BD, Schaedler TA (2017) Geometric characterization of additively manufactured polymer derived ceramics. Addit Manuf 18:95–102
373. Li Z, Chen Z, Liu J, Fu Y, Liu C, Wang P, Jiang M, Lao C (2020) Additive manufacturing of lightweight and high-strength polymer-derived SiOC ceramics. Virtual Phys Prototyp 15:163–177
374. Agrafiotis CC, Mavroidis I, Konstandopoulos AG, Hoffschmidt B, Stobbe P, Romero M, Fernandez-Quero V (2007) Evaluation of porous silicon carbide monolithic honeycombs as volumetric receivers/collectors of concentrated solar radiation. Sol Energy Mater Sol Cells 91:474–488
375. Fu Y, Chen Z, Xu G, Wei Y, Lao C (2019) Preparation and stereolithography 3D printing of ultralight and ultrastrong ZrOC porous ceramics. J Alloys Compd 789:867–873
376. Xiao J, Liu D, Cheng H, Jia Y, Zhou S, Zu M (2020) Carbon nanotubes as light absorbers in digital light processing three-dimensional printing of SiCN ceramics from preceramic polysilazane. Ceram Int 46:19393–19400
377. Konstantinou G, Kakkava E, Hagelüken L et al (2020) Additive micro-manufacturing of crack-free PDCs by two-photon polymerization of a single, low-shrinkage preceramic resin. Addit Manuf 35:101343
378. Brigo L, Schmidt JEM, Gandin A, Michieli N, Colombo P, Brusatin G (2018) 3D nanofabrication of SiOC ceramic structures. Adv Sci 5:1800937
379. Gyak KW, Vishwakarma NK, Hwang YH, Kim J, Yun HS, Kim DP (2019) 3D-printed monolithic SiCN ceramic microreactors from a photocurable preceramic resin for the high temperature ammonia cracking process. React Chem Eng 4:1393–1399
380. Baldwin LA, Rueschhoff LM, Deneault JR, Cissel KS, Nikolaev P, Cinibulk MK, Koerner H, Dalton MJ, Dickerson MB (2018) Synthesis of a two-component carbosilane system for the advanced manufacturing of polymer-derived ceramics. Chem Mater 30:7527–7534
381. De Hazan Y, Penner D (2017) SiC and SiOC ceramic articles produced by stereolithography of acrylate modified polycarbosilane systems. J Eur Ceram Soc 37:5205–5212
382. Huang K, Elsayed H, Franchin G, Colombo P (2020) 3D printing of polymer-derived SiOC with hierarchical and tunable porosity. Addit Manuf 36:101549
383. Wei L, Li J, Zhang S, Li B, Liu Y, Wang F, Dong S (2020) Fabrication of SiOC ceramic with cellular structure via UV-Assisted direct ink writing. Ceram Int 46:3637–3643
384. Xiong H, Chen H, Zhao L, Huang Y, Zhou K, Zhang D (2019) SiCw/SiCp reinforced 3D-SiC ceramics using direct ink writing of polycarbosilane-based solution: microstructure, composition and mechanical properties. J Eur Ceram Soc 39:2648–2657
385. Xiao J, Jia Y, Liu D, Cheng H (2020) Three-dimensional printing of SiCN ceramic matrix composites from preceramic polysilazane by digital light processing. Ceram Int 46:25802–25807

386. Chen J, Wang Y, Pei X, Bao C, Huang Z, He L, Huang Q (2020) Preparation and stereolithography of SiC ceramic precursor with high photosensitivity and ceramic yield. Ceram Int 46:13066–13072
387. Chen H, Wang X, Xue F, Huang Y, Zhou K, Zhang D (2018) 3D printing of SiC ceramic: direct ink writing with a solution of preceramic polymers. J Eur Ceram Soc 38:5294–5300
388. Narang SB, Bahel S (2010) Low loss dielectric ceramics for microwave applications: a review. J Ceram Process Res 11:316–321
389. Goat CA, Whatmore RW (1999) The effect of grinding conditions on lead zirconate titanate machinability. J Eur Ceram Soc 19:1311–1313
390. Lejeune M, Chartier T, Dossou-Yovo C, Noguera R (2009) Ink-jet printing of ceramic micro-pillar arrays. J Eur Ceram Soc 29:905–911
391. Chen Z, Song X, Lei L et al (2016) 3D printing of piezoelectric element for energy focusing and ultrasonic sensing. Nano Energy 27:78–86
392. Wang W, Sun J, Guo B, Chen X, Ananth KP, Bai J (2020) Fabrication of piezoelectric nano-ceramics via stereolithography of low viscous and non-aqueous suspensions. J Eur Ceram Soc 40:682–688
393. Qi F, Chen N, Wang Q (2017) Preparation of PA11/$BaTiO_3$ nanocomposite powders with improved processability, dielectric and piezoelectric properties for use in selective laser sintering. Mater Des 131:135–143
394. Chen W, Wang F, Yan K, Zhang Y, Wu D (2019) Micro-stereolithography of KNN-based lead-free piezoceramics. Ceram Int 45:4880–4885
395. Chen Y, Bao X, Wong CM, Cheng J, Wu H, Song H, Ji X, Wu S (2018) PZT ceramics fabricated based on stereolithography for an ultrasound transducer array application. Ceram Int 44:22725–22730
396. Dufaud O, Corbel S (2002) Stereolithography of PZT ceramic suspensions. Rapid Prototyp J 8:83–90
397. Guo D, Li LT, Cai K, Gui ZL, Nan CW (2004) Rapid prototyping of piezoelectric ceramics via selective laser sintering and gelcasting. J Am Ceram Soc 87:17–22
398. Jin L, Qiao J, Hou L et al (2018) A strategy for obtaining high electrostrictive properties and its application in barium stannate titanate lead-free ferroelectrics. Ceram Int 44:21816–21824
399. Shi R, Pu Y, Wang W, Shi Y, Li J, Guo X, Yang M (2019) Flash sintering of barium titanate. Ceram Int 45:7085–7089
400. Gaytan SM, Cadena MA, Karim H, Delfin D, Lin Y, Espalin D, MacDonald E, Wicker RB (2015) Fabrication of barium titanate by binder jetting additive manufacturing technology. Ceram Int 41:6610–6619
401. Rödel J, Webber KG, Dittmer R, Jo W, Kimura M, Damjanovic D (2015) Transferring lead-free piezoelectric ceramics into application. J Eur Ceram Soc 35:1659–1681
402. Khomyakova E, Wenner S, Bakken K, Schultheiß J, Grande T, Glaum J, Einarsrud MA (2020) On the formation mechanism of $Ba_{0.85}Ca_{0.15}Zr_{0.1}Ti_{0.9}O_3$ thin films by aqueous chemical solution deposition. J Eur Ceram Soc 40:5376–5383
403. Bai Y, Matousek A, Tofel P, Nan B, Bijalwan V, Kral M, Hughes H, Button TW, Pooladvand H (2015) Phase transitions and dielectric, ferroelectric and piezoelectric properties of $Bi_{0.5}(Na_{0.82}K_{0.18})_{0.5}TiO_3$-doped $(Ba_{0.85}Ca_{0.15})(Zr_{0.1}Ti_{0.9})O_3$ ceramics. In: 2015 joint IEEE international symposium on the applications of ferroelectric, pp 268–271
404. Nan B, Galindo-Rosales FJ, Ferreira JMF (2020) 3D printing vertically: direct ink writing free-standing pillar arrays. Mater Today 35:16–24
405. Villafuerte-Castrejón ME, Morán E, Reyes-Montero A, Vivar-Ocampo R, Peña-Jiménez JA, Rea-López SO, Pardo L (2016) Towards lead-free piezoceramics: facing a synthesis challenge. Materials 9:21
406. Li Y, Li L, Li B (2015) Direct ink writing of three-dimensional (K, Na)NbO_3-based piezoelectric ceramics. Materials 8:1729–1737
407. Wen Y, Xun S, Haoye M, Baichuan S, Peng C, Xuejian L, Kaihong Z, Xuan Y, Jiang P, Shibi L (2017) 3D printed porous ceramic scaffolds for bone tissue engineering: a review. Biomater Sci 5:1690–1698

408. Poitout DG (2004) Biomechanics and biomaterials in orthopedics. Biomech Biomater Orthop Springer
409. Stępak B, Dzienny P, Franke V, Kunicki P, Gotszalk T, Antończak A (2018) Femtosecond laser-induced ripple patterns for homogenous nanostructuring of pyrolytic carbon heart valve implant. Appl Surf Sci 436:682–689
410. Petersen R (2016) Carbon fiber biocompatibility for implants. Fibers 4:1
411. Fan J, Lin T, Hu F, Yu Y, Ibrahim M, Zheng R, Huang S, Ma J (2017) Effect of sintering temperature on microstructure and mechanical properties of zirconia-toughened alumina machinable dental ceramics. Ceram Int 43:3647–3653
412. Brown TS, Van Citters DW, Berry DJ, Abdel MP (2017) The use of highly crosslinked polyethylene in total knee arthroplasty. Bone Jt J 99B:996–1002
413. Khaskhoussi A, Calabrese L, Bouaziz J, Proverbio E (2017) Effect of TiO_2 addition on microstructure of zirconia/alumina sintered ceramics. Ceram Int 43:10392–10402
414. Ohtsuki C, Kamitakahara M, Miyazaki T (2009) Bioactive ceramic-based materials with designed reactivity for bone tissue regeneration. J R Soc Interface 6:349–360
415. Gmeiner R, Deisinger U, Schönherr J, Lechner B, Detsch R, Boccaccini AR, Stampfl J (2015) Additive manufacturing of bioactive glasses and silicate bioceramics. J Ceram Sci Technol 6:75–86
416. Hench LL, Splinter RJ, Allen WC, Greenlee TK (1971) Bonding mechanisms at the interface of ceramic prosthetic materials. J Biomed Mater Res 5:117–141
417. Elsayed H, Zocca A, Schmidt J, Günster J, Colombo P, Bernardo E (2018) Bioactive glass-ceramic scaffolds by additive manufacturing and sinter-crystallization of fine glass powders. J Mater Res 33:1960–1971
418. Ong JL, Chan DCN (2000) Hydroxyapatite and their use as coatings in dental implants: a review. Crit Rev Biomed Eng 28:667–707
419. Fihri A, Len C, Varma RS, Solhy A (2017) Hydroxyapatite: a review of syntheses, structure and applications in heterogeneous catalysis. Coord Chem Rev 347:48–76
420. Buchanan F (2008) Degradation rate of bioresorbable materials. Woodhead Publishing Limited
421. Zhang H, Zhou L, Zhang W (2014) Control of scaffold degradation in tissue engineering: a review. Tissue Eng—Part B Rev 20:492–502
422. Trombetta R, Inzana JA, Schwarz EM, Kates SL, Awad HA (2017) 3D printing of calcium phosphate ceramics for bone tissue engineering and drug delivery. Ann Biomed Eng 45:23–44
423. Kitamura M, Ohtsuki C, Ogata SI, Kamitakahara M, Tanihara M (2004) Microstructure and bioresorbable properties of α-TCP ceramic porous body fabricated by direct casting method. Mater Trans 45:983–988
424. Wang WJ, Qian C, Hu M, Xu T, Wang YG, Sun J (2020) Optimisation of scanning parameters in stereolithography for dental zirconia ceramic fabrication. Adv Appl Ceram 119:244–251
425. Wei Y, Zhao D, Cao Q et al (2020) Stereolithography-based additive manufacturing of high-performance osteoinductive calcium phosphate ceramics by a digital light-processing system. ACS Biomater Sci Eng 6:1787–1797
426. Zhou T, Zhang L, Yao Q, Ma Y, Hou C, Sun B, Shao C, Gao P, Chen H (2020) SLA 3D printing of high quality spine shaped β-TCP bioceramics for the hard tissue repair applications. Ceram Int 46:7609–7614
427. Zhou Z, Lennon A, Buchanan F, McCarthy HO, Dunne N (2020) Binder jetting additive manufacturing of hydroxyapatite powders: effects of adhesives on geometrical accuracy and green compressive strength. Addit Manuf 36:101645
428. Marques CF, Perera FH, Marote A, Ferreira S, Vieira SI, Olhero S, Miranda P, Ferreira JMF (2017) Biphasic calcium phosphate scaffolds fabricated by direct write assembly: mechanical, anti-microbial and osteoblastic properties. J Eur Ceram Soc 37:359–368
429. Liu J, Hu H, Li P, Shuai C, Peng S (2013) Fabrication and characterization of porous 45S5 glass scaffolds via direct selective laser sintering. Mater Manuf Process 28:610–615
430. Liu X, Rahaman MN, Liu Y, Bal BS, Bonewald LF (2013) Enhanced bone regeneration in rat calvarial defects implanted with surface-modified and BMP-loaded bioactive glass (13–93) scaffolds. Acta Biomater 9:7506–7517

431. Kolan KCR (2015) Selective laser sintering of bioactive glass scaffolds and their biological assessment for bone repair, Thesis
432. Zhang H, Yang Y, Hu K, Liu B, Liu M, Huang Z (2020) Stereolithography-based additive manufacturing of lightweight and high-strength Cf/SiC ceramics. Addit Manuf 34:101199
433. Lu Z, Xia Y, Miao K, Li S, Zhu L, Nan H, Cao J, Li D (2019) Microstructure control of highly oriented short carbon fibres in SiC matrix composites fabricated by direct ink writing. Ceram Int 45:17262–17267
434. Fu H, Zhu W, Xu Z, Chen P, Yan C, Zhou K, Shi Y (2019) Effect of silicon addition on the microstructure, mechanical and thermal properties of C_f/SiC composite prepared via selective laser sintering. J Alloys Compd 792:1045–1053
435. Yunus DE, He R, Shi W, Kaya O, Liu Y (2017) Short fiber reinforced 3d printed ceramic composite with shear induced alignment. Ceram Int 43:11766–11772
436. Zhu W, Fu H, Xu Z, Liu R, Jiang P, Shao X, Shi Y, Yan C (2018) Fabrication and characterization of carbon fiber reinforced SiC ceramic matrix composites based on 3D printing technology. J Eur Ceram Soc 38:4604–4613
437. Lu ZL, Lu F, Cao JW, Li DC (2014) Manufacturing properties of turbine blades of carbon fiber-reinforced SiC composite based on stereolithography. Mater Manuf Process 29:201–209
438. O'Masta MR, Stonkevitch E, Porter KA, Bui PP, Eckel ZC, Schaedler TA (2020) Additive manufacturing of polymer-derived ceramic matrix composites. J Am Ceram Soc 103:6712–6723
439. Xing H, Zou B, Wang X, Hu Y, Huang C, Xue K (2020) Fabrication and characterization of SiC whiskers toughened Al_2O_3 paste for stereolithography 3D printing applications. J Alloys Compd 828:154347
440. Zhu Q, Dong X, Hu J, Yang J, Zhang X, Ding Y, Dong S (2020) High strength aligned SiC nanowire reinforced SiC porous ceramics fabricated by 3D printing and chemical vapor infiltration. Ceram Int 46:6978–6983
441. Lv X, Ye F, Cheng L, Fan S, Liu Y (2019) Fabrication of SiC whisker-reinforced SiC ceramic matrix composites based on 3D printing and chemical vapor infiltration technology. J Eur Ceram Soc 39:3380–3386
442. Li S, Zhang Y, Zhao T, Han W, Duan W, Wang L, Dou R, Wang G (2020) Additive manufacturing of SiBCN/Si_3N_4w composites from preceramic polymers by digital light processing. RSC Adv 10:5681–5689
443. Liu C, Ding J (2020) Carbon nanotubes reinforced alumina matrix nanocomposites for conductive ceramics by additive manufacturing. Procedia Manuf 48:763–769
444. You X, Yang J, Huang K, Wang M, Zhang X, Dong S (2019) Multifunctional silicon carbide matrix composites optimized by three-dimensional graphene scaffolds. Carbon 155:215–222
445. Román-Manso B, Figueiredo FM, Achiaga B, Barea R, Pérez-Coll D, Morelos-Gómez A, Terrones M, Osendi MI, Belmonte M, Miranzo P (2016) Electrically functional 3D-architectured graphene/SiC composites. Carbon 100:318–328
446. Shuai C, Gao C, Feng P, Peng S (2014) Graphene-reinforced mechanical properties of calcium silicate scaffolds by laser sintering. RSC Adv 4:12782–12788
447. Sun Q, Liu J, Cheng H, Mou Y, Liu J, Peng Y, Chen M (2019) Fabrication of 3D structures via direct ink writing of kaolin/graphene oxide composite suspensions at ambient temperature. Ceram Int 45:18972–18979
448. Azhari A, Toyserkani E, Villain C (2015) Additive manufacturing of graphene-hydroxyapatite nanocomposite structures. Int J Appl Ceram Technol 12:8–17
449. Zheng T, Wang W, Sun J, Liu J, Bai J (2020) Development and evaluation of Al_2O_3–ZrO_2 composite processed by digital light 3D printing. Ceram Int 46:8682–8688
450. Xing H, Zou B, Liu X, Wang X, Chen Q, Fu X, Li Y (2020) Effect of particle size distribution on the preparation of ZTA ceramic paste applying for stereolithography 3D printing. Powder Technol 359:314–322
451. Wu Z, Liu W, Wu H, Huang R, He R, Jiang Q, Chen Y, Ji X, Tian Z, Wu S (2018) Research into the mechanical properties, sintering mechanism and microstructure evolution of Al_2O_3-ZrO_2 composites fabricated by a stereolithography-based 3D printing method. Mater Chem Phys 207:1–10

452. Wu H, Liu W, He R, Wu Z, Jiang Q, Song X, Chen Y, Cheng L, Wu S (2017) Fabrication of dense zirconia-toughened alumina ceramics through a stereolithography-based additive manufacturing. Ceram Int 43:968–972
453. Zhang J, Huang D, Liu S et al (2019) Zirconia toughened hydroxyapatite biocomposite formed by a DLP 3D printing process for potential bone tissue engineering. Mater Sci Eng C 105:110054
454. Verga F, Borlaf M, Conti L, Florio K, Vetterli M, Graule T, Schmid M, Wegener K (2020) Laser-based powder bed fusion of alumina toughened zirconia. Addit Manuf 31:100959
455. Schwarzer E, Holtzhausen S, Scheithauer U, Ortmann C, Oberbach T, Moritz T, Michaelis A (2019) Process development for additive manufacturing of functionally graded alumina toughened zirconia components intended for medical implant application. J Eur Ceram Soc 39:522–530
456. Wu H, Liu W, Lin L, Li L, Li Y, Tian Z, Zhao Z, Ji X, Xie Z, Wu S (2019) Preparation of alumina-toughened zirconia via 3D printing and liquid precursor infiltration: manipulation of the microstructure, the mechanical properties and the low temperature aging behavior. J Mater Sci 54:7447–7459
457. Pusch J, Wohlmann B (2018) Carbon fibers. Inorg Compos Fibers Prod Prop Appl Elsevier 31–51
458. Narottam PB (2005) Handbook of ceramic composites. Springer
459. Silvestroni L, Sciti D, Melandri C, Guicciardi S (2010) Toughened ZrB_2-based ceramics through SiC whisker or SiC chopped fiber additions. J Eur Ceram Soc 30:2155–2164
460. Pienti L, Sciti D, Silvestroni L, Guicciardi S (2013) Effect of milling on the mechanical properties of chopped SiC fiber-reinforced ZrB_2. Materials 6:1980–1993
461. Nekahi S, Sadegh Moghanlou F, Vajdi M, Ahmadi Z, Motallebzadeh A, Shahedi Asl M (2019) Microstructural, thermal and mechanical characterization of TiB_2–SiC composites doped with short carbon fibers. Int J Refract Met Hard Mater 82:129–135
462. Huang S, Zhou W, Luo F, Wei P, Zhu D (2014) Mechanical and dielectric properties of short carbon fiber reinforced Al_2O_3 composites with MgO additive. Ceram Int 40:2785–2791
463. Yang F, Zhang X, Han J, Du S (2008) Mechanical properties of short carbon fiber reinforced ZrB_2-SiC ceramic matrix composites. Mater Lett 62:2925–2927
464. Suemasu H, Kondo A, Itatani K, Nozue A (2001) Probabilistic approach to the toughening mechanism in short-fiber-reinforced ceramic-matrix composites. Compos Sci Technol 61:281–288
465. Rice RW (2008) Ceramic matrix composite toughening mechanisms: an update, pp 589–607
466. Polozov I, Razumov N, Masaylo D, Silin A, Lebedeva Y, Popovich A (2020) Fabrication of silicon carbide fiber-reinforced silicon carbide matrix composites using binder jetting additive manufacturing from irregularly-shaped and spherical powders. Materials 13:1766
467. Sun X, Zhu H, Li J, Huang J, Xie Z (2018) High entropy alloy FeCoNiCu matrix composites reinforced with in-situ TiC particles and graphite whiskers. Mater Chem Phys 220:449–459
468. Nevarez-Rascon A, Aguilar-Elguezabal A, Orrantia E, Bocanegra-Bernal MH (2011) Compressive strength, hardness and fracture toughness of Al_2O_3 whiskers reinforced ZTA and ATZ nanocomposites: Weibull analysis. Int J Refract Met Hard Mater 29:333–340
469. Nebol'sin VA, Shchetinin AA, Dolgachev AA, Korneeva V V. (2005) Effect of the nature of the metal solvent on the vapor-liquid-solid growth rate of silicon whiskers. Inorg Mater 41:1256–1259
470. Mizuhara Y, Noguchi M, Ishihara T, Takita Y (1995) Preparation of silicon nitride whiskers from diatomaceous earth: I, reaction conditions. J Am Ceram Soc 78:109–113
471. Xiong H, Zhao L, Chen H, Wang X, Zhou K, Zhang D (2019) 3D SiC containing uniformly dispersed, aligned SiC whiskers: printability, microstructure and mechanical properties. J Alloys Compd 809:151824
472. Meyyappan M, Srivastava D (2002) Carbon nanotubes. Handb Nanosci Eng Technol
473. Kinloch IA, Suhr J, Lou J, Young RJ, Ajayan PM (2018) Composites with carbon nanotubes and graphene: an outlook. Science 362:547–553

474. Ahmad K, Wei P, Wan C (2014) Thermal conductivities of alumina-based multiwall carbon nanotube ceramic composites. J Mater Sci 49:6048–6055
475. Yang Y, Ramirez C, Wang X, Guo Z, Tokranov A, Zhao R, Szlufarska I, Lou J, Sheldon BW (2017) Impact of carbon nanotube defects on fracture mechanisms in ceramic nanocomposites. Carbon 115:402–408
476. Qi F, Chen N, Wang Q (2018) Dielectric and piezoelectric properties in selective laser sintered polyamide11/$BaTiO_3$/CNT ternary nanocomposites. Mater Des 143:72–80
477. Li Y, Feng Z, Huang L, Essa K, Bilotti E, Zhang H, Peijs T, Hao L (2019) Additive manufacturing high performance graphene-based composites: a review. Compos Part A Appl Sci Manuf 124:105483
478. Nieto A, Bisht A, Lahiri D, Zhang C, Agarwal A (2017) Graphene reinforced metal and ceramic matrix composites: a review. Int Mater Rev 62:241–302
479. Porwal H, Grasso S, Reece MJ (2013) Review of graphene-ceramic matrix composites. Adv Appl Ceram 112:443–454
480. Ahmad I, Yazdani B, Zhu Y (2014) Recent advances on carbon nanotubes and graphene reinforced ceramics nanocomposites. Nanomaterials 5:90–114
481. Jin Y, Chen N, Li Y, Wang Q (2020) The selective laser sintering of a polyamide 11/$BaTiO_3$/graphene ternary piezoelectric nanocomposite. RSC Adv 10:20405–20413
482. Román-Manso B, Moyano JJ, Pérez-Coll D, Belmonte M, Miranzo P, Osendi MI (2018) Polymer-derived ceramic/graphene oxide architected composite with high electrical conductivity and enhanced thermal resistance. J Eur Ceram Soc 38:2265–2271
483. Yu WH, Sing SL, Chua CK, Kuo CN, Tian XL (2019) Particle-reinforced metal matrix nanocomposites fabricated by selective laser melting: a state of the art review. Prog Mater Sci 104:330–379
484. deSeBde MMCC, Elias CN, Duailibi Filho J, de Oliveira LG (2004) Mechanical properties of alumina-zirconia composites for ceramic abutments. Mater Res 7:643–649
485. De Aza AH, Chevalier J, Fantozzi G, Schehl M, Torrecillas R (2002) Crack growth resistance of alumina, zirconia and zirconia toughened alumina ceramics for joint prostheses. Biomater 23:937–945
486. Stanciuc AM, Sprecher CM, Adrien J, Roiban LI, Alini M, Gremillard L, Peroglio M (2018) Robocast zirconia-toughened alumina scaffolds: processing, structural characterisation and interaction with human primary osteoblasts. J Eur Ceram Soc 38:845–853
487. Gheisari R, Chamberlain H, Chi-Tangyie G et al (2020) Multi-material additive manufacturing of low sintering temperature $Bi_2Mo_2O_9$ ceramics with Ag floating electrodes by selective laser burnout. Virtual Phys Prototyp 15:133–147

The manufacturer's authorised representative in the EU is Springer Nature Customer Service Centre GmbH, Europaplatz 3, 69115 Heidelberg, Germany. If you have any concerns regarding our products, please contact ProductSafety@springernature.com

Printed and bound by CPI Group (UK) Ltd, Croydon, CR0 4YY
30/09/2026
02225242-0012